NDE Handbook

NDE Handbook
**Non-destructive examination methods
for condition monitoring**

Edited by
Knud G. Bøving
The Danish Maintenance Society
(Den Danske Vedligeholdsforening – DDV)

Butterworths
London Boston Singapore Sydney Toronto Wellington

 PART OF REED INTERNATIONAL P.L.C.

All rights reserved. No part of this publication may be reproduced in any material form (including photocopying or storing it in any medium by electronic means and whether or not transiently or incidentally to some other use of this publication) without the written permission of the copyright owner except in accordance with the provisions of the Copyright, Designs and Patents Act 1988 or under the terms of a licence issued by the Copyright Licensing Agency Ltd, 33–34 Alfred Place, London, England WC1E 7DP. Applications for the copyright owner's written permission to reproduce any part of this publication should be addressed to the Publishers.

Warning: The doing of an unauthorised act in relation to a copyright work may result in both a civil claim for damages and criminal prosecution.

This book is sold subject to the Standard Conditions of Sale of Net Books and may not be re-sold in the UK below the net price given by the Publishers in their current price list.

First published in Danish by Danish Technical Publishers A/S, Copenhagen, 1987
English translation by Frank Bason, Silkeborg SolData, Denmark, first published by Butterworths, 1989

© Teknisk Forlag A/S (Danish Technical Press), 1987 and 1989

British Library Cataloguing in Publication Data

Bøving, Knud G.
  NDE handbook.
  1. Non-destructive testing
  I. Title
  620.1′127

ISBN 0-408-04392-X

Library of Congress Cataloging-in-Publication Data

NDE ståbi.   English.
  NDE handbook.
  Translation of: NDE ståbi.
  Includes bibliographical references.
  1. Non-destructive testing—Handbooks, manuals, etc.
  I. Bøving, Knud G.   II. Danske vedligeholdsforening.
  III. Title.
  TA417.2.N465T3 1989   620.1′127   89-22163

ISBN 0-408-04392-X

Printed and bound by Hartnolls Ltd., Bodmin, Cornwall

Preface

The idea for the NDE Handbook arose in the Standing Committee on Condition Monitoring of the Danish Maintenance Society (Den Danske Vedligeholdsforening – DDV) in recognition of the general lack of information concerning the many NDE methods, their applications and limitations in connection with condition monitoring.

The Committee therefore decided to provide this information by publishing a handbook in the series of handbooks published by Teknisk Forlag A/S (Danish Technical Press). Subsequently, because of the general usefulness of NDE techniques and the unique collection of readily accessible information represented by the Danish version of the NDE Handbook, Butterworths commissioned a translation of the book into English.

The Council of Technology in Denmark supported project management for the Danish version under contract no. 85.787 without which it is doubtful whether the project could ever have seen the light of day.

The Standing Committee has acted as the project supervisory group, and the book has been edited by committee members Kaj Tranholm Olesen, Frank Jensen and K.G. Bøving, who also acted as project manager and coordinator.

Altogether 33 authors from all over Denmark have contributed with descriptions of their specialities. A total of 85 different NDE methods are presented in the current volume in a total of 38 sections.

The Standing Committee and the Editors wish to thank all those who have contributed to the creation of this handbook. We believe that it has been of great benefit to many industries in the Scandinavian countries, and now with its publication in English, we hope that a much wider audience will be able to share this knowledge and to benefit industry in many other countries.

Copenhagen, January 9th, 1989

Knud G. Bøving
Chairman of the Standing Committee

Contents

Preface v
Introduction 1
Cross reference table 4
Authors 6
  1. Acoustic cross correlation 9
  2. Acoustic emission 17
  3. Coating thickness 24
  4. Dye penetrant examination 32
  5. Eddy current testing 42
  6. Emission spectroscopy 60
  7. Endoscopy 72
  8. ER probe 81
  9. Ferrography 88
 10. Hardness testing 103
 11. Hydrogen cell 120
 12. Isotope techniques 128
 13. Laser distance measurements 144
 14. Leak testing 148
 15. LPR probe 159
 16. Magnetic plugs 165
 17. Magnetic particle examination 171
 18. Mechanical calibration 189
 19. NDE method combination 195
 20. NDE methods under development 202
 21. Noise measurements 223
 22. Optical pattern recognition 231
 23. P scan 240
 24. Pinhole detection 251
 25. Pressure testing 258
 26. Radiography 263
 27. Replica technique 276
 28. Spectrometric oil analysis program (SOAP) 286
 29. Strain gauge technology 295
 30. Stroboscopy 302
 31. Test coupons 306
 32. Thermography 310
 33. Ultrasonic leak detection 321
 34. Ultrasonics 326
 35. Vibration monitoring 340
 36. Visual inspection 379
 37. X-ray crawlers 397
 38. X-ray diffraction 404
Index 413

Introduction

Manufacturing control

The development of non-destructive examination techniques began years ago when it became more and more common to join materials by means of welding, and the need arose to check the quality of the weld joints.

Non-destructive examination tools such as X-rays (radiography), ultrasonics, dye penetrant methods, magnetic particle techniques, etc. were employed to detect defects which had occurred in connection with or as a consequence of manufacturing processes. These testing methods are collectively referred to as **NDT** – non-destructive testing.

As the demand for manufacturing controls increased, the choice of method and procedures as well as acceptance and classification criteria became well-defined and are today fixed to a large degree by means of national and international standards.

Condition monitoring

With respect to condition monitoring, where non-destructive examination techniques are used to check the condition of equipment, the situation is quite different.

In this case knowledge of the processes of material deterioration in a given environment for a given material in a given construction is decisive for the choice of the testing method or testing methods which will be used in connection with checking equipment condition. Furthermore, the conditions for acceptance and classification are rarely particularly well-defined because they depend among other things upon the construction, the degree of wear and the actual operating conditions.

The rejection criteria must therefore in many cases be specified on the basis of an interaction among several parameters which must be identified qualitatively and quantitatively on the basis of various methods of condition monitoring.

During the condition monitoring phase it will therefore often be necessary to combine several testing methods to describe the condition of a material or a component. It is thus impossible in the maintenance-condition monitoring phase to choose a control method or methods and fix acceptance and rejection criteria before a clear basis of operational and material experience has been established concerning the deterioration mechanisms which are present and what effect they can have upon the material of interest.

As a consequence of the above it is only possible in the condition monitoring phase to establish standard operating procedures for the non-destructive examination techniques for certain types of failure modes, while it will not generally be possible to establish standards for the choice of method, or the acceptance and rejection criteria.

In order to distinguish between the manufacturing control methods, which as stated are termed NDT, and the monitoring methods, which in addition to the methods traditionally used for manufacturing control also include many methods especially developed for condition monitoring, The Danish Maintenance Society has decided to term the condition monitoring methods NDE - non destructive examination. Furthermore, a specific training program for NDE technicians has been developed, and implementation of the program will take place early in 1990.

Scope of the NDE Handbook

The methods of condition monitoring described in the NDE Handbook can be applied in the mechanical engineering and processing industry in the widest possible sense. Some methods can be used in construction and electrotechnical fields as well. In some cases also these applications will be discussed in the respective Chapters.

Organization of the NDE Handbook

The various NDE methods and equipment are listed alphabetically in the Contents, so the methods will appear in alphabetical order. For the sake of the user who is not already familiar with the designations for the various methods, but who needs to find an NDE method for detection of a certain type of defect or condition (e.g. leaks), a cross reference table is presented right after this introduction, providing an overview of the NDE methods and their main areas of application.

An author index is included after the cross reference table, and a subject index is available at the end of the book.

Condition Monitoring
– by all NDE methods – from the simplest to the most advanced • Individually and in all combinations

Systematic Maintenance, manual or EDP-operated, is executed for
▼ Refineries
▼ Gas treatment plants
▼ Offshore installations
▼ Industrial plants of any kind
▼ Power stations
▼ Vehicles • Aircraft • Vessels • Ships • Trains • Etc.

MOCS
Maintenance **O**ptimising **C**omputerised **S**ystem
The predictive program for corrosion control based on NDE corrosion monitoring.
Designed for extensive piping systems, tubes, pressure vessels, tanks, etc. Provides automatic planning of inspections and repair, analytic reports, and graphical condition overviews for decision purposes. Documented cost/benefit factors up to 1:24. A third generation system developed by the Danish Welding Institute under practical conditions since 1972.

P-scan
Projection image **scan**ning
A highly developed ultrasonic image technique for weld inspection and corrosion mapping including data processing and storage.
The picture shows corrosion mapping of a storage tank where a self-supporting P-scan device is climbing vertically up the tank wall.
The P-scan equipment has been developed by the Danish Welding Institute.

Eddy Current
The illustration shows mechanised eddy current inspection of heat exchanger tubes. Cracks as well as corrosion are localized and recorded.
The method can be used on ferromagnetic as well as on non ferromagnetic materials.
Up to 1000 tubes are examined in one day.
Probes and pistol have been developed by the Danish Welding Institute.

▼ Systematic Maintenance – Quality Assurance in the production phase
▼ Condition monitoring of components and materials
▼ Integrated NDE-methods
▼ Predictive NDE-programs
▼ Inspection
▼ Education and training
▼ Research and development

The Danish Welding Institute   Copenhagen

▼ An independent non-commercial institute under the auspices of the Danish Academy of Technical Sciences

Member of
▼ The International Institute of Welding
▼ The Welding Institute (UK)
▼ Eurotest
▼ The Danish Maintenance Society

SVEJSE centralen
The Danish Welding Institute
Park Allé 345
DK-2605 Copenhagen Brondby · Denmark
World-wide Services
200 experts at your disposal
Get more information
☎ +45 42 96 88 00
Telex 3 33 88 svc dk
Telefax +45 42 96 26 36

Cross reference: applications for NDE-methods

NDE METHOD	Cracks	Wear	Fractures	Corrosion	Erosion	Leaks	Mat'l Analysis	Mat'l Condition	Stress	Deformation	Mat'l Thickness	Deposit Thickness	Phys.Restrictions	Other	REMARKS
1 Acoustic cross correlation						X									Locating buried pipes
2 Acoustic emission	X	X				X	X		X					X	Internal structural noise
3 Coating thickness												X		X	Magnetic methods and eddy currents. Ferrite content of ferritic-austenitic steels
4 Dye penetrant examination	X	X				X									Including the chalk, water, alcohol methods
5 Eddy current testing	X	X	X	X	X	X					X	X		X	Heat exchanger tubes, wire rope, surface checks, sorting
6 Emission spectroscopy (Metascope)							X								Low and high alloy steels. Including X-ray fluorescence
7 Endoscopy	X	X	X	X	X	X						X	X		Inspec'n of internal surfaces
8 ER-probe				X											Average corrosion rates
9 Ferrography		X													Lubricated mechanical systems
10 Hardness testing								X							Brinel, Vickers, Rockwell B,C & N, Rockwell superficial, Knoop Shore, Scleroscope, Equotip, UCI
11 Hydrogen cell				X											Average corrosion rates.
12 Isotope techniques		X				X		X			X	X	X	X	Tracer tech., Ball test, radiometry, collim. photon scatt.
13 Laser distance measurements (optocator)		X								X				X	Topography, symmetry
14 Leak testing						X								X	Liquid penetrant, ultrasonics pressure change, foam, tracers, sulphur diffusion, ozalide paper, halogen
15 LPR-probe, polarization resistance				X											Instantaneous corrosion rate
16 Magnetic plugs		X													Lubricated mechanical systems
17 Magnetic particle examination	X													X	Weld defects, laminations - only ferromagnetic materials
18 Mechanical calibration		X	X	X							X	X		X	Physical dimensions
19 NDE method combination	X	X	X	X	X	X	X	X	X	X	X	X	X	X	Check of entire component condition. Predictive programmes
20 NDE meth. under Dev.	(X)						(X)	(X)	(X)				(X)		
20.1 SPATE									X						Stress pattern analysis by thermal emission

Cross reference: applications for NDE-methods

NDE METHOD	Cracks	Wear	Fractures	Corrosion	Erosion	Leaks	Mat'l Analysis	Mat'l Condition	Stress	Deformation	Mat'l Thickness	Deposit Thickness	Phys.Restrictions	Other	REMARKS
20.2 Pulsed video thermography (PVT)								X						X	Composite materials. Glued metals, delamination, coatings
20.3 Moiré contour mapping (MCM)										X				X	Topography
20.4 Holographic interferometry (HI)							X							X	Lack of adhesion, material defects, thin samples
20.5 Computerized tomography (CT)	X													X	Annual rings, knots, moisture, concrete column cross sections
20.6 Positron annihilation (PA)							X							X	Voids in metals. Fatigue in titanium alloys
21 Noise measurements														X	Noise level, bearing checks
22 Pattern recognition	X	X	X	X	X						X	X	X		
23 P-scan	X	X	X	X	X					X				X	Weld inspec., stress corrosion corrosion topography, creep defects. Full documentation
24 Pinhole detection														X	Coatings, high/low voltage
25 Pressure testing	X		X			X				X					Including vacuum testing. See also leak detection.
26 Radiography	X	X	X	X	X	X					X	X	X	X	Check of joints, geometry, laminations, reinforced concrete and corrosion/erosion
27 Replica technique	X	X	X					X		X				X	Surface microstructure, crack type, wear grooves, topography
28 Spectrometric oil analysis program		X													Lubricated mechanical systems
29 Strain gauge techn.									X	X					Weight, pressure, oscillation
30 Stroboscopy	X	X	X											X	Visual condition monitoring Rotation direction and rate
31 Test coupons				X	X										Average corrosion rate
32 Thermography	X			X		X					X			X	Surface temp., bearing pressure, moisture, energy loss
33 Ultrason. leak det.						X								X	Electrical discharge, flow
34 Ultrasonics	X	X	X	X	X	X	X	X	X	X					Including sound attenuation
35 Vibration monitoring	X	X	X											X	Machinery incl. bearings, gears turbines, centrifuges, etc
36 Visual inspection	X	X	X	X	X	X	X			X		X	X		Spark pattern & chem. analysis
37 X-ray crawlers														X	Checking welds inside pipes
38 X-ray diffraction									X						Measurement resid. stresses

Authors

A. Lindegaard-Andersen
Professor, Dr. Techn.
Technical Physics Laboratory
Danish Technical University
+45-42 88 24 88

Martin Rørbye Angelo,
M.Sc. (Elec. Eng.), M.B.A.
Brüel & Kjær
+45-42 80 05 00

Hans Arup, Director
Korrosionscentralen (The Danish Corrosion Centre)
+45-42 63 11 00

Knud G. Bøving, Manager of Quality Assurance and Maintenance, B.Sc.
SVEJSEcentralen (The Danish Welding Institute)
+45-42 96 88 00

Kirsten M. Dorph, M.Sc., Ph.D.
Danish National Testing Board
+45-31 85 10 66

Thomas Dresler, Managing Director
Præcisionsteknik A/S
+45-42 84 99 33

Alli Fabrin, B.Sc. (Mech. Eng.)
Sekura Industries A/S
+45-86 42 66 44

Axel Feddersen, M.Sc. (Mech. Eng.)
SVEJSEcentralen (The Danish Welding Institute)
+45-42 96 88 00

Kjeld Grønfeldt, B.Sc. (Mech. Eng.)
Dantest
+45-31 54 08 30

Hardy Hansen, B.Sc. (Mech. Eng.)
SVEJSEcentralen (The Danish Welding Institute)
+45-42 96 88 00

J. Vagn Hansen, M.Sc. (Mech. Eng.)
Korrosionscentralen (The Danish Corrosion Centre)
+45-42 63 11 00

Ole L. Høppermann, M.Sc. (Chem. Eng.)
Nordisk Materialkontrol a/s
+45-31 64 15 44 (private)

Lars Tofte Johansen, B.Sc. (Mech. Eng.)
SVEJSEcentralen (The Danish Welding Institute)
+45-42 96 88 00

Anders Korsbæk, B.Sc. (Mech. Eng.)
Dansk Olie og Naturgas A/S
+45-42 15 31 22

Peter Krarup, M.Sc. (Elec. Eng.)
SVEJSEcentralen (The Danish Welding Institute)
+45-42 96 88 00

B. W. Kristensen, Research Engineer
Aalborg Portland
+45-98 16 77 77

Finn Kristensen, B.Sc. (Chem. Eng.)
Dantest
+45-31 54 08 30

Gert Lassen, B.Sc. (Mech. Eng.)
Air Materiel Command, R.D.A.F.
+45-44 68 09 00

Per Bo Ludwigsen, M.Sc. (Head of Metallurgy Dept.)
Korrosionscentralen (The Danish Corrosion Centre)
+45-42 63 11 00

Lene B. Mikkelsen, B.Sc. (Chem. Eng.)
Aalborg Portland
+45-98 16 77 77

Jørgen Møller, M.Sc. (Chem. Eng.)
Korrosionscentralen (The Danish Corrosion Centre)
+45-42 63 11 00

Erik Mørch, M.Sc. (Chem. Eng.)
Isotopcentralen (Danish Isotope Centre)
+45-31 21 41 31

Jesper Nilausen, B.Sc. (Elec. Eng.)
Rugmarken 5
DK-3550 Slangerup
+45-42 33 37 54

Kaj Tranholm Olesen, Section Manager
Aalborg Portland
+45-98 16 77 77

Henning Pamperin, Managing Director
Præcisionsteknik A/S
+45-42 84 99 33

Thomas Fich Pedersen, M.Sc. (Mech. Eng.)
Korrosionscentralen (The Danish Corrosion Centre)
+45-42 63 11 00

Knud Erik Poulsen, M.Sc.
The Jutland Technological Institute
+45-86 14 24 00

Søren Poulsen, M.Sc. (Elec. Eng.)
Vision Automation A/S
+45-31 62 83 33

Ebbe Rislund, B.Sc. (Chem. Eng.)
Korrosionscentralen (The Danish Corrosion Centre)
+45-42 63 11 00

Michael Svan, B.Sc. (Mech. Eng.)
SVEJSEcentralen (The Danish Welding Institute)
+45-42 96 88 00

Steen Teller, M.Sc. (Elec. Eng.), Ph.D.
Isotopcentralen (Danish Isotope Centre)
+45-31 21 41 31

Norman Thomsen, M.Sc. (Mech. Eng.)
SVEJSEcentralen (The Danish Welding Institute)
+45-42 96 88 00

Leif Ødegaard, M.Sc.
Ødegaard & Danneskiold-Samsøe a/s
+45-31 26 60 11

The Danish Maintenance Society
Den Danske Vedligeholdsforening (DDV)
Danish Technical University
+45-42 87 40 00

Chapter 1

Acoustic cross correlation

NDE principle

Acoustic cross correlation involves the detection of leaks in buried main water pipes or district heating pipes by means of a specially designed measuring instrument which operates on the basis of the flow noise generated at the point of the break.

A leak will cause noise at the point of the break. The sound is propagated on either side of the leak. The sound can either be measured directly in the liquid, directly on the pipe or on connected valves and taps.

The sound signal is detected by means of transducers, for example accelerometers mounted directly on the pipe or on connected valves and taps. Hydrophones can be used as transducers placed within the piping system itself.

The leakage sound signal is converted by the transducers to an electrical signal which via amplifiers is sent to the analysing instrument.

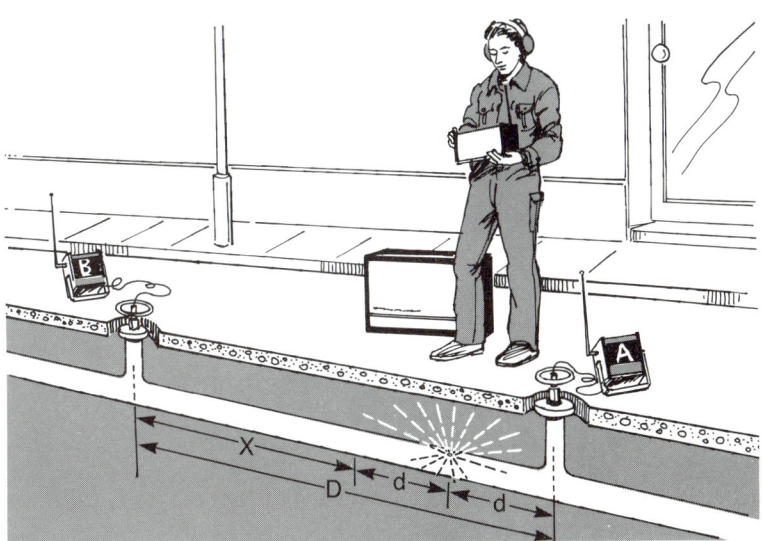

Figure 1.1. Sketch of a piping system with a leak L between two transducers A and B separated by a distance D from one another.

The method requires that the sound of the leak is detected simultaneously at two points of the piping system, for the measurement principle involved is the determination of the difference in the propagation time of the signal between the two measuring points (see Figure 1.1). The time difference is determined by means of the cross correlation function between the signals detected.

When a signal propagates from a point A to a point B, the cross correlation function between the signals detected at A and B, respectively, will have a maximum corresponding to the difference in propagation time.

When the speed of sound is known, the point of the leak can be determined by means of simple geometric considerations. The speed of sound in the piping system can be either calculated or measured.

The method can only be used for piping systems which are filled and operational.

Development of the method

The cross correlation method is a general measuring technique in acoustics. The method can for example be used to determine the time difference between reflected signals or to reveal sound signals which are partly hidden in noise.

The use of the cross correlation method for leak detection eliminates most of the disadvantages which characterize simpler methods, for example sound detection of leak noise at the ground surface.

The measuring method of cross correlation is objective. This means that the result is both independent of the experience of the measurement team and also independent of traffic noise, which is one of the problems with the simple method of listening at the ground surface.

Calculation method

Leak detection by means of the cross correlation method differs from all other leak detection methods in that the method does not reveal the position of greatest sound intensity but instead determines the time difference of the propagation of the sound signal from the leak in the piping system to the two transducers.

Knowing the propagation speed in the piping system, the distance between the transducers and the measured time difference, the position of the leak can be ascertained.

The propagation velocity can be determined theoretically by means of the equation

$$c = \sqrt{\frac{K}{\varrho(1 + \frac{K}{E} \cdot \frac{d}{t})}}$$

where

* K is the compression modulus of the liquid in N/m^2
* E is the elastic modulus of the piping material in N/m^2
* d is the internal diameter of the piping in metres
* t is the thickness of the piping walls in metres
* ϱ is the density of the liquid in kg/m^3

The principle involved in the measurement of the sound signal from the leak is indicated in Figure 1.1.

By simultaneously detecting the leak sound signal by the transducers at A and B, the time difference t_d is determined, where t_d is the time it takes the leak sound signal to propagate from C to A, i.e. the distance x.

By simple consideration of the figure it can be seen that

$$D = 2d + x$$

which can be rewritten by replacing x by the product of the time difference t_d and the propagation speed c, whereby one obtains:

$$d = (D - c\,t_d)/2$$

This expression indicates the distance d between the leak and the nearest measuring point. If t_d is negative, then the leak is nearest to position B, and if t_d is positive, the leak is nearest to A.

In case the leak is located outside the measurement section A-B, for example beyond A, then the measured time difference t_d will correspond to the propagation time for the distance D. In this case the propagation speed c can be determined directly by means of the expression:

$$c = D/t_d$$

Thus, by appropriate placement of the transducers it is possible to measure the propagation speed in the piping section under examination. Furthermore, it is possible to determine on which side of the measurement positions the leak is located. A positive time difference means that the leak noise has reached position A before position B, which indicates that the leak is beyond position A.

Areas of application

The method can be used on most types of pipes, i.e. steel pipes, cast iron pipes, asbestos pipes and PVC pipes.

The distance between the transducers can be up to 1000 metres. The distance limitations for some of the instruments used will be partly due to the transducer sensitivity, and partly due to the range of time differences which the analysis instrument is able to indicate.

By using the cross correlation method the accuracy of the leakage determination is enhanced substantially compared with other simpler methods. Tests have shown that when cross correlation is used for leak detection, 93% of the leaks detected are located within ±2 metres, and 78% of the leaks are located within ±1 metre. Due to the physical size of modern excavation equipment, greater accuracy is unnecessary.

The development of the microprocessor has made it possible today to produce small, transportable instruments. Presently there are many different types of instruments with widely varying technical capabilities. Thus it is possible to transfer signals from transducers via radio, avoiding the need for long cable connections.

Practical examples

In the following an example of leak determination in a mains water supply by means of the correlation method is provided, where the leak was characterized as impossible to locate using conventional listening.

The water main pipe in question was an 80 mm (3.15") cast iron pipe. A 38 mm (1.5") branch pipe with a valve was located opposite each residence. Figure 1.2 provides a sketch of the piping system.

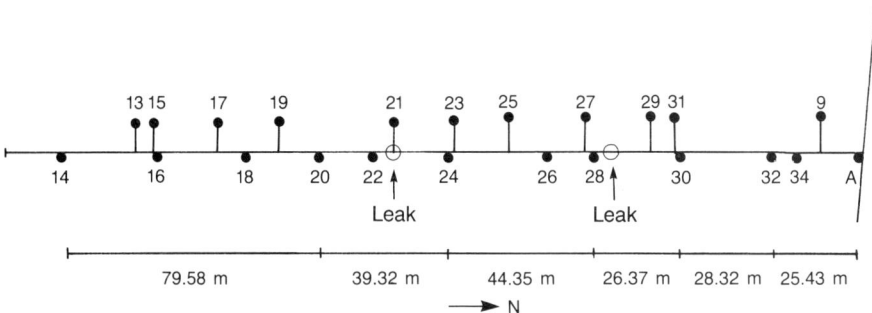

Figure 1.2. Sketch of the water piping system. Valves are marked with the symbol •.

Transducers, in this case accelerometers, were placed on valve A and on taps No. 24 and No. 14, respectively. The distances between the transducers were 124.47 metres and 243.47 metres, respectively. The correlation functions determined between the noise signals recorded simultaneously in position A and in position 2 and position 14, respectively, are provided in Figure 1.3 and Figure 1.4.

On the basis of the theoretical propagation speed of sound c in an 80 mm cast iron pipe of 1272 m/s and a distance D of 124.47 metres between position A and position 24, the maximum difference in the propagation time t_{max} could be computed to 97.5 ms.

From Figure 1.3 it can be ascertained that three characteristic maxima in the correlation function occur:

$t_1 = -97.53$ ms
$t_2 = -19.53$ ms
$t_3 = 101.97$ ms

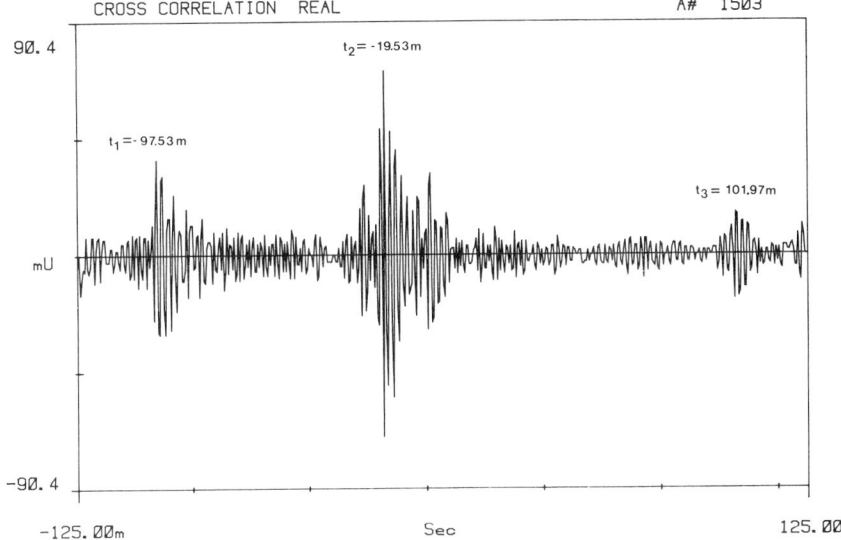

Figure 1.3. The cross correlation function between signals recorded simultaneously in position A and position 24.

It is apparent that $t_1 = t_{max}$, from which it can be concluded that a leak is located outside the measurement section, and that it is located south of position 24. The actual propagation speed c could be computed to 1276 m/s, which is in good accord with the theoretically computed value.

Furthermore, it could be determined that a leak between position A and position 24 corresponding to the time difference t_2 of −19.53 ms. The location of the leak could be computed to be 74.8 metres south of position A. The last maximum corresponding to t_3 of 101.97 ms was due to a reflection of the leak sound signal by the perpendicular water main at position A.

In Figure 1.4 two maxima in the correlation function are clearly seen at the time of observation:

$t_d = -30.07$ ms
$t_d = 74.46$ ms

The maximum path time difference t_{max} could in this case be computed to 190.8 ms. This implies that there are two leaks between position A and position 14. The locations of the leaks could be computed to 140.9 metres south of position A and

74.2 metres south of position A, respectively.

Subsequent excavation revealed leaks at 74.0 metres and 141.1 metres south of position A, which is an extremely good correspondence.

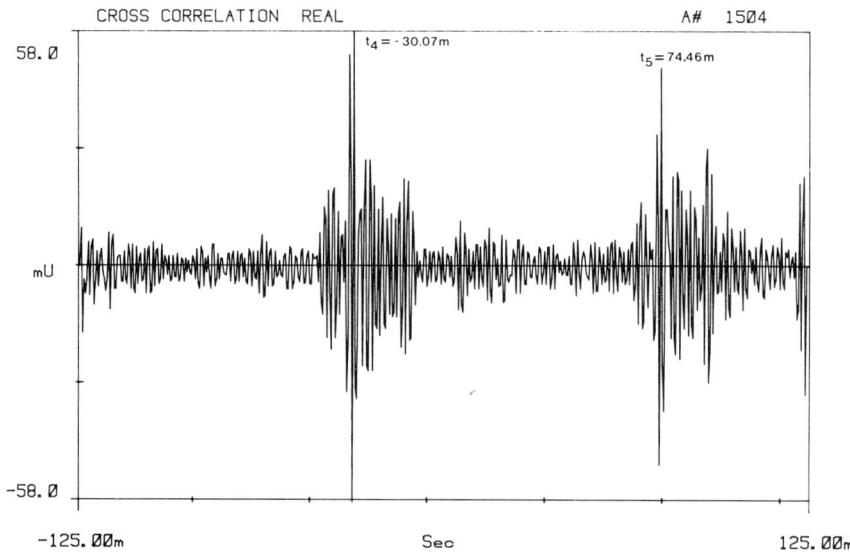

Figure 1.4. The cross correlation function between signals recorded simultaneously in position A and position 14.

Training required/desired

No specific operator training is required for personnel to use the correlation method for leak detection. It is, however, expected that the operator is familiar with the instrument and the principle involved in this measuring technique. More modern instruments today contain built-in microprocessors which take care of all calculations concerning the location of the leak and the propagation speed of sound in the piping system. The operator using these instruments only needs to evaluate the nature of the maxima which occur in the correlation function. The location of the leak is computed automatically. In Figure 1.5 an example of a portable instrument is shown. It was especially developed for leakage detection using the cross correlation method.

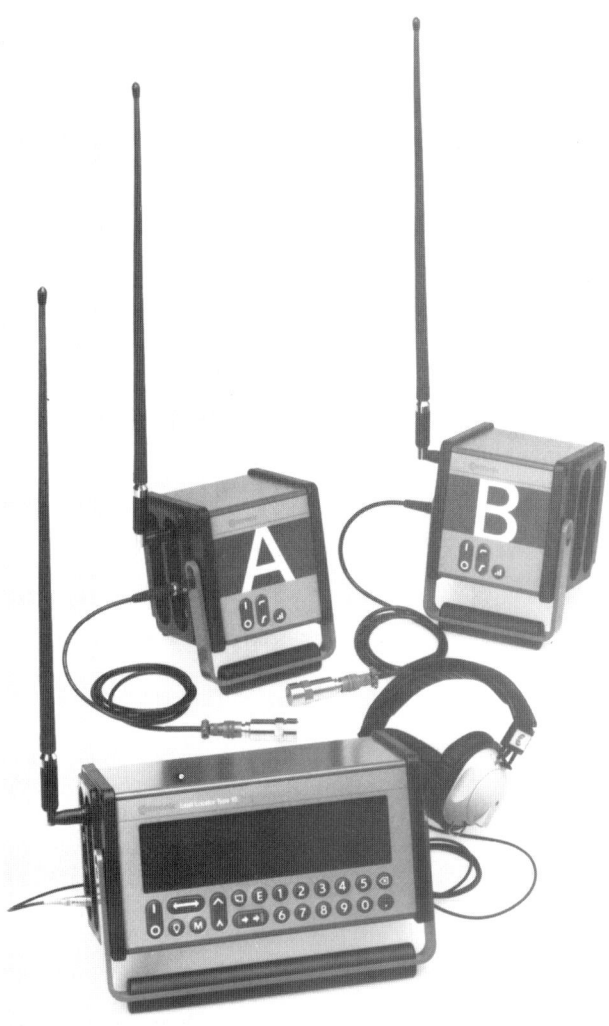

Figure 1.5. Portable leak locator consisting of two transmission units and one correlator, developed and manufactured by Caltronic A/S, Copenhagen, Denmark.

Leif Ødegaard

References

1) *Wasserlecksuchen - Neue Möglichkeiten durch Korrelationsmesstechnik.* Peter Sewerin. Gwf - Wasser/abwasser 1982 H-4, pp. 184-189.

2) *Detection of Water Leakage Point using the Cross Correlation Method.* Kageo Akizuki. Proceedings of IMEKO symposium on flow measurements and control in Industry. Tokyo, Japan 1979, pp. 199-204.

3) *A New Method of Leak Detection in Distribution Systems under Pressure: Acoustic Correlation.* Arnac Pascal. Compagnie Générale de Laux, Paris.

Chapter 2

Acoustic emission

NDE principle

Acoustic emission (AE) has become the designation for the NDE principle which consists of listening to a material.

In certain materials changes of condition can occur which cause the emission of sound waves due to thermal, mechanical or other effects. These changes of condition can be fractures, crack formation, crack growth or metallurgical changes such as plastic deformation, dislocation diffusion and changes in crystal structure.

During such changes in condition energy is released which propagates in the surrounding material as elastic vibrations. These vibrations or sound waves can be detected by placing a sensor on the surface of the material. AE sensors (or transducers) are in principle high frequency microphones which are glued or affixed in some other manner to the clean surface so that the best possible acoustic contact is achieved (see Figure 2.1).

AE technology is thus used for production control of the material and condition monitoring of constructions consisting of materials for which the ability to emit sound is present. Because not all materials emit sound waves from defects under load, and since the propagation of sound is damped depending on the material, it is necessary to evaluate the possibility of using the AE method in each individual case. It may also be necessary to supplement this evaluation with laboratory experiments performed on the material of interest with the relevant defects. Furthermore, it is necessary to take the geometry of the object of interest and possible sensor placements into account.

The AE method can only detect changes in materials. Thus defects already present, e.g. a crack, cannot be revealed unless the size of the crack increases under load. AE is therefore used as a supplement to other NDE methods, where critical regions with defects detected by means of other NDE techniques can be monitored using AE. And vice-versa, AE can indicate regions where other NDE examinations are required.

Since the AE technique consists of listening, clearly undesired sounds will disturb AE measurements. Acoustic or mechanical noise can arise from:

– friction with other parts of the construction

– loose particles in or on the construction

– mechanical equipment such as pumps, etc., connected to the construction

Electrical noise can also interfere with AE measurements, as AE signals are usually very weak. Even though signal amplifiers are installed near the AE sensors, the signals may be disturbed by noise fields in the construction or due to the often very long measurement cables.

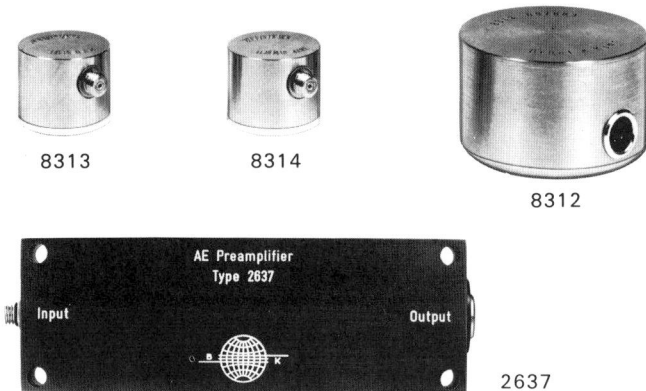

Figure 2.1. AE sensors and amplifiers (courtesy Brüel & Kjær).

If the noise sources mentioned above cannot be avoided, it is possible to filter out some of them by means of appropriate signal handling. In Figure 2.1 various AE sensors and a pre-amplifier are illustrated.

Development of the method

Historically speaking, the seismological equipment which was produced for the purpose of measuring oscillations of the earth as a consequence of earthquakes, can be considered to be the first AE instruments.

The first experiments with AE measurements of materials such as metals and wood were carried out in the early 1950s by the German Joseph Kaiser. Using sensitive electronic instruments he could hear sounds from many different metals subject to deformation. He also discovered that a material which had been previously stressed first exhibited AE again when the previous stress was exceeded (the so-called "Kaiser Effect").

Some years after Kaiser's discovery American researchers began to examine the connection between the metallurgy of metals and AE more closely, but it was not before the early 1960s that intensive research was initiated in many laboratories all over the world. At the same time the evolution of technology had improved AE instrumentation. Around 1970 the use of AE began outside of laboratories, where pressure vessels, pipelines, etc. were equipped with AE equipment during pressure testing. Also concrete constructions and bridges were monitored using AE during stress testing.

By using modern electronics various types of AE apparatus have been produced, so that the processing of AE signals can be carried out according to a variety of criteria. The introduction of computerized systems has made it possible to collect and to handle large amounts of data.

The future prospects for the use of the AE method as an NDE tool are quite bright, as the material research which is ongoing in laboratories all over the world will enhance knowledge of the mechanisms which cause AE. These results can be used to carry out AE measurements on more materials and to expand knowledge of the ability of defects to exhibit AE.

Areas of application

AE technology has been used with many different metals, but primarily with steel, aluminium, etc. Furthermore concrete, wood, plastic, glass fibre and many types of laminated and fibre reinforced materials have been examined.

AE examinations are carried out on many types of components from large oil storage tanks to small electronic components. AE examinations can be made in various ways:

* *Production monitoring*, where a component is stressed according to a specified testing procedure during AE monitoring. The loading is scheduled to achieve the optimum AE activity, e.g. by step-wise stressing, so that AE can be measured both during the increasing stress and during constant stress.

* *Condition monitoring*, where components are subjected to a loading procedure, as indicated above, after a predetermined period of operation. In such cases it is often useful to be able to compare with previous AE measurements so that the development of any defects can be traced.

* *On-line inspection*, where components are examined during operation, preferably at given intervals, to reveal irregularities. As above a "fingerprint" of AE measurements is a valuable aid in the analysis of future measurements.

* *Monitoring*, where AE equipment is permanently mounted on the component with a view to issuing an alarm in case critical AE signals are detected.

In the above mentioned types of examination the size and shape of components are critical to the choice of AE instrumentation. Because the acoustic emission from a defect is damped during propagation in the material and also by flanges, etc., it is necessary to determine the maximum permissible distance between AE sensors required to achieve certain detection of those defects which are relevant to the examination in question. In the case of small components, or if only a small critical region of a larger construction is tested, two to four sensors are appropriate, while the supporting structure of an off-shore platform may require 100 sensors (with correspondingly complex electronic equipment for signal handling), if all nodes are to be encompassed by the AE measurement.

In connection with all of the methods mentioned above the measurement principle which is best suited to reveal the defects of interest is used. Appropriate AE equipment has been developed for each of the various applications. This equipment can carry out more or less complex analyses of the signals arriving from the AE sensors. The following types are normally used for AE analysis, listed in increasing order of complexity:

- registration of signal amplitude as a function of stress and time
- registration of the number of signals which exceed a reference threshold
- computation of the energy of the signals detected
- registration of the frequency content of the signals
- localization of the signal source by triangulation
- correlation analysis, where each signal detected is compared with previously recorded signals

Some analysis equipment can perform combinations of the above mentioned possibilities. For example registration of signal amplitude is often carried out in connection with localization, so both the magnitude and the position of the AE source in a construction can be ascertained. An AE setup with localization equipment is shown in Figure 2.2. Many of these types of instruments are available on the world market, and many NDE organizations have produced special equipment for use in their own studies.

Figure 2.2. AE setup with localization equipment

Other applications

The listening technique which characterizes AE methods can also be used to solve other problems which are not directly related to the emission of AE signals from internal defects. These applications are:

* *Leak detection.* When a liquid or a gas flows through a hole in a material, sound waves will be emitted (due to turbulent flow), depending upon the size of the hole and pressure conditions. By mounting AE sensors on the construction and comparing the signals received, the location of the leak can be determined. This technique is used with buried pipes, where a number of AE sensors are mounted directly on the pipe, e.g. at 50 metre intervals.

* *Searching for loose particles.* Loose particles, such as flakes of rust and broken pieces of a construction, which are forced through the construction by a liquid will emit sound waves due to friction with internal surfaces. These sounds can be detected by AE sensors. This method is in use in power plants in a number of countries to avoid damage to pumps, etc.

Standards

ASTM, NORDTEST and EWGAE (European Working Group on Acoustic Emission) work to prepare standards for the execution of AE inspections as well as for the calibration and control of AE instrumentation.

Practical applications

AE examination has been applied, for example, in the following areas:

- Checking the condition of pressure vessels and pipes to detect possible cracks during pressure testing.

- "On-line inspection" of various components under pressure such as pressure pistons, containers and pipes.

- Monitoring of components on off-shore installations.

- Research programmes designed to reveal the ability of materials to emit AE.

Figure 2.3a. Leak detection with AE equipment.

Figure 2.3b. AE sensors mounted on an uneven surface.

Training required/desired

The use of AE technology requires knowledge of both the AE characteristics of materials as well as ultrasound technology. It should therefore be carried out by specialists with knowledge of the equipment.

Peter Krarup

References

The majority of AE literature is in English and is available as many thousand articles concerning special AE applications.

1) *Journal of Acoustic Emission*, Acoustic Emission Group, Los Angeles, USA.

2) Special technical publications from ASTM (American Society for Testing and Materials).

Chapter 3

Coating thickness

NDE principle

The thickness of coatings on metallic substrates can be measured by means of simple methods when the coatings and the substrates have significantly different magnetic or electrical characteristics.

Magnetic methods

The thickness of non-ferromagnetic coatings on ferromagnetic substrates is easiest to measure with instruments based on magnetic or electromagnetic principles.

Magnetic attraction
In its simplest form the attractive force between a bar magnet placed at right angles to the surface of the coated object and the ferromagnetic base material is measured. The principle here is that the force is roughly proportional to the magnetic flux density between the magnet and the substrate and that the flux density decreases when the magnet is moved away (see Figure 3.1). The attractive force is measured by means of a specially calibrated spring balance, or even better by means of a spiral spring, which, placed around the middle of a balanced, counter-sink supported balance arm, is placed under tension by an angular turn around the balance arm support. The angular turn is used as a measure of the coating thickness, see Figure 3.2. The apparatus shown is popularly referred to as a "banana".

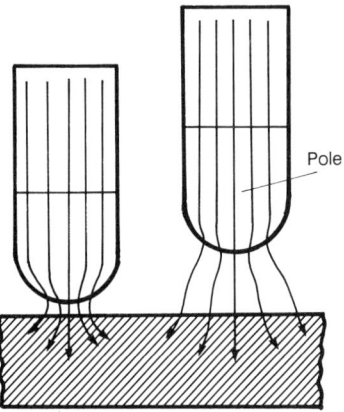

Figure 3.1. The magnetic flux density in the transition cross section at the pole of a permanent magnet of high permeability.

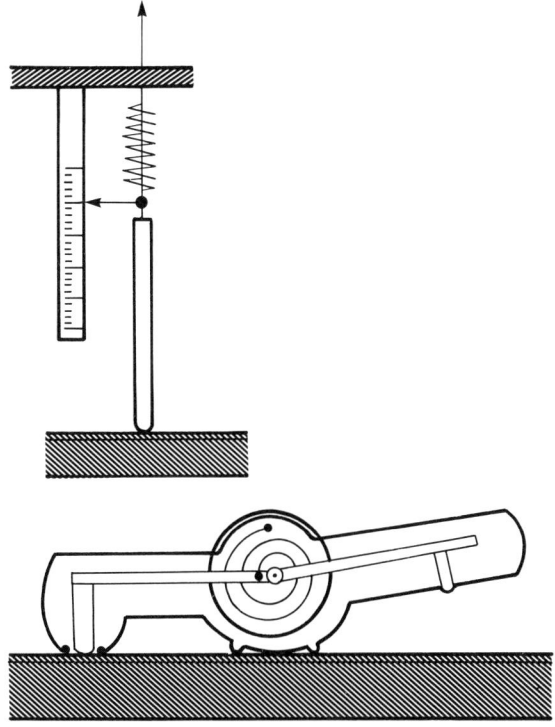

Figure 3.2. Magnetic attraction. The principle (above). The practical implementation (below).

Magnetic induction
Measurements which are more accurate and more reproducible can be achieved by means of methods employing electromagnetic induction.

The non-ferromagnetic layer creates a gap in the magnetic circuit between the base material and the measuring probe. The inductance of a coil with an iron core or in a transformer core varies with the size of the gap (see Figure 3.3).

Changes in the inductance can be detected and by means of an appropriate electronic transformation be converted to a measure of the coating thickness which can be read from a panel meter or a digital display.

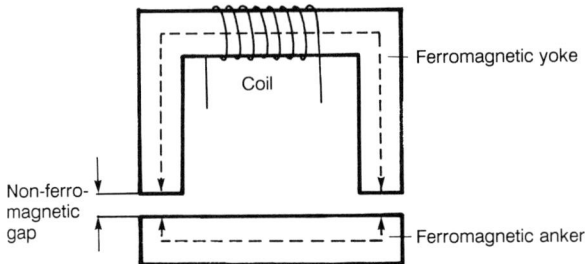

Figure 3.3. A magnetic induction circuit. The inductance of the coil is dependent upon the size of the gap.

Instruments constructed according to this principle are sold as one or two poled devices depending upon whether the measuring probe has one or two protruding poles; see Figures 3.4 and 3.5.

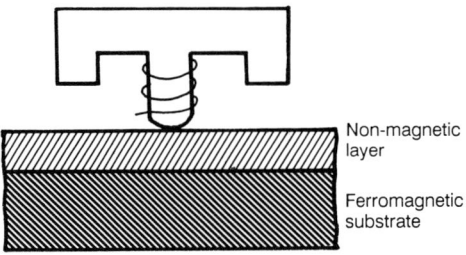

Figure 3.4. "Single poled" measuring probe.

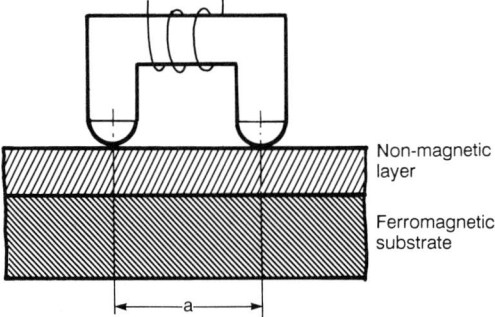

Figure 3.5. Bipolar measuring probe.

Magnetostatics
There are or have been pieces of apparatus on the market which operate according to magnetostatic principles. In some of these the field direction is recorded in an

extra air gap in a magnetic circuit by means of a magnetic dipole suspended in the air gap. In others a semiconductor is inserted into the circuit. The resistance change in the semiconductor, which is sensitive to the strength of a magnetic field, is measured.

Eddy currents

The measurement of nonconducting coatings on non-ferromagnetic metals, e.g. paint layers, plastic layers anodized on aluminium, can be carried out by means of eddy currents. The impedance of a coil which is supplied by an alternating current is dependent upon the distance of the coil from a metal in which it induces eddy currents and can thus be used as a measure of the thickness of an insulating layer which lies between the coil and the metal. The eddy current method is regarded as less suitable for measurements when steel is the base material.

The eddy current method is not treated separately in the following because the same general rules apply to apparatus of this type as for devices for measuring magnetic induction. They resemble one another so much in their exterior construction that they are sometimes combined in the same piece of equipment with a common power supply, adjustment dials and readout device.

Development of the method

It has always been essential that coatings designed to protect against corrosion or decorative coatings have a certain thickness in order to fulfil their mission. There has therefore been a need for many years for equipment which permitted rapid and relatively accurate measurement of the coating thickness on metals.

With steel as the most important construction material it was natural that attention should be directed from the very first at the magnetic attractive force, which clearly is diminished with distance when a magnet is moved away from a steel object. The spring-loaded magnet as thickness measurement device almost invented itself.

Unfortunately the accuracy of the device was not impressive, so as semiconductor technology advanced, and lightweight, stable amplifiers were developed which could record very small changes in electromagnetic circuits, a series of continuously improved devices appeared, all using principles of magnetic induction. As early as the beginning of the 1960s equipment appeared which was as good as anything available today, but they were quite clumsy and not particularly inexpensive. Developments since that time have been in the direction of handier equipment, and in recent years the tendency has been to provide them with memory, statistical calculation programs and printer or read-out options.

Areas of application

The magnetic methods are used mainly for the measurement of coating thicknesses of paints, plastics and layers of non-magnetic metals on "soft" steel, which in general would be steels with hardnesses up to 550-600 HV.

Equipment of this type can also be used to measure the thicknesses of less permeable coatings on more magnetically permeable substrates, e.g. a layer of nickel on steel.

Finally, with appropriate calibration, they can be used to measure the ferrite content of ferritic-austenitic stainless steel, including welding joints of sufficient extent in austenitic stainless steel.

Equipment which utilize the magnetic force normally have a limited range, for the sensitivity decreases rapidly with increasing coating thickness. Apparatus is available each covering a particular measuring range, e.g. 0-100 μm, 100-600 μm, 0.5-6 mm.

Other equipment normally has several measuring ranges, with the decreasing sensitivity compensated for by means of amplifier technology. For particularly thick coatings (i.e. above 2-3 mm), special equipment is required which incorporates bipolar probes with some distance between the poles. In this manner the sensitivity to magnetic effects outside the measuring circuit can be reduced. Equipment is on the market which can measure thicknesses of up to at least 25 mm.

Measurement accuracy differs for the various methods. One can count on a measuring accuracy of about 10% when measuring by means of magnetic attraction, while techniques using magnetic induction yield a measurement accuracy of 3-4%.

Measurement errors

Accuracy is, however, contingent upon correction for various external factors which influence the measurements:

1) The permeability of the steel
2) The conductivity of the steel (high measuring frequency)
3) The strength/hardness of the steel
4) The thickness of the steel
5) The curvature (convex/concave) of the object
6) Distance to edge or support
7) External magnetic fields and AC fields
8) (Temperature)

It is important to check factors 4, 5 and 6 at once when a coating thickness measuring instrument has been acquired. It is also very useful to become familiar with the temperature sensitivity of the measuring probe and/or the associated electronics. It is recommended that one try placing the apparatus and probe in a refrigerator and then observe the reliability of the basic adjustment during the warm-up period. In the best equipment based on magnetic induction the compensation of the adjustment for temperature drift is exceptionally good.

Adjustment of equipment

Most equipment for coating thickness measurement is supplied with a limited selection of calibration foils which, depending on the manufacturer and the price, can be of varying quality.

Equipment which uses magnetic attraction has only one calibration option: one can adjust the tension of the spring. This is often limited and inconvenient.

Normally such equipment does not get out of adjustment, but letting the apparatus lie near electrical AC conductors with high current levels can cause partial demagnetization of the magnet with substantial errors in measurements as a consequence. The equipment should therefore always be checked before use.

Equipment which is based on the principle of magnetic induction is provided with two potentiometers or pushbuttons for adjusting to a high and low scale value. These adjustments should be carried out prior to and during use.

It is advisable as a general rule to always check the readout of the thickness measurement device in the neighbourhood of the specified or measured thickness, for one can not always be sure that the readout of the device is correct over the entire readout range. For example, some apparatus develops a systematic readout error as the measuring probe's pole(s) is(are) worn down.

Specialized equipment

In recent years coating thickness measuring devices have appeared on the market with built-in memory so that a large number of measurements can be recorded. This equipment can also be provided with a simple statistics program which can compute the mean and the standard deviation and read out the greatest and the least thickness measured. Finally, equipment of this sort is equipped with a printer or computer output so that all values and calculations can be printed out, and the results can be presented in the form of a histogram.

Standard operating procedures

All the equipment is small and light and thus practical to use in almost any situation. But the level of documentation, experience has shown, is quite strongly dependent upon the weather. When a substantial need for control and documentation exists, one should seriously consider the use of measuring instruments with built-in memory. As an alternative one might supplement the measuring equipment with a memo tape recorder. Using the tape recorder the measurements can also be localized geographically, a feature which the devices equipped with memory do not yet offer.

Figure 3.6. Coating thickness measurement of hot dip galvanizing on a support structure.

Practical applications

In Figure 3.6 the thickness measurement of hot dip galvanizing on a support structure is shown, and in Figure 3.7 a similar check of glass fibre reinforced polyester on oil storage tanks is being carried out.

Training required/desired

There is no particular training requirement in connection with coating thickness measurement with the equipment discussed, but the operator should be aware of its limitations so that he or she does not collect data with systematic errors.

Jørgen Møller

Figure 3.7. The coating thickness of the fiberglass reinforced polyester on oil tanks is being measured.

References

1) *ISO 2360 Measurement of coating thickness. Eddy current method.*

2) *DS/R 454, Recommendations for the corrosion protection of steel structures,* 1. edition, February 1982. Translated edition, June 1984.

3) *ISO 2178, 1982, Non-magnetic coating on magnetic substrates – Measurement of coating thickness – magnetic method.*

Chapter 4

Dye penetrant examination
(Penetrant flaw detection)

NDE principle

Examination with penetrants is a non-destructive examination method used to reveal defects which reach the surface of non-porous materials. Defects such as cracks, porosities, cleavages and leaks in steel, cast iron, plastics, ceramics, etc.

The examination (see Figure 4.1) is carried out in such a manner that the penetrating liquid (the *penetrant*), which is dyed or fluorescent, is applied to the cleaned surface of the component.

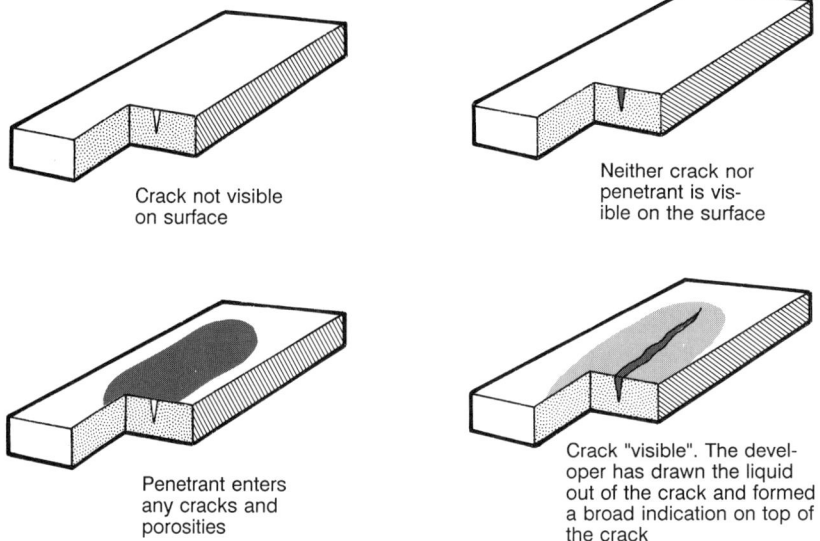

Figure 4.1. The procedure used when performing an examination with a penetrant.
1. *Pre-clean, remove grease and dry the component.*
2. *Penetrant is applied to the component and acts for a brief period.*
3. *Excess penetrant is completely removed from the surface.*
4. *A developer is applied and dried off. Inspect for indication of defects.*

The penetrant must be allowed to act for a period of time depending, among other things, upon the temperature and the component under examination. Excess penetrant is carefully removed from the surface of the component, after which a developing liquid is applied and dried off. The developer acts like a blotter, drawing the penetrant out of the defect.

After a short time indications appear in the developer which are wider than the defect and which, therefore, can be seen directly or under ultraviolet light due to the enhancement of contrast which results between the penetrant and the developer.

Development of the method

Even many years ago the use of a suspension of crushed chalk in alcohol – or for hot surfaces water – has been employed. The surface to be examined for cracks, etc. was cleaned, swabbed with petroleum and then cleaned again. Finally the mixture of chalk and alcohol was painted on the surface. When the suspension medium had evaporated, the dry chalk was left, and any cracks could be seen as dark stripes of petroleum on the white chalked surface.

It is this method which today has been developed to become the *dye penetrant* examination, where petroleum has been replaced by a capillary colouring agent, and the chalk/alcohol mixture has been replaced by the developer. Both are available in aerosol cans and in other forms.

Furthermore, special cleaning liquids, also available in aerosol cans, have been developed. A dye penetrant set of this type is shown in Figure 4.2. One advantage of having the liquids stored in aerosol cans is that one can be reasonably certain that fresh, clean liquids will always be available.

The technique does not require any particularly big investment, and the cans are easy to transport and do not require much room during the examination. However, considerable care must be exercised by the operator in order to perform all steps of the examination correctly.

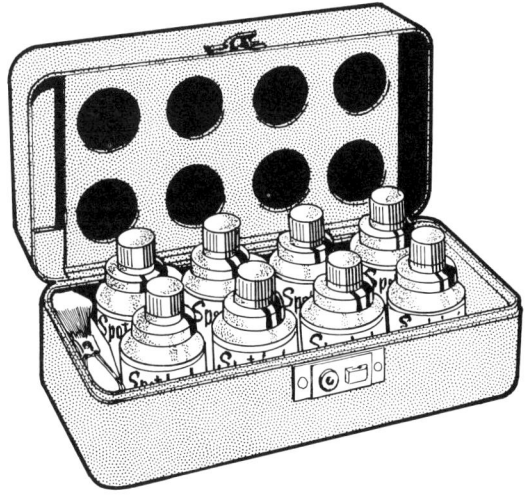

Figure 4.2. Capillary liquid test set.

Dye penetrant examination

The penetrant method is quite slow when one must examine large surfaces, while it can be very quick and easy to use for checking the assembly line production of small parts. It is even possible to automate the method and to have the liquids in baths and to allow the items on the assembly line to pass through the various baths. The process can be made additionally effective by means of ultrasonic baths. Only the inspection for defects must be performed manually, and even this part of the process can be carried out automatically by means of optical pattern recognition.

Areas of application

Almost all materials which are not porous (e.g. unglazed pottery) can be examined by means of dye penetrant liquids.

A rough surface, as for example of cast iron, reduces the possibility of an effective examination, for the rough surface is more difficult to clean and to interpret for excess penetrant. The method can be used on all items of all sizes and shapes, provided they do not have too many odd angles and depressions which are so large that they are difficult to clean. Similarly, components which are screwed or pressed together, where natural joints can lead to misinterpretations, are not suited for testing. Furthermore, groups of pores can conceal more serious defects.

In addition the method can be used in all orientations, for the liquid can be drawn up as well as down into a crack. Also, one should check before testing that the component will not be damaged by the test liquids and will not, for example, be dissolved or corroded. For stainless steel which contains nickel and for austenitic stainless steel the sulphur content must be less than 1% and the chlorine content must be less than 1% in the test liquids.

Because the penetrant method works by allowing a liquid to enter a defect on the surface of a component and be drawn out to the surface afterwards, it is only possible to find defects which are open onto the surface. Foreign matter such as paint or oil must not be allowed to enter the defect. In this case the penetrant is prevented from entering.

Surface coatings, layers due to heating, oil, grease and hammered or sand-blasted surfaces exclude the use of dye penetrant methods. To a certain extent one can use methods such as magnetic particle flaw detection and eddy current techniques instead. On the other hand magnetic particle flaw detection, *magnetoflux*, can only be used on ferromagnetic materials and eddy current techniques can only be used on electrically conducting materials.

Modern dye penetrant systems are extremely sensitive to open surface defects. Under favourable conditions and with sufficient care, it should be possible to indicate cracks which are only about 2 μm wide. Penetrants are normally developed for use from about 15 to 40°C, but they can be used as low at 2°C. At low temperatures the penetration time must be increased by a factor 2 to 3 times the normal, and at higher temperatures (above 50°C) it is reduced to just a few minutes. The method should be used at room temperature, i.e. at about 20°C, but never below 2°C. Special types are available for use with high temperatures. According to the manufacturers they can be used at temperatures of up to 250°C.

Penetrant types

Penetrants are classified into two main categories:

Dye penetrants: The liquids are coloured so that they provide good contrast against the developer. The liquids are as a rule red with white developer. Observation is performed in ordinary daylight or good indoor illumination.

Fluorescent penetrants: The liquids contain additives to give fluorescence under ultraviolet light ("black light"). Inspection is performed protected from visible light. The indications are easy to see in the dark, because any developed penetrant emits light under the action of ultraviolet light.

Furthermore, combined types are available which are both coloured and fluorescent.

Depending upon the type of liquid, a further classification into three main groups is possible:

Water washable penetrants: The liquid contains an emulsifier which allows excess surface penetrant to be removed using water.

Post-emulsifiable penetrant: After the liquid has been applied, an emulsifier must be applied to the excess surface penetrant to make it water soluble.

Solvent removable penetrant: The penetrant can only be removed fully from the surface by means of an appropriate organic solvent.

When selecting a method (type of penetrant) there are a number of factors to be considered before proceeding. Here a few of these can be mentioned: the size and type of defects which one is looking for, whether the surface structure of the object is rough or smooth, the extent of the examination, environmental and safety requirements, costs and space available. The selection of a penetrant depends for the most part on the required sensitivity and the extent of the examination and the size and shape of the object.

If the required sensitivity is very high, i.e. if one wants to detect the smallest possible defects which can be revealed, then fluorescent penetrant should be chosen, for in practice it is easier to detect something which is light against a dark background than to observe a small coloured area on a white background. Fluorescent penetrants are also well suited to small objects, objects with a rough surface and to threads. In addition control and removal of excess liquid is more dependable with ultraviolet light.

When it is a case of spot testing of objects which can not be shielded from visible light, a red penetrant in aerosol cans is often selected.

The next thing to be decided is the selection of a method for the removal of the penetrant. The water washable types have primarily the advantage of requiring fewer steps than the post-emulsifying and are therefore faster to use and cheaper.

Developer types

The developer can be classified into three main groups:

- *Dry powder developers*
- *Water-based wet developers*
- *Non-water based wet developers*

The latter two types can be further divided into two sub-categories. In one case the developer particles are in suspension, and in the other they are dissolved in the liquid medium.

In most cases wet developers are used. On occasion it can be advantageous to use dry developers, as e.g. on rough surfaces and for objects with concave corners and threads, where there is a tendency for the wet developer to form a layer which is too thick. The dry developer is then used in connection with a fluorescent penetrant. Among the non-water soluble wet developers organic solvent or alcohol based developers can be mentioned. These are used almost exclusively in aerosol cans and are therefore particularly easy to transport and to use for spot examinations. The developer particles are here suspended in a liquid and therefore need to be mixed thoroughly before use. In order to make this mixing more effective, the aerosol can contains a glass marble. The aerosol can must be shaken vigorously, and one should be able to hear the glass marble move freely inside the can.

One should avoid using penetrants, emulsifiers, cleaning liquids and developers produced by different manufacturers. On the contrary, it is important to make sure that these items will work well together (see Figure 4.3), and one should follow the instructions provided by the supplier and demand clear, detailed instructions about how the particular system is intended to be used.

Inspection

Inspection for defect indications, when coloured dye penetrants are used, must be carried out in sufficiently bright daylight or artificial illumination. Most norms specify that the area to be observed must be illuminated with a luminous intensity of at least 500 lux, corresponding roughly to the light from a new 80 W lamp one metre away.

On the other hand, if fluorescent penetrants are used, the room in which the inspection is to be made should be darkened, i.e. the illumination level should be no more than 10 lux. The eye must be allowed an accommodation time of at least 5 minutes to adjust itself to the darkness before the inspection begins. The ultraviolet lamp reaches its full intensity after about 15 minutes. There should then be a light intensity of about 500-800 $\mu W/cm^2$. The ultraviolet radiation from the lamp is normally within the wavelength range from 320 and 400 nm. Ultraviolet radiation at these wavelengths is not harmful to human skin or eyes.

Ultraviolet lamps with cracks in the filter glass should not be used, for in such cases damaging radiation may be present. When inspecting indications of defects it is often advantageous to use a magnifying glass. However, one should not use eyeglasses which have been specially treated so that they darken when they are exposed to bright illumination. When ultraviolet lamps are used, the eyes can be protected with contrast glasses with absorbing glass to avoid a temporary impairment in the observer's vision.

Penetrants of different manufacturers and type are dissolved differently by the developer. If too much colour is dissolved in the developer and it spreads out too much, it can be difficult to locate faults precisely and to make a judgement about the type of defect. It is therefore an advantage to repeat the inspection at intervals during the development process.

If one is in doubt about whether a particular indication which is observed is relevant or not, then it may be necessary to repeat the examination. A relevant indication comes from a real defect, while a false indication may be due to inadequate cleansing of the penetrant before development. A relevant indication will look just the same when the examination is repeated, while a false indication will not appear again.

When repeating an inspection it is necessary to repeat every step in the process and to be particularly careful about cleaning before new developer is applied. If one elects to use a different type of penetrant for the re-examination, one must be very careful that the original penetrant has been completely removed from the defects. In particular it should be noted that remnants of a coloured penetrant can react with a fluorescent penetrant and cause a suppression of the fluorescent effect.

Safety and environmental requirements must of course be observed. The examination materials should be used with great care. Highly volatile organic solvents and additives can to some degree be poisonous and very flammable. One must always be sure to provide good ventilation during the examination.

A detailed description of the procedure for a dye penetrant examination is not provided here. Only some of the considerations which must be taken into account to achieve a satisfactory result are described. A flow chart overview of typical steps in such an examination with various types of liquid is shown in Figure 4.3.

Reference components

There are several types of reference components with artificially generated cracks of known dimensions which can be used to check whether the system of penetrants used works satisfactorily. They can also be used if one is going to work with low temperatures.

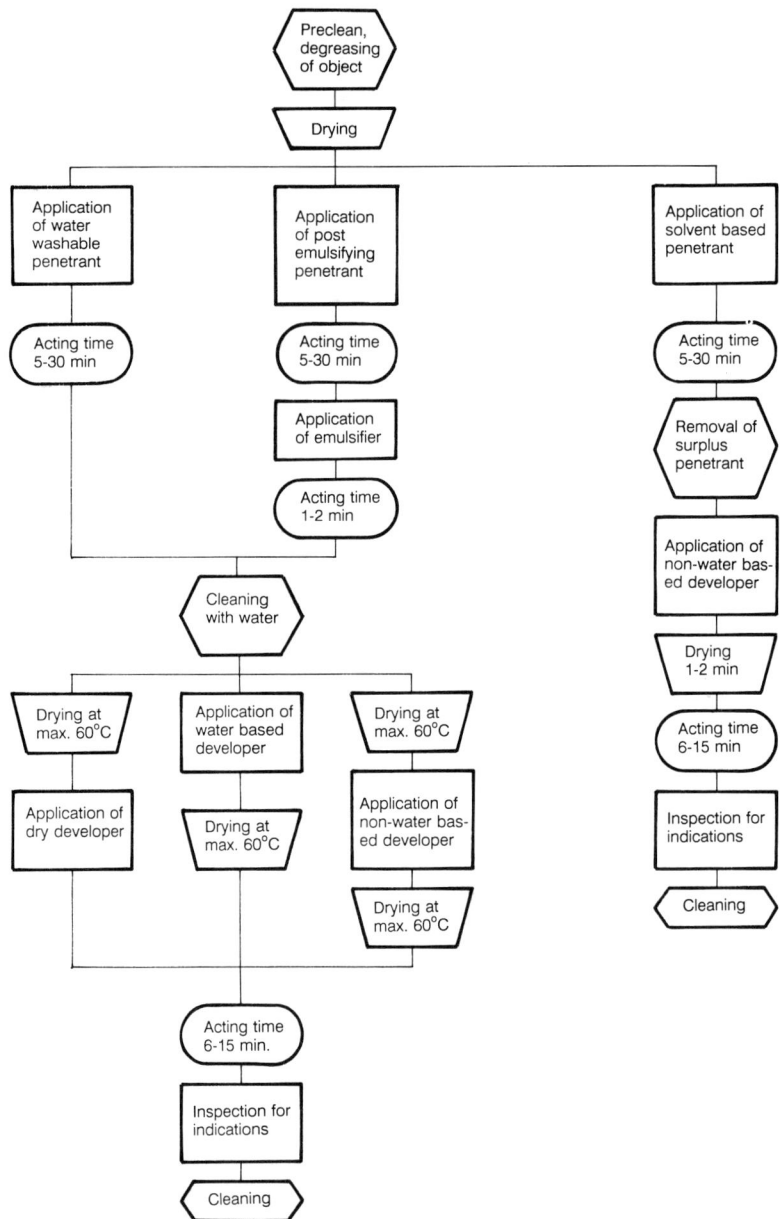

Figure 4.3. Overview flow chart for the dye penetrant examination process.

Standards

Within each set of norms there are standards which deal with dye penetrant examination. In Denmark the standard is DS/ISO 3879: *Welding. Recommended standard operating procedures for dye penetrant examinations*. In Great Britain the standard is BS 4416: *Penetrant testing of welded or brazed joints in metals*.

Practical applications

The method is used not only for the detection of surface defects but also to find leaks as e.g. shown in Figure 4.4, which illustrates the testing of an open tank. Leak testing can in fact take some time. In a case where a 1.5 m² diameter frying pan from an institutional kitchen was to be tested, it turned out that after repeated applications of capillary liquid and several hours of waiting there were apparently no leaks. Not before the next morning had some of the liquid penetrated through the several centimetre thick bottom.

Figure 4.5 shows a brake drum with a crack. Figure 4.6 shows a drive shaft with a crack, and Figure 4.7 shows a circular crack in a flywheel.

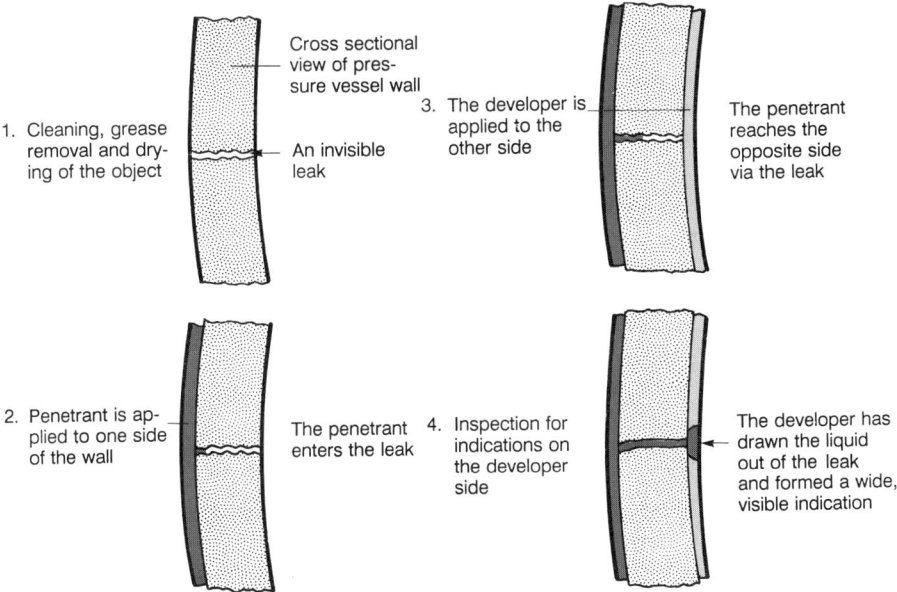

Figure 4.4. Leakage testing using dye penetrant.
1) Cleaning, grease removal and drying of the component.
2) Penetration liquid is applied to one side of the component.
3) The developer is applied to the other side.
4) Inspection for indications on the developer side.

Dye penetrant examination

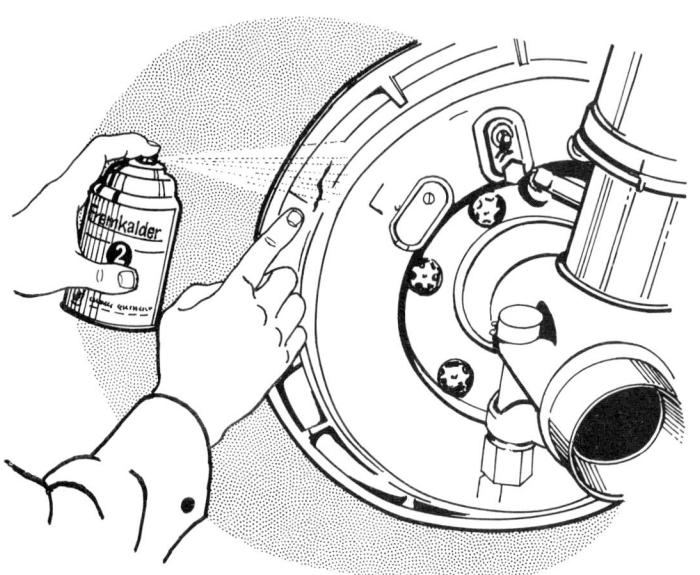

Figure 4.5. A brake drum with a crack.

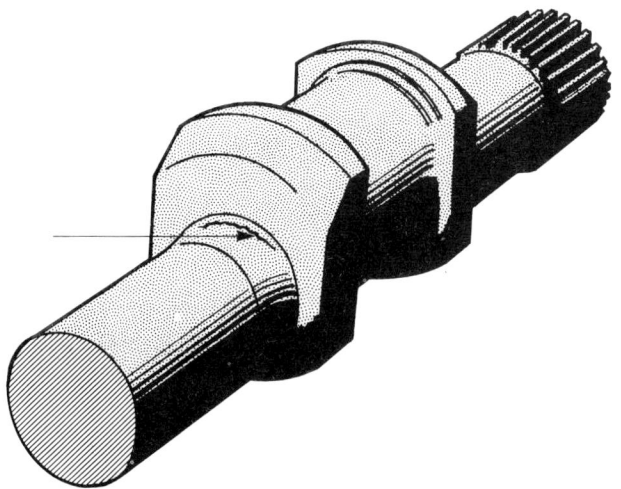

Figure 4.6. Driveshaft with a crack.

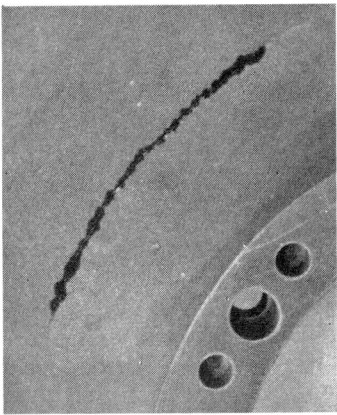

Figure 4.7. Circular crack in a flywheel.

Training required/desired

The method is generally regarded as being so simple that the NDE operator should be able to train himself. If the operator is to have an all-round background, however, a day or two of training to learn more about the details of the method is necessary. In addition some experience is required to be able to make proper judgments about the indications. If the operator is to acquire a certificate attesting to his or her qualifications, then he or she can be trained at special courses which are periodically held for this purpose.

Alli Fabrin

References

1) *Programmed Instruction Handbook Non-destructive Testing, PI-4-2 Liquid Penetrant Testing.* General Dynamics, Convair Division.

2) *Oförstörande provning. Sandvikens handbok.* (Non-destructive testing Handbook available from Sandviken AB in Sweden.)

Chapter 5
Eddy current testing

NDE principle

Eddy current testing is an electromagnetic testing technique which can be used on all electrically conducting materials.

The term *inductive testing* is sometimes used as an alternative designation for eddy current testing.

Among the uses for the method are the following:

- Crack detection
- Corrosion and material thickness measurements
- Sorting material
- Identification of heat-affected areas
- Coating thickness measurement
- Electrical conductivity measurement
- Metal detection

The test equipment consists of three components: a generator, a test coil and recording equipment, e.g. a galvanometer or an oscilloscope.

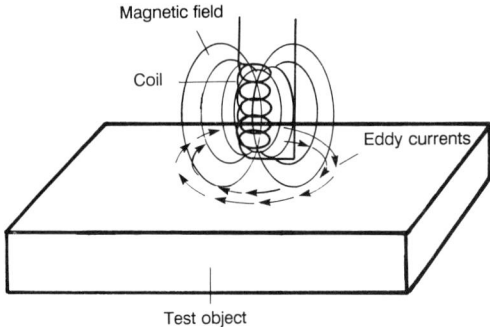

Figure 5.1. Induction of eddy currents.

Eddy currents

An alternating current is produced in the generator and sent through the test coil. When this current passes through the test coil, a magnetic field, *the primary field*, arises surrounding the coil. This magnetic field induces eddy currents in the test object when the test coil is held above it (see Figure 5.1). At the same time eddy currents in the test object will induce a magnetic field, *the secondary field*, which is opposite to the primary field. The intensity of the secondary field compared with the primary field is dependent, among other factors, upon the electrical and magnetic properties of the test object.

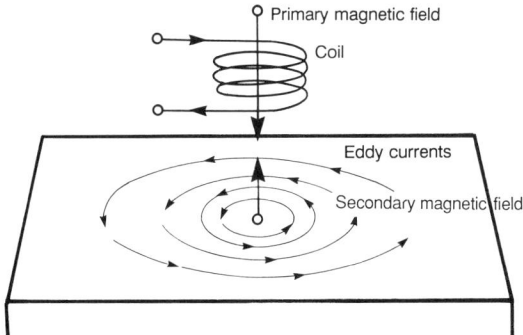

Figure 5.2. The primary magnetic field H_p and the secondary magnetic field H_s.

If a galvanometer is attached to the test coil, then it will show a standard reading which is characteristic for the state of the test object. If the galvanometer is zeroed, it will remain zeroed as the test coil is moved across the test object, provided that the eddy currents can act unhindered in the object. If the currents change value, the magnitude of the secondary field will also change. This will then change the intensity of the primary field in the test coil and thereby change the current through it, producing a deviation in the reading on the galvanometer.

The value of the eddy currents is affected among other things by changes in the electrical and magnetic properties of the test object: cracks or cavities in the test object and the distance from the test coil to the object. It is these properties which are utilized to perform eddy current measurements.

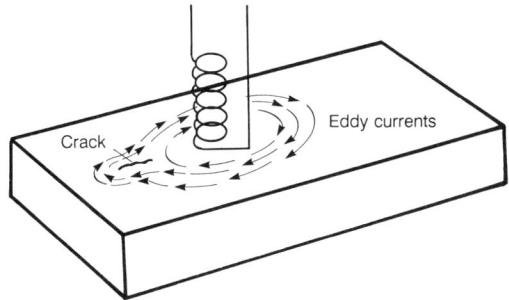

Figure 5.3. Changes in the magnetic field during passage of a crack.

Eddy current testing

Advantages to eddy current testing

Eddy current testing has a number of advantages compared with other methods used to detect cracks. Among these are the following:

- The method is quick to use, for pre- and post-treatment of the test object is not required.

- The method does not require direct contact between the test coil and the test object, which means among other things that the method can be used over painted surfaces. It is also possible to use the method under water.

- The method is inexpensive to use, for there are no materials to be consumed.

When checking a series of items, e.g. on an assembly line, the method can be automated.

Development of the method

During the period 1775-1900 scientists like Coulomb, Ampère, Faraday, Ørsted, Arago, Maxwell and Kelvin discovered most of the basic facts about electricity and magnetism which we possess today. It is this knowledge which forms the basis of the principles of electromagnetic non-destructive examination which are used today.

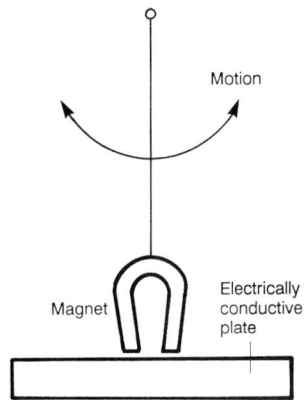

Figure 5.4. Arago's pendulum.

In 1824 Arago discovered that the oscillations of a magnetized pendulum were strongly damped when it came close to an unmagnetized, electrically conducting material. In 1820 Ørsted discovered the magnetic field which is present around an electrical conductor through which a current passes. The same year Ampère observed that two currents flowing in opposite directions through two parallel conductors which are close to one another will cause the magnetic fields to cancel. Faraday discovered the principles of electromagnetic induction in 1831. Maxwell gathered these and other observations in a two volume work which was published in 1873. Maxwell's equations for magnetism and electromagnetism are still used as the foundation for research in these fields.

Quite a few years passed before this knowledge was applied in connection with non-destructive testing. In the 1930s and 40s substantial progress was made in this area. One of the prime movers in the field was the German Dr. Förster. He carried out many significant experiments during this period and formulated theories for eddy current testing and also constructed completely new types of equipment for these procedures.

Since Dr Förster constructed his first equipment there has been a dramatic evolution within the field of eddy current testing. In particular the developments during the time period 1975-85 have been phenomenal, both with respect to equipment and with respect to applications of the method. In the mid-1980s the first generation of microprocessor-based equipment began to appear on the market. It is possible to store calibrated data, reference and defect indications.

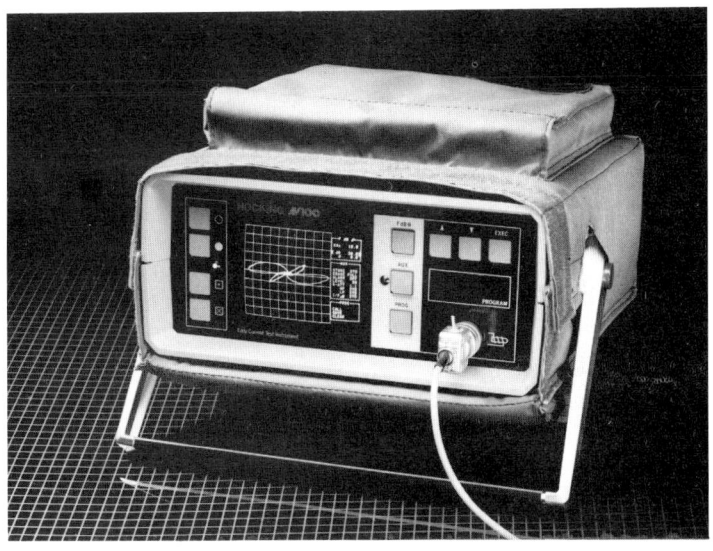

Figure 5.5. Microprocessor-based equipment with a wide range of application possibilities.

A wide range of different types of equipment are available depending upon the area of application. The equipment differs primarily with respect to physical design, for the principles of operation are identical for the various devices. There is also a tendency to produce equipment with a wide range of facilities which can be used for many different types of examination.

The test coil is placed in a measuring probe which determines the application of the equipment. Probes can be classified into three categories:

- *Surface probes* which are used for crack detection and corrosion measurements on machine components, etc.
- *Internal bobbin probes* which are used to check heat exchanger tubing.
- *Encircling probes* which are used primarily for production control of wire, etc.

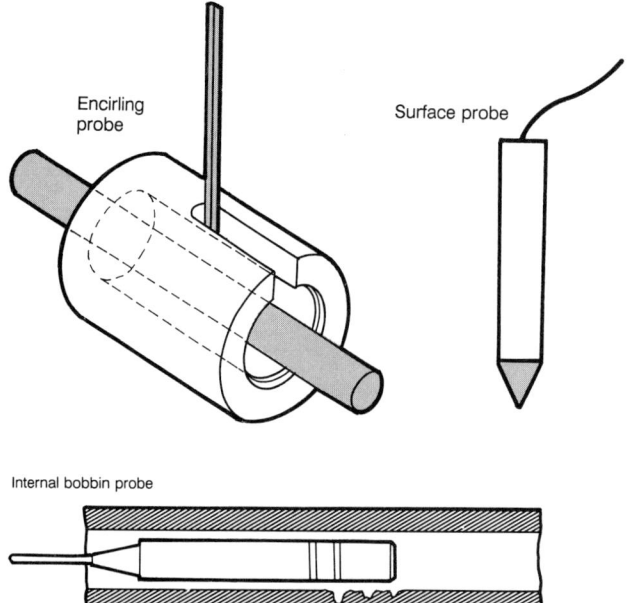

Figure 5.6. Various types of probes.

Surface probes

Surface probes are available in a wide range of shapes designed for specific types of examination. In general one distinguishes between high- and low-frequency surface probes.

High-frequency probes are used to detect defects on the surface and just below the surface of the test object. On the other hand *low-frequency probes* are used to detect corrosion or cracks deeper down in the material of the test object.

Thus it is the frequency of the alternating current which determines how deeply the eddy currents penetrate a given material and thereby where defects can be detected in the test object. Figure 5.7 shows a diagram of the penetration depth of eddy currents for a wide range of materials as a function of the frequency and the conductivity of the material indicated in % IACS (International Annealed Copper Standard).

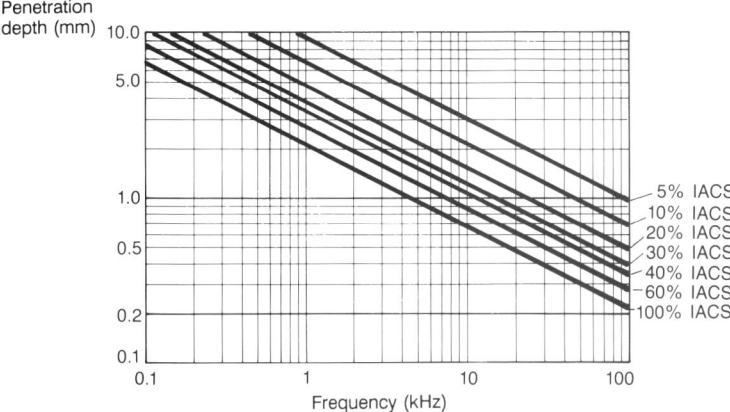

Figure 5.7. Penetration depth as a function of frequency and conductivity.

The examination itself is carried out by the inspector passing the measuring probe over the surface of the object after the equipment has been calibrated by means of a calibration block containing known and well-defined cracks.

Any cracks or other defects in the test object will affect the eddy currents, whereby the instrument readout will show a deflection.

Internal bobbin probes

Internal bobbin probes are primarily used for the examination of tubes in heat exchangers. The examination of heat exchanger tubes is one of the traditional areas of application for eddy current testing. The method has been used for safety checks of heat exchangers for many years. Within the past decade use of the method has increased dramatically, a phenomenon which is due, among other things, to a reduction in the time required to perform the inspection due to the technological evolution of the equipment used. At the same time the philosophy of preventive maintenance has gained acceptance, and regular condition monitoring has become more and more common.

The examination of a heat exchanger is undertaken by sending a measuring probe through each tube at a constant speed. While it moves it will be affected by any defects in or on the tube.

In order to reduce the time required for the examination, the probe is sent into the tube at high speed, and immediately after it comes out at the opposite end it is sent back again at the desired examination speed, normally 0.5-1.0 m/s.

While the probe is drawn back and the examination is made, a chart recorder or a tape recorder is run to record the results of the measurement automatically. These can be evaluated later, since it is not possible to judge the state of the tube in detail in real time due to the high examination speed. However it is possible for the inspector to register large defects which require additional special examination.

The examination is normally conducted by two inspectors who can examine, depending upon the conditions in each case, up to 1000 tubes 6 metres in length daily.

Figure 5.8. Eddy current examination of a heat exchanger.

The following overview diagram illustrates the equipment used:

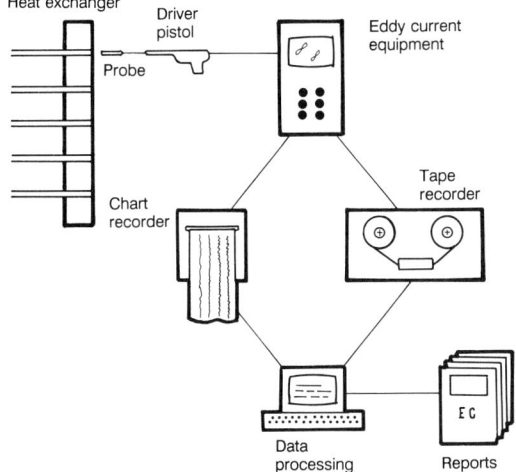

Figure 5.9. Overview diagram, eddy current examination of a heat exchanger.

Since the measurements are compared with indications from a calibration tube containing artificially generated defects, it can be desirable to verify evaluations of some of the defects found. This can for example be accomplished by removing some tubes and then cutting them up to perform visual and mechanical evaluations of the defects. It is not usually possible to remove tubes from a heat exchanger. In such cases a visual evaluation of the defect can be carried out by means of a video endoscope.

Measurements of the thickness of the tube walls can be verified by means of a small ultrasonic head, whereby it is possible to undertake measurements of the wall thickness at certain points in the tube.

A rapid evolution of the equipment used to perform eddy current testing of tubes has occurred. Thus one of the greatest limitations of the method has been overcome. The method has been limited up to now to non-magnetic tubes, but equipment is currently available on the market which can be used to examine magnetic tubes by saturating the tube magnetically with a direct current, so that the material reacts as non-magnetic material does when the measurement is performed.

Another new development is the rotating probe which can be pulled through the tubes. These probes measure just a little region of the circumference of the tube, and thus they can locate very small defects and determine their exact location on the circumference. Furthermore any eccentricity of the reduction in wall thickness can be detected.

Encircling probes

Encircling probes are used mainly for automatic production control, where the product passes through a coil or passes a surface probe, e.g. during the production of tubes, wires and similar products.

The principle of the examination itself does not differ from that previously discussed. Equipment for production monitoring is normally connected to alarm systems which are activated by defects in the product, so that the apparatus can be used automatically on the production line. The defective region is often marked e.g. by means of spray paint.

Areas of application

Surface probes

Crack detection with high-frequency equipment can be carried out on all electrically conducting objects which have a reasonably smooth surface. The method can detect cracks which are open to the surface as well as cracks which lie just below it.

Eddy current examinations for cracks are often undertaken as routine examinations of components where there is a risk that fatigue cracks will be formed. It is also used for production monitoring and for the examination of components which have been exposed to excessive stress.

The method is widely used in the aviation industry to examine aircraft wheels, landing gear components, fittings, turbine blades, etc. In addition the method is becoming more and more widely used in the machine industry and as a condition monitoring method at power plants, processing plants and on large steel constructions.

Low-frequency equipment can be used to examine non-magnetic, electrically conductive materials.

Corrosion measurement or crack detection with low-frequency equipment is mainly used in the aviation industry where constructions consisting of many layers of plates are quite common. Because eddy currents penetrate deeply into the material at low frequencies, it is possible to detect corrosion or cracks in the underlying plates.

The examination of ferritic welding joints is one of the very new areas of application for eddy current testing. In the discussion of high-frequency equipment, it was mentioned that the application is limited to reasonably smooth surfaces. It has, therefore, not previously been possible to use the method on ferritic welding joints due to the normally rough and uneven character of the surfaces of such joints. But by means of a special new probe design it is now possible to reduce the "noise" from the roughness in the welding joints to a level where it is possible to achieve a satisfactory ratio between noise and the indications from cracks in the welding joints.

The experience gained from this application demonstrates that the reliability of the method for the detection of cracks in welding joints is just as good as for magnetic powder testing.

This is true of the use of the method for condition monitoring of coated welding joints. Here the use of eddy current testing is significantly quicker and less expensive than magnetic powder testing, which requires a cleaning of the coating before the examination and the application of a new coating of surface protection afterwards.

The areas of application for the method are typically condition monitoring of large steel constructions such as cranes, ships and offshore installations.

In connection with the sorting of materials, the measurement of electrical conductivity and the detection of regions which have been heated, the fact that the intensity of eddy currents depends upon the electrical conductivity of the material is utilized.

Figure 5.10 shows a conductivity curve which would appear on the oscilloscope of eddy current equipment when measuring an array of objects of different metals. As a rule equipment used for the measurement of conductivity permits the conductivity to be read directly as a numerical value on the eddy current readout.

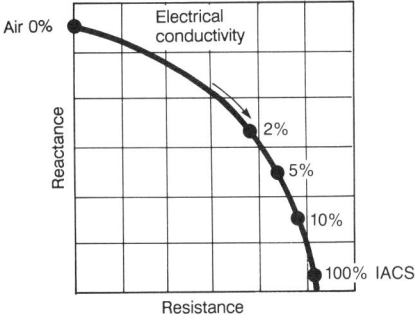

Figure 5.10. Conductivity curve for materials with various conductivities.

It is thus possible to measure the electrical conductivities of various metals as well as to sort out metals which have characteristics which deviate from a known reference or as a check of material delivered before it is worked.

It is also possible to find areas which have been overheated, for the structure of the material is altered, and thus the metallic characteristics change due to heating. For example it may be a question of parts of a construction which have been exposed to a fire or an aircraft wheel rim which has been overheated due to hard braking during a landing.

The intensity of the eddy currents in an object is also dependent upon the distance of the coil from the object. This can be used for the measurement of the thickness of *non-metallic coatings*. Variations in the distance between the coil and the object provides indications which are normally regarded as "noise" in the measurements.

This noise is normally referred to as "lift-off". Figure 5.11 shows examples of lift-off curves for various metals, as they would appear on the oscilloscope when the probe is moved from the air to direct contact with the object. The placement of the graph is thus related to the distance from the coil to the object.

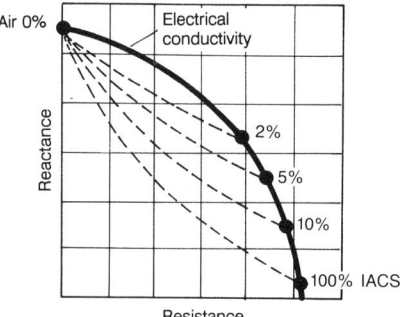

Figure 5.11. Lift-off curves for various metals.

Internal bobbin probes

Eddy current examinations of heat exchangers can be used to detect extensive corrosion or local defects as for example cracks, pitting or holes in the tubes. Such regular checking of a heat exchanger can also provide information about the processes which take place inside it.

In addition to the use of eddy current testing for condition monitoring, it is often used to detect leaks in leaky heat exchangers. Internal bobbin probes usually have diameters in the range 5-50 mm.

Encircling probes

Encircling probes are used mainly for automatic production control, e.g. in connection with the production of pipes, wires and similar items.

Another use of encircling probes is control of wires on cranes, ski-lifts, elevators, towing cables, etc.

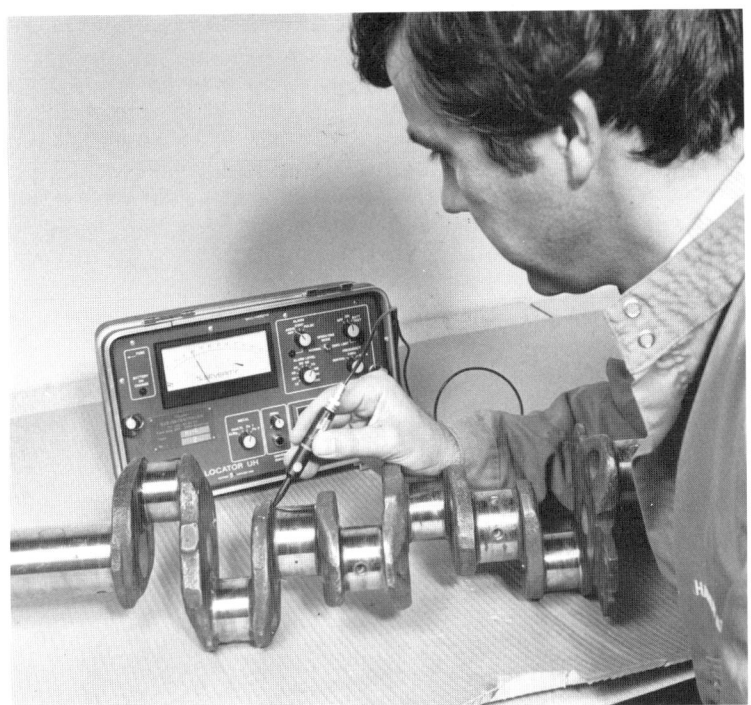

Figure 5.12. Eddy current examination of a driveshaft for cracks. The apparatus is equipped with a panel meter readout.

Practical applications

Surface probes

The examination for cracks (see Figure 5.12) of the driveshaft is carried out with a high-frequency pencil point probe so that it is possible to reach all the nooks and crannies on the object.

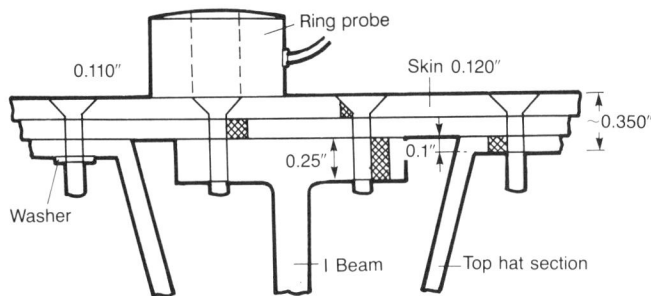

Figure 5.13. Examination of a rivet joint.

Eddy current testing

Figure 5.13 shows the examination of the region around a rivet in an aircraft wing. For this purpose a low-frequency ring probe is used. It is shaped in this manner so that the entire region around the rivet can be examined at once by centring the probe over the rivet.

Figures 5.14 and 5.15 show cracks in a weld and the corresponding defect indication on the eddy current equipment. The defect indication is visible above the horizontal line, while the indications under it are due to the nature of the welding joint.

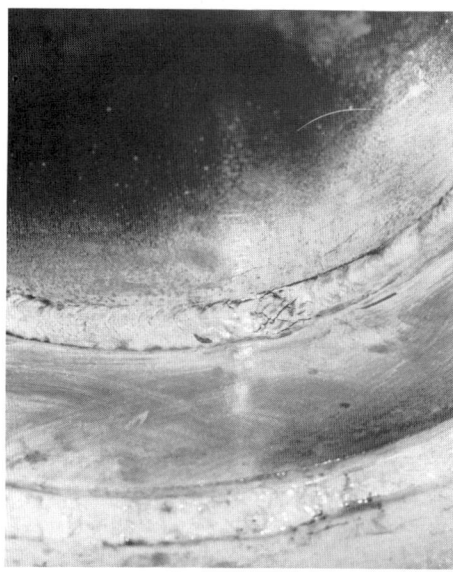

Figure 5.14. Cracks in a weld.

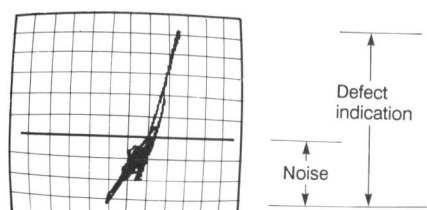

Figure 5.15. Defect indication.

Internal bobbin probes

When the internal bobbin probe is drawn through a tube, the average thickness of the wall of the tube adjacent to the probe is measured as well as any local defects, e.g. cracks and pitting. Furthermore, support plates and tube sheets are also indicated so that it is possible to indicate the linear placement of the defects in the tube with respect to them.

Figure 5.16 shows an example of a typical chart recording with a lengthwise cross section of the tube illustrated beside it.

At the upper tube sheet indications are seen due to the expansion and superimposed on these are indications due to local reductions in the wall thickness. Immediately below the tube sheet there is a slightly conical reduction in the wall thickness. Down through the pipe there are eight indications due to support plates. Between support plates 3 and 4 some local corrosion is indicated, between the 6th and the 7th support plates a conic reduction in the wall thickness begins, where the thickness of the tube wall is reduced from 2.80 mm to 2.62 mm just above the tube sheet in the bottom of the heat exchanger.

The chart recorder registration for each tube thus contains a considerable amount of information which can be used in connection with maintenance. For example if the measurement is undertaken at periodic intervals it is possible to compute the remaining lifetime for the heat exchanger. Furthermore, even minor process changes will often be revealed, for they can cause considerable variation from the known or calculated corrosion rate.

Sometimes up to eight different wall thicknesses are measured per tube, corresponding to about 20,000 values for an ordinary, medium-sized heat exchanger. In order to handle all these data a portable computer is used, making it possible in a brief period of time to create drawings which, with the aid of special symbols, can provide a clear overview of the state of the heat exchanger.

The overview drawing shown in Figure 5.17 shows clearly that about half of the tubes are worn, while there is only minor wear in a few tubes on the other side. For this heat exchanger a consequence of the measurements was that the entry distribution plates were altered, providing a significant extension of the expected lifetime of the heat exchanger.

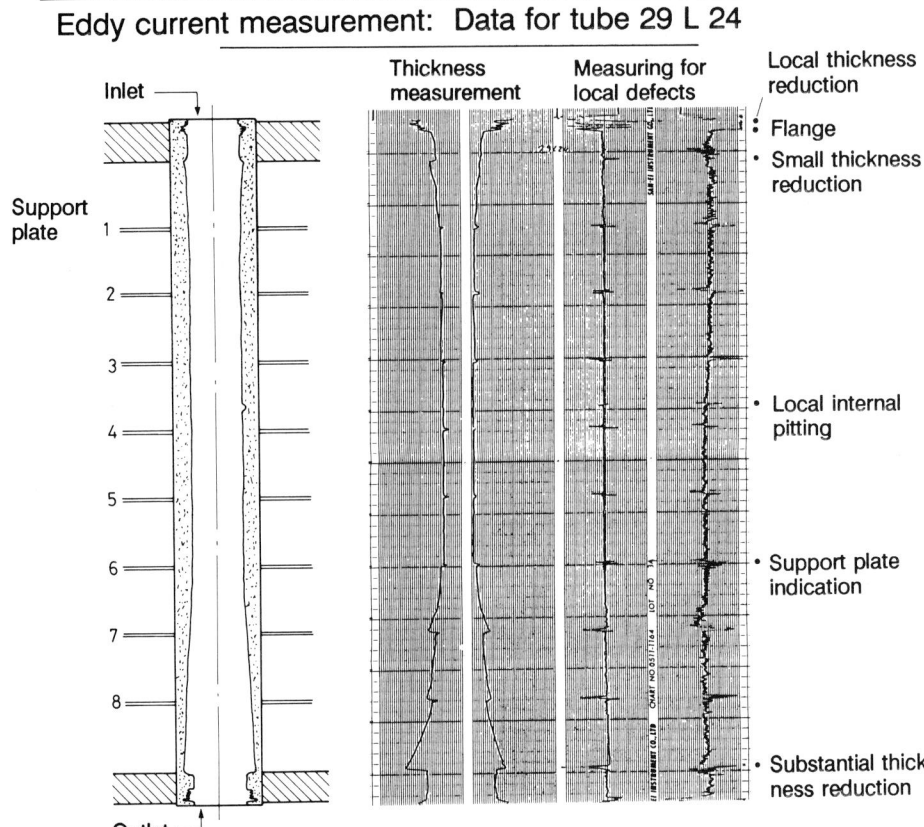

Figure 5.16. An example of a chart recording of an eddy current measurement of a heat exchanger tube.

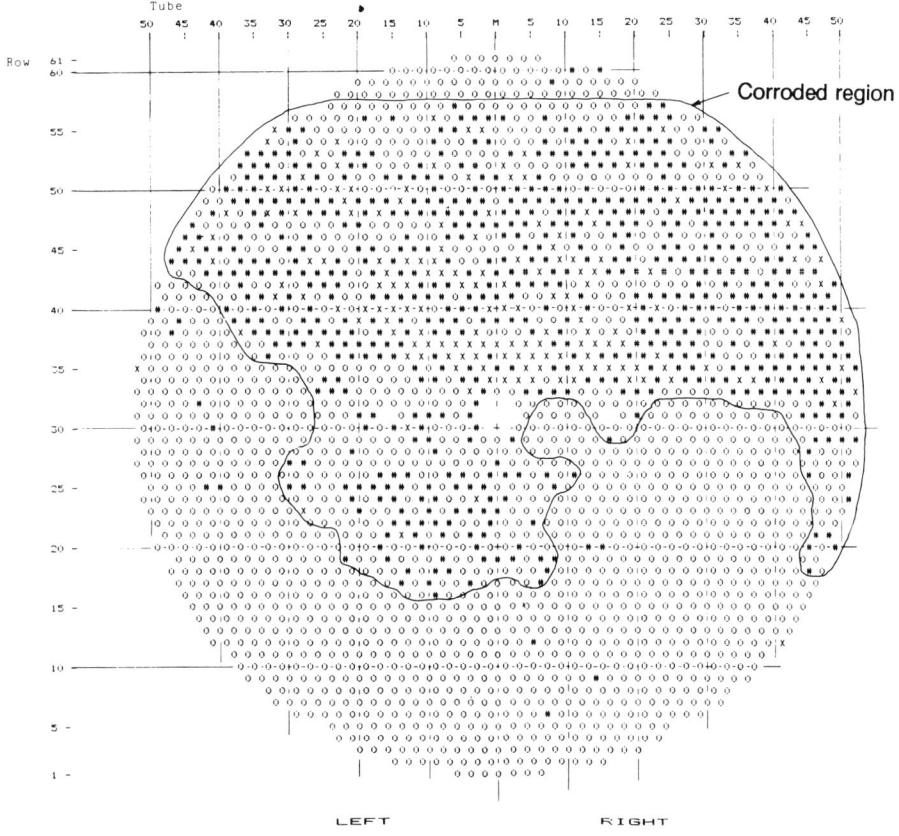

```
0 :    3.65 < WT              nominal material thickness
# :    3.60 < WT ≤ 3.65 mm    modest thickness reduction
X :           WT ≤ 3.60 mm    substantial thickness reduction
```

Figure 5.17. An overview drawing of heat exchanger tubing.

Encircling probes

The examination of wires is carried out by placing a coil around the wire. The coil registers any defects in the wire when the wire runs through the coil or the coil is passed over the wire.

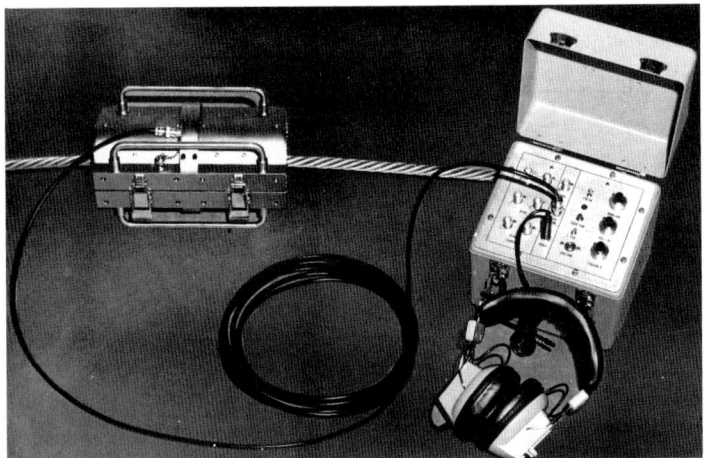

Figure 5.18. Equipment for eddy current examination of wire.

In this fashion it is possible to reveal corrosion as well as broken strands of wire.

Training required/desired

Eddy current examinations of heat exchangers are normally carried out by companies specializing in inspections, a situation which in part is due to the time-consuming and expensive training required for an inspector. In addition periodic refresher courses are needed.

The equipment which is used for this type of testing costs in the order of £50,000-£100,000 and must often be specialized to specific tasks.

It will, however, be feasible to have one's own inspectors trained to carry out simpler, routine tasks in an industrial operation. For example crack detection can be carried out provided the inspections are undertaken often enough to maintain the inspectors' skills in the use of the equipment. A week-long course of instruction would be adequate to provide the necessary background to perform this type of work.

Surface probe equipment can be acquired for about £2,000 while an adequate universal instrument costs about £7,000.

Equipment for automatic production control will normally be delivered and installed by the manufacturer's own specialists. It can then be operated by the company's own personnel after an appropriate period of instruction. Prices for this type of equipment vary widely depending upon the particular application.

Requirements today are becoming more and more strict, in that some inspections must be carried out by certified inspectors. In the area of eddy current testing the guidelines of the American Society for Non-destructive Testing are normally used.

A number of norms and standards for eddy current testing have been published. It would be beyond the scope of this book to name all of them here, for they are often prepared for specific types of examination. Norms and standards are published by, among others, the following organizations:

ASNT	American Society for Non-destructive Testing
DIN	Deutsche Norm
SS	Svensk Standard
BS	British Standard

Michael Svan

References

1) *Eddy Current Testing Programmed Instruction Handbooks*, Volumes I-II. General Dynamics/Convair Division, 1967. Self-study materials regarding eddy current testing.

2) *Introduction to Electromagnetic Non-destructive Test Methods,* H.L. Libby, 1971. John Wiley & Sons, Inc. New York. Theoretical treatment of the electromagnetic testing methods.

3) *ASNT Level III Study Guide, Eddy Current Method,* 1983, American Society for Non-destructive Testing, Inc. Columbus, Ohio. A treatment of the information required to pass a level III certification in eddy current testing in the ASNT certification system.

4) *Various NDT journals*. New applications for eddy current testing are discussed regularly in NDT journals.

5) *Non-destructive Testing Handbook, Vol. 4, Electromagnetic Testing*, American Society for Non-destructive Testing, 1986.a

Chapter 6
Emission spectroscopy

NDE principle

Emission spectroscopy is a technique of qualitative and semi-quantitative analysis of the chemical composition of steel, used predominantly for type identification and sorting.

The equipment includes an analysis unit and power supply (see Figure 6.1). The analysis is achieved by means of an arc which is ignited between a vibrating tungsten electrode and the sample surface. The light which is emitted from the arc is diffracted in a prism system into spectral lines which can be observed through the upper ocular, see Figure 6.2.

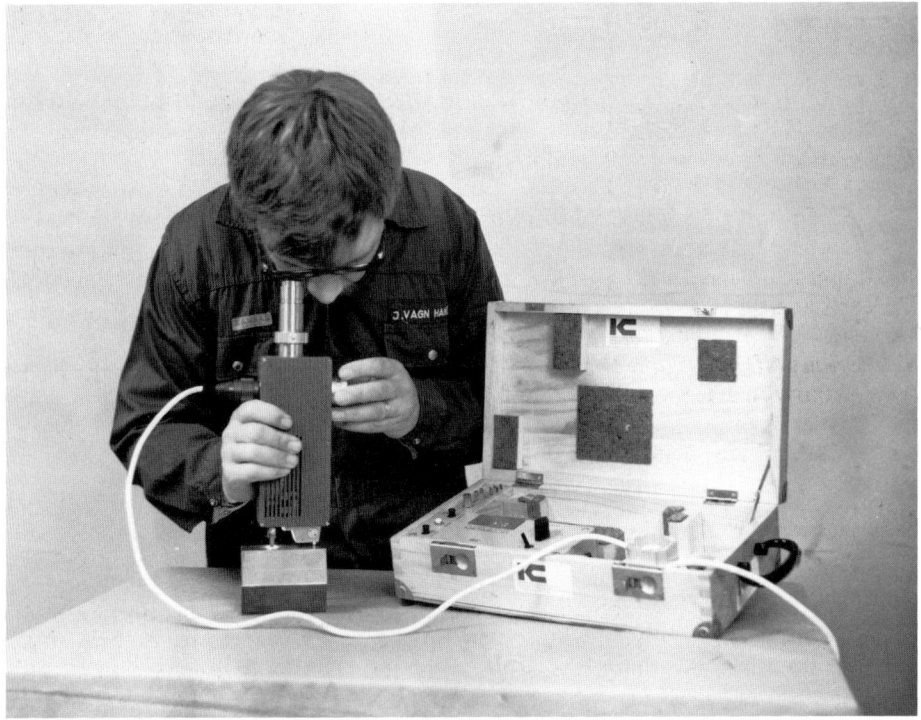

Figure 6.1. Metascope analysis.

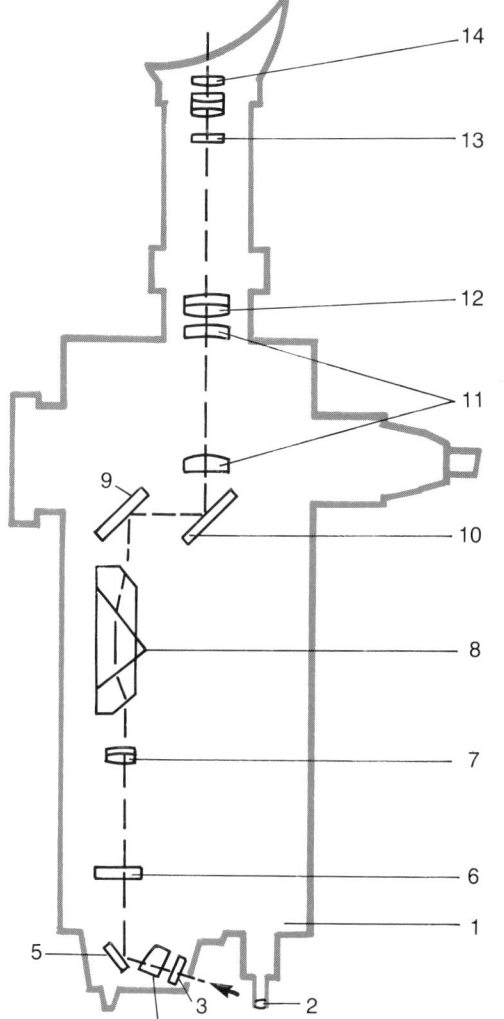

1. Vibration generator
2. Vibrating Tungsten electrode
3. Protective glass
4. Convex lens
5. Illumination mirror
6. Slit iris
7. Collimator
8. Amici Straight-vision prism
9. Mirror
10. Revolving mirror
11. Cylindrical lens system
12. Objective
13. Glass disk with arrow mark
14. Ocular

Figure 6.2: Typical optical path in Metascope analysis equipment.

The spectrum consists of lines for iron which are used as reference lines and of lines which are characteristic of the other elements in the sample. A number of characteristic lines are identifiable directly on the calibrated scale, see Figures 3 and 4. Other lines must be found by the operator based on the known lines and a table of spectral lines in the manual. Samples with known quantities of the element of interest can be a big help when searching for spectral lines for which the scale value is not known.

Emission spectroscopy

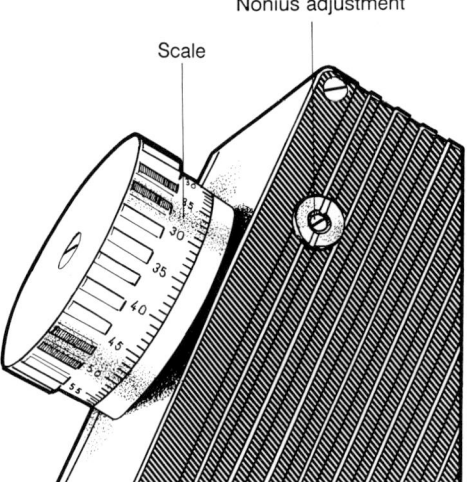

Figure 6.3. Closeup of the scale adjustment dial.

A measure for the quantity of the individual elements is obtained by comparison of the intensities of pairs of lines. The same light intensity in the line pair indicates a given concentration level. Line pairs are tabulated in the manual (see e.g. Figure 6.5). Detection thresholds and tabulated concentration levels, line pairs, are shown in Figure 6.6.

The evaluation of light intensities is most often carried out by the operator directly, but it can be backed up by means of photographs taken with a small image camera, available as an accessory. The equipment can only identify alloyed steel containing the characteristic elements with detection limits above those indicated in Figure 6.6. Among the important alloy elements which can *not* be analysed are carbon (C), sulphur (S), phosphorus (P) and nitrogen (N). Tungsten (W) can not normally be detected simply because the standard electrode is composed of this material. A molybdenum (Mo) electrode is available as an accessory.

ELEMENT	NANO-METRES	SCALE	ELEMENT	NANO-METRES	SCALE
CO 1	486.79	29.3	MN 5	475.40	32.6
CO 2	484.03	30.1	MO 3	553.30	15.2
CO 3	481.35	30.8	MO 4	550.65	15.6
			MO 5	441.23	45.3
CR 2	534.58	18.5			
CR 3	529.83	19.4	NA	589.30	10.0
CR 5,6,7	520.60	21.2	NI 1	547.69	16.1
CR 12	425.44	53.2	NI 2	503.54	25.0
			NI 3	471.44	33.9
CU 3	521.82	21.0			
CU 4	515.32	22.3	PB 2	560.88	14.0
CU 5	510.55	23.4	PB 3	500.54	25.7
			PB 5	405.78	6.1
FE 51	561.56	13.9			
FE 54	558.68	14.3	SN 1	563.17	13.6
FE 56	557.28	14.6	SN 2	452.47	40.6
FE 142	495.75	26.9			
FE 147	492.05	27.9	TI 3	499.95	25.9
FE 151	489.12	28.7	TI 5	498.91	26.1
FE 153	487.17	29.2			
FE 196	452.86	40.4	V 5	488.16	28.9
FE 222	441.51	45.2	V 6	487.55	29.1
FE 224	440.48	45.6	V 10	439.00	46.3
FE 228	438.35	46.6	V 12	437.92	46.8
MG 1	518.36	21.7	W 1	573.51	12.1
MG 2	517.27	21.9	W 2	551.47	15.5
MG 3	516.73	22.0	W 5	430.21	50.6
MN 1	482.35	30.6	ZN 1	636.24	4.8
MN 2	478.34	31.7	ZN 2	481.05	30.9
MN 3	476.64	32.2	ZN 3	472.22	33.6
MN 4	476.24	32.4	ZN 4	468.01	35.0

Figure 6.4. Wavelengths of calibrated spectral lines and corresponding scale values.

Chromium	
R%	Line Pair
0.1	Cr 12 = W 7
0.3	Cr 2 = Fe 87
0.8	Cr 2 = Fe 79
1.0	Cr 3 = Fe 90
1.2	Cr 1 = Fe 90
1.5	Cr 4 = Fe 98
2.4	Cr 12 = Fe 251
5.0	Cr 12 = Fe 249
7.5	Cr 2 = Fe 89
12	Cr 9 = Fe 154
18	Cr 8 = Fe 112
25	Cr 4 = Fe 100

Figure 6.5. Line pairs for semi-quantitative analysis of chromium (Cr).

Emission spectroscopy

Element	Detection Limit in Steel Alloys	Tabulated Analysis Concentrations, Line Pairs, %
Al	<1.0	
Co	2.5	3-5
Cr	0.05	0.1-0.3-0.8-1.0-1.2-1.5-2.4-5.0-7.5-12-18-25
Cu	0.2	0.2-0.25- <1.0
Mg	0.01	0.03-0.08-0.12
Mn	0.2	0.2-0.3-0.4-0.6-0.8-1.0-1.7-13
Mo	0.2	0.1-0.15-0.5-1.0-1.3-2.0-3.0
Na	calibration line	
Ni	1.8	1.5-3.2-4.4-13.0-19.0
Si	>5.0	
Ti	0.1	>0.2-0.6
V	0.05	0.1- >0.2-0.7-1.0-2.1- >3.2
W	1.5	2.0-6.0-9.0-16-25

Figure 6.6. Detection thresholds and tabulated concentration levels for line pairs.

The equipment is, as is apparent from Figure 6.1, very compact. The weight is only about 7.5 kg (16½ lb), of which 5 kg (11 lb) is due to the power supply.

A thorough analysis, e.g. as a basis for a materials certificate, can not be carried out due to the limitations indicated, but an experienced operator can by means of a systematic (and time-consuming) analysis provide a reasonably accurate estimate of the type of steel alloy in an unknown sample. If a number of possible alternatives can be provided in advance, then the analysis can be performed more rapidly and efficiently, as the analysis can be simplified into a matter of sorting the samples.

The equipment can sort steel samples with significant differences in concentration of one or more elements. Known reference samples are of great value in this work, but one must be aware of the danger of contamination of the electrode, which must therefore be cleaned between analysis of the sample and the reference.

A typical job could be verification of the specification or certificate on incoming materials or components or as a check during renovation or in case of damage.

Development of the method

The analysis of the chemical composition of metals has traditionally been carried out by means of wet chemical quantitative analysis in the laboratory. During the past 20-30 years instrumentation has been developed which can quickly provide an analysis of 10-20 of the most important alloys in a simple process. The most common methods are optical emission spectroscopy and X-ray fluorescence analysis.

Optical emission spectrometry is the fundamental principle upon which the Metascope is based, but to achieve the analytical precision of the laboratory, the electrical discharge must take place in a protective gas atmosphere and the optical path must be in a vacuum. In this manner it is also possible to analyse materials containing the lighter elements such as carbon, sodium and boron. The light intensities are recorded automatically and the signals are handled in a microprocessor unit. The level of precision for analysing steel samples by *optical emission spectroscopy* can be determined by doubling the standard deviation, see Figure 6.7.

By co-analysing certified reference samples the actual results can be improved by correction and the exact accuracy can be quantified.

The emission spectrometer can be equipped for copper and light alloys as well. The newest generation of optical emission spectrometers (the Danish Corrosion Centre acquired one in 1989) has an additional feature: *glow discharge equipment*. Glow discharge analysis enables the user to analyse the composition of thin surface layers and to measure concentration gradients by continuously analysing as the surface of a sample is gradually "burned off".

Element	Approximate Concentration, %	Precision, % (95% Confidence), 2s
C	0.06	±0.003
	0.15	0.006
	0.50	0.015
	0.90	0.020
Mn	0.35	0.01
	0.60	0.02
	1.00	0.03
	1.50	0.04
P	0.006	0.001
	0.04	0.002
S	0.005	0.002
	0.04	0.003
	0.06	0.005
Si	0.02	0.005
	0.30	0.008
	0.50	0.015
Ni	0.03	0.003
	0.10	0.004
	0.70	0.01
	1.60	0.04
Cr	0.04	0.003
	0.30	0.01
	0.80	0.02
Sn	0.003	0.0008
	0.02	0.002
	0.05	0.003
V	0.01	0.002
	0.03	0.003
	0.25	0.010
Mo	0.03	0.004
	0.30	0.012
Cu	0.02	0.001
	0.15	0.004
Ti	0.02	0.003
	0.20	0.012
Al	0.006	0.003
	0.02	0.004
	0.07	0.005
Cb/Nb	0.02	0.002
	0.07	0.004
B	0.001	0.0002
	0.07	0.004
Zr	0.05	0.003
Pb	0.01	0.002
Se	0.02	0.003

Figure 6.7. Recommended precision requirements for steel using an optical emission spectrometer. ASTM E 1009-84. Precision equals twice the standard deviation.

X-ray fluorescent analysis is based upon irradiation of a sample with X-rays. Due to the irradiation every element emits radiation at certain characteristic wavelengths (see Figure 6.8a). By analysing the wavelengths after diffraction by means of a crystal lattice, thus determining the amount of radiation at each wavelength, it is possible to compute the composition of the sample (see Figure 6.8b). The method can only be used to identify elements with an atomic number greater than 8. This means for example that carbon and boron can not be identified.

A common characteristic of the laboratory equipment is, however, that it is necessary to remove a sample or a component for analysis.

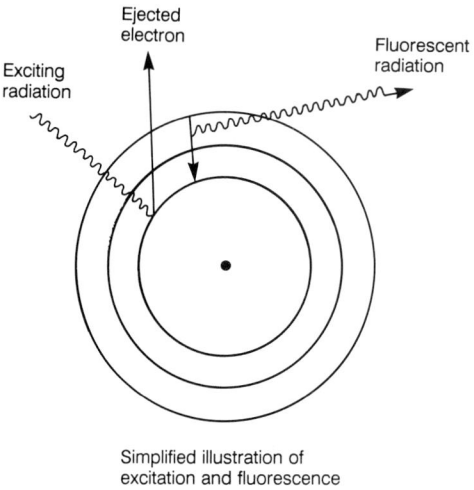

Figure 6.8a. Simplified illustration of excitation and fluorescence.

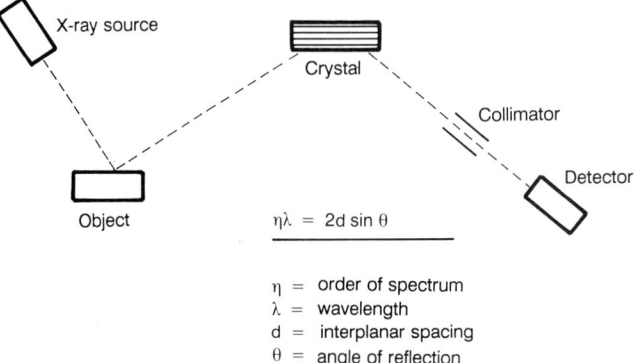

$\eta\lambda = 2d \sin \theta$

η = order of spectrum
λ = wavelength
d = interplanar spacing
θ = angle of reflection

Figure 6.8b: Simplified illustration of X-ray diffraction technique.

Figure 6.9. Analysis using a transportable Baird Spectromobile emission spectrometer.

Transportable emission spectrometers

Optical emission spectroscopy is used not only in the simple Metascope equipment but also in a type of transportable equipment which operates in a manner similar to the Quantovac. In contrast to the analysis with the Quantovac, however, the analysis is carried out by means of a gun with a 5 or 10 metre cable and for some equipment without gas protection. The optical path via the cable to the detectors is not through vacuum. Equipment of this type is supplied among others by Baird and by ARL, as e.g. the Baird Spectromobile illustrated in Figure 6.9. The short term standard deviations for the results of measurements of the same sample are shown in Figure 6.10. In addition to the increased standard deviation in the results, the decisive difference compared to the Quantovac is the relatively large standard

Emission spectroscopy

deviation for the carbon analysis. Carbon analysis with reasonable accuracy can only be carried out on non-alloyed steel. Subsequent experiments performed by one manufacturer have demonstrated that the deviation can be narrowed by placing an argon nozzle around the analysing electrode. Even with this equipment carbon analysis is not presently dependable for high- or low-alloyed steel.

```
Tolerance table of analyses performed on BAIRD Spectromobile MS 3
according to supplier's information, October 1985.

Area of analysis

Element       Low alloy       Stainless       High speed      Mn steel
              steel           steel           tool steel

C             0.05-1.2
Si            0.01-1.0        0.10- 2.0                       0.1  - 1.5
Mn            0.01-2.0        0.50- 2.5       0.1  - 1.2      2.0 -20.0
Ni            0.01-4.5        1.0 -30.0       0.10- 4.5       0.01- 4.0
Cr            0.01-1.0        2.0 -30.0       0.50-15.0       0.01- 4.0
Mo            0.01-1.0        0.10- 4.50      0.10-10.0       0.01- 3.0
V             0.01-0.50       0.01- 5.0       0.10- 5.0       0.01-0.50
Ti            0.01-0.30       0.01- 2.0
Nb            0.01-1.0        0.01- 5.0
Cu            0.01-0.60       0.01- 0.50                      0.01- 1.0
Al            0.01-0.50       0.01- 2.0                       0.01-0.20
W             0.05-0.50                       0.10-20.0
Co                            0.01- 5.0       0.10-20.0
```

Standard deviation of measurements
C ±0.05% absolute

Others, expressed as relative standard deviation, RSD

```
Concentrations    RSD
>1%               2- 5%
0.1 -1.0%         4- 8%
0.01-0.1%         10-20%
```

Figure 6.10. Tolerance values for the Baird type transportable emission spectrometer.

In addition to general analysis facilities the equipment also offers the capability of sorting, i.e. comparison with precoded standards and with the option of programming a rejection level with optical or acoustic alarms at the rejection level.

Portable X-ray fluorescence analysis equipment

Portable versions of X-ray fluorescence analysis equipment is also available, e.g. Texas Nuclear: Alloy Analyser. The equipment which has a total weight of about 6 kg (13.2 lb), of which 2.3 kg (5 lb) is due to the test head, is shown in Figure 6.11. Detection limits and standard deviations of individual results are shown in Figure 6.12. The deviations can be reduced by increasing the analysis time by a factor of 3 or 4.

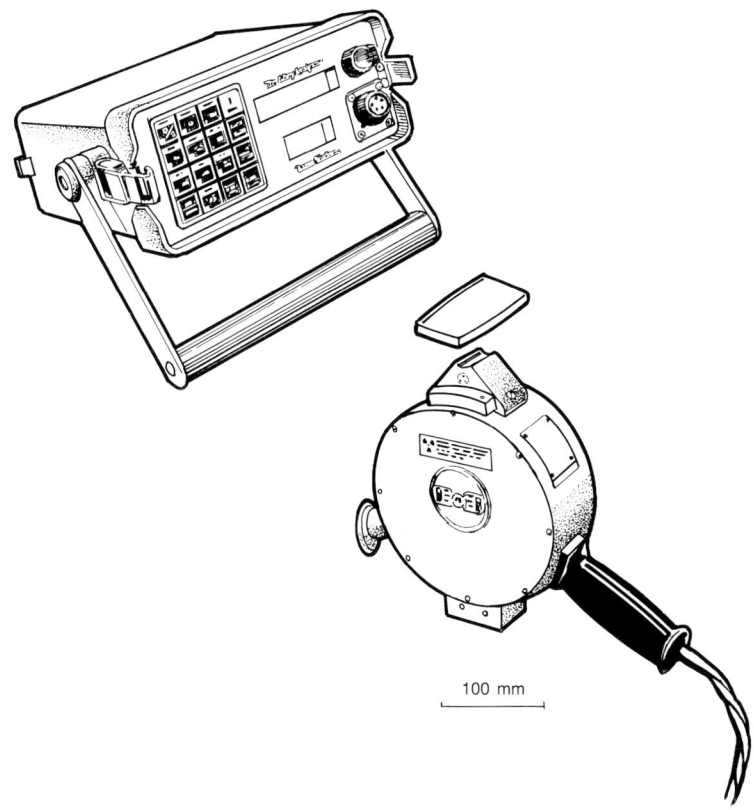

Figure 6.11. Texas Nuclear: Alloy Analyser portable fluorescent analysis equipment.

The equipment has analysis options for individual elements as well as the capability to classify about 100 different standard qualities. The equipment can be used for the analysis of copper alloys as well as steel.

Standard analysis, the precision of single elements can in some instances be increased by increasing the time of the analysis.

Element	SD, absolute ±%, also indicates detection limit		
	Low alloy steel	Stainless steel	Nickel alloys
Mn	0.3	0.3	0.095
Mi	0.9	0.6	0.22
Cr	0.1	0.3	0.35
Mo	0.02	0.025	0.04
V	0.02	0.02	0.04
Ti	0.02	0.03	0.04
Nb	0.02	0.03	0.05
Cu	0.07	0.5	0.4
W	0.04	0.04	0.05
Co	0.35	0.5	0.6
F	0.55	0.55	0.05

Figure 6.12. Tolerance table for analyses performed by a Texax Nuclear: Alloy Analyser according to manufacturers information dated 1982.

Emission spectroscopy

In conclusion it can be mentioned that new equipment for EDAX (Energy Dispersive Analysis by X-ray) either in the form of X-ray equipment or as a Scanning Electron Microscope (SEM) has the capability to identify and semi-quantitatively analyse small shavings and particles. Even samples in diamond impregnated abrasive tape can provide semiquantitative analysis.

Areas of application

Metascope analysis can be undertaken on all types of steel which contain alloy elements above a certain minimum, i.e. low- and high-alloyed steel, see Figure 6.6. The analysis can be carried out on a cleaned surface, or where surface layers with a different composition have been removed. In hardened samples a minimum thickness must be specified to ensure that cleaning and surface removal of arcburns to about 0.5 mm will not bring the thickness of the material below a permissible minimum.

The equipment is used, as already mentioned, primarily for sorting. Furthermore analysis can be used to check the chemical composition of samples or components for which the removal of samples is not permissible, or impractical.

The limited accuracy of the analysis, see Figure 6.6, means that it is not possible to use the technique for certification of materials as such. It is however possible to ascertain the likelihood or to accomplish reasonable verification, particularly if known reference samples are used. As a rule this will be adequate documentation for a customer/builder/inspector or supervisory authority, provided that there is no reason to suspect that an obscure steel quality has been used. By obscure steel quality in this case we mean a steel quality with the correct concentration of alloy elements but with an incorrect content of, say, carbon or high concentrations of impurities such as sulphur and phosphorus.

Within the areas of application outlined here, the more advanced transportable emission spectrometers and X-ray fluorescence analysis equipment offer better analysis options. But particularly compared with emission spectrometers the Metascope has an advantage due to its size in connection with examinations of large, complex systems. A decisive factor when choosing equipment is, however, often the price, which is substantially higher for both the large emission spectrometers and X-ray fluorescence equipment.

Practical applications

Metascope analysis has been used for sorting, e.g. in connection with the auditing of refineries, where entry checks were made on new piping. Preserved parts of the plant were also checked. In certain cases such auditing has revealed that a repair had been done using incorrect – but readily available – steel qualities.

In a machine workshop which has specialized in items which are fabricated of special tool steel, confusion of the raw material can cause substantial liability suits and the loss of customers and reputation. An entry check of the items received from the wholesaler was therefore desired. On this basis an appropriate marking of the items can then be carried out. With a set of reference samples the Metascope would be well-suited to the task.

During the construction of a power plant, for example, the traceability of pipes can be lost because stamps are transferred without approval (the stamp of the supervisory authority), or the markings are cut or polished away while the pipe is being fitted. As a rule certificates are available which according to the builder correspond to the pipes of interest. By means of analyses based on known reference samples, the claims of the builder can be substantiated or rejected with adequate certainty.

Training required/desired

An operator should go through an adequate training period to provide knowledge of how to handle the equipment and to provide adequate experience to recognize the capabilities and the limitations of the equipment. The operator should also have adequate knowledge of which alternative methods of analysis are available which could solve a given analysis problem.

It is imperative for achievement of accurate results with the Metascope that the operator gain adequate experience and maintain his or her routine with the instrument.

J. Vagn Hansen

References

1) *Kemisk analyse af metalliske materialer (Chemical analysis of metalic materials).* Finn Kristensen. DMS Winter Meeting, 1986.

2) *Ikke-destruktiv metallurgi (Non-destructive metallurgy).* J. Vagn Hansen. DMS Winter Meeting, 1986.

3) *Materials identification in the field.* B. Ostrofsky, **Materials Evaluation**, August 1978.

Chapter 7

Endoscopy

NDE principle

An endoscope is an optical instrument which is used for visual inspection of interior surfaces in tubes, holes or other hard-to-reach places, see Figure 7.1.

Figure 7.1. An endoscope can be used for the visual inspection of hard-to-reach locations. (Reproduced with the permission of Polack A/S.)

Endoscopes can be grouped into two main categories, namely the stiff *borescopes* and the flexible *fibrescopes*.

It should be noted that other terms may be used for endoscopes: intrascope, motoscope, technoscope, autoscope, varioscope, etc.

Stiff borescopes

The stiff borescopes can be compared to a periscope, where the objective is placed quite near the object or detail to be examined, while the ocular is placed at the desired distance from the objective.

The objective and the ocular are connected by means of one or more removable extension tubes. The borescope's length can thus be varied as required.

The image is transferred from the objective to the ocular by means of an optical system which is built into the tube.

The illumination of the surface to be observed is achieved by means of a 6-15 V halogen lamp of from 20-150 W which is contained in the objective tube.

The above mentioned borescopes are available in diameters in the range 9-40 mm and in lengths in the range 0.5-30 metres.

In short borescopes the light may also be transferred by means of a bundle of optical fibres which thus serve as a light guide. The light guide surrounds the lens system itself and is in principle equivalent to that of the fibrescopes which are discussed later under "flexible fibrescopes".

The above-mentioned short borescopes cannot be extended and must therefore be ordered with the desired length. They are available in diameters in the range 1.7-18 mm and in lengths in the range 50-3500 mm.

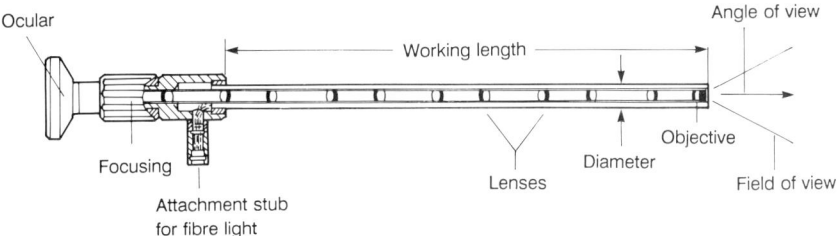

Figure 7.2. Borescope with lenses and optical fibre light guide.

Flexible fibrescopes

Flexible fibrescopes can, in contrast to the stiff borescopes, be inserted into curved pipes and cavities. The light in the fibrescopes is transmitted via ultrathin optical fibres with a diameter as small as 7 μm (0,007 mm).

A fibrescope contains up to 120,000 fibres, each consisting of two layers of glass, namely a core of glass surrounded by an outer layer of glass of differing refractive index. When the light strikes the interface between the core material and the outer layer, a nearly perfect internal reflection occurs. In this way the light is guided through each individual fibre, see Figure 7.3.

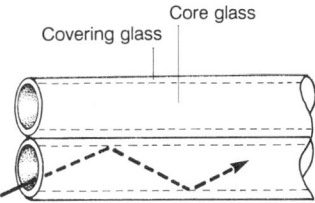

Figure 7.3. The ray of light is subject to total internal reflection during its passage through the optical fibre.

The fibres are separated into two bundles, one for transmitting light, the other for transmitting the image. The optical fibres are placed like a cloak outermost, and their function is to transmit the light from a cold light generator to the object which is to be perceived through the fibrescope. It terminates in one or more "projectors" surrounding the objective so that the object can be seen without awkward shadows. The individual threads of glass are not well-ordered, for their mission is exclusively the transmission of light to the object. The image-forming bundle on the other hand is well-ordered. Thus every one of the 120,000 threads of glass has exactly the same relative position in both ends, i.e. both at the objective and at the ocular. The image is thus composed of 120,000 points of light (pixels) just like the image on a video monitor.

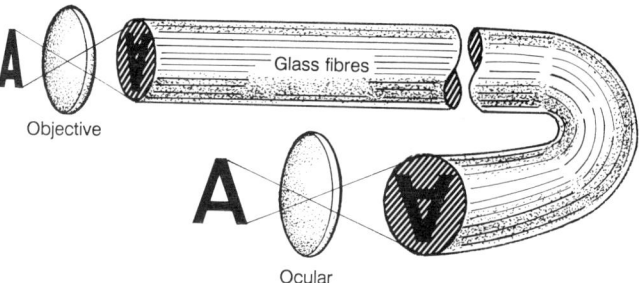

Figure 7.4. Image transfer through a flexible bundle of fibres.

Both borescopes and fibrescopes can be supplied with distance adjustments just like an ordinary camera, so that the image can be focused sharply at any distance from a few millimetres up to infinity. When adjusting to very small distances the image is magnified, so that even very small details can be inspected.

The flexible fibrescope can be supplied in diameters ranging from about 1.7 to 13.5 mm and in lengths from about 300 to 6000 mm.

Video endoscope

The video endoscope represents a further development of the borescopes and fibrescopes discussed previously. Video endoscope equipment operates on the principle that a small electronic sensor detects the image and transmits the signals recorded to a video processor from which the signal is sent to a colour TV monitor. The electronic sensor is placed at the end of a flexible fibre optics cable. The cable can be obtained in lengths up to 30 metres and with a diameter down to 6 mm.

In addition to providing a large, sharp colour image – the equipment is capable of automatic focus – it is possible to type in a text in the image field, e.g. to indicate the location of the examination, the object, etc. The date and the time for the examination are shown automatically on the screen. It is also possible to freeze the image on the monitor so that details can be studied at leisure.

Figure 7.5. Video endoscope equipment with video tape recorder.

Documentation

A large selection of accessories is available to aid in the documentation of examinations using the borescopes, fibrescopes and video endoscope equipment mentioned above, e.g. photographic, TV or video equipment. A video tape recorder can be directly attached to the video endoscope.

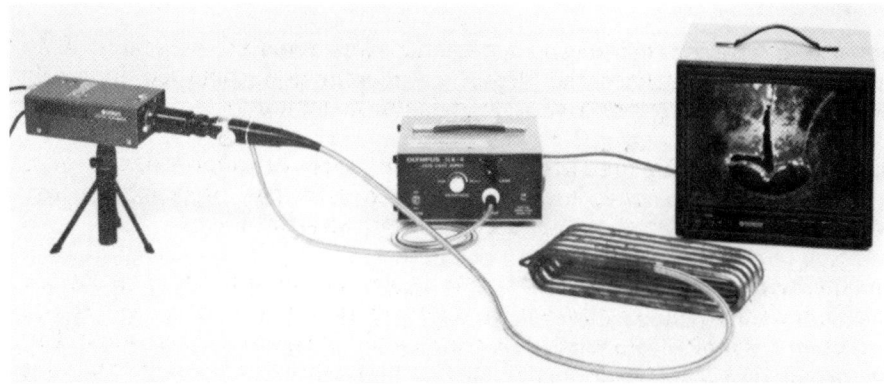

Figure 7.6. Fibrescope with cold light generator, TV camera and monitor.

The rapid evolution of video technology has – in the space of just a few years – resulted in lightweight and sensitive TV cameras which have the capability to document subjects in motion with simultaneous activation of a video tape on which comments can be recorded in real time.

Prices

Endoscope equipment costs from £1,500 to £20,000, depending upon whether the equipment is a borescope, fibrescope or a video endoscope.

The equipment of most well-known manufacturers is based on modular systems which can be expanded as the need arises.

Development of the method

Visual inspection is perhaps the oldest form of non-destructive examination. It is still one of the most important. A wealth of data is acquired with a visual inspection including information about the condition of the surface, the form, colour, occurrence of defects, etc.

The use of endoscopes compensates for the inability of the eye to see around corners and for the limited resolution of the eye.

The etymology of the word *endoscope* is that it derives from the Greek and can be loosely translated as "internal vision".

The first endoscopes – which made it possible to see inside hollow organs or cavities from outside the body – were developed for medical purposes.

Based on these first steps in the development process, technical instruments called borescopes were derived. These borescopes are and were stiff instruments provided with lens optics and a miniature incandescent lamp for illumination.

In the next generation the incandescent lamp was replaced by optical fibre illumination. In this manner the degree of illumination could be increased significantly thus allowing better photographic documentation.

A decisive stage of development was achieved with the construction of flexible glass fibre optics. In this manner it became possible to carry out inspections of very inaccessible surfaces, even those not accessible in a straight line.

As an alternative to the transmission of images via optical fibres as in the fibrescope, the image can now be acquired by means of a small electronic sensor which transmits the signals to a video processor and then on to a TV monitor. This means that it is possible to use much longer cables with the endoscope than was previously possible.

Areas of application

Endoscopes can be used for a wide range of applications. If internal visual inspection of pipes, boilers, cylinders, motors, reactors, heat exchangers, turbines, and other products with narrow, inaccessible cavities and/or channels is to be performed, then the endoscope is an important, if not an indispensable instrument.

Many jobs make special demands upon the endoscopy equipment. A glance at a few equipment catalogues shows a wide assortment of both traditional as well as more specialized equipment, of which the following will be described briefly here.

- Most fibrescopes are explosion-proof and watertight, some can handle up to 3 bars (about 3 atmospheres of pressure). They can be used directly in liquid-filled containers and piping systems without the risk of causing an explosion, short-circuit or excessive heating, for a cold light source is used without any electrical potential in the probe.

- Some types of endoscopes are provided with ultraviolet (UV) illumination sources with quartz glass as the light conductor, so that surfaces which have been treated with fluorescent material can be inspected for cracks, porosity, etc. with greater sensitivity than if white light were used.

- Some types of endoscopes are provided with additional channels through which air or liquid can be sent to clean the areas which are inspected. Other types can be equipped with pincers so that e.g. lost objects can be retrieved.

- Yet other types of endoscopes are provided with optical measuring gratings so that it is possible to provide an accurate length determination through the ocular, such as the length of a crack.

- Endoscopes can normally be provided with accessories to change the viewing angle from straight ahead, 45°, 90° or 110° (slightly backwards) and viewing fields of from 10° to 80°. Some endoscopes are designed so that the viewing angle can be varied during the inspection by means of a movable prism which is located at the tip of the optical path.

- The flexible fibrescopes can normally be manoeuvred into every position by means of a handle and then locked in the desired orientation.

Precautions

As is the case for all optical instruments, endoscopes are quite sensitive to external effects. Therefore the following precautions should be taken:

- Lenses and oculars may only be cleaned with a soft cloth.

- Never try to take an endoscope apart.

- An endoscope should be protected from shocks and should not be stepped on.

- Never bend a fibre optics cable too sharply.

- Never twist a fibre optics cable more than 360°.

- Never dip an endoscope in a liquid for which it was not designed.

- Never use a fibrescope at temperatures which are higher than allowed (usually about 100°C).

- When using high power built-in lamps, take the evolution of heat into consideration.

Practical applications

As mentioned previously, endoscopes can be used for a wide variety of purposes. Figures 7.7 and 7.8 provide examples of practical applications where the endoscope can be used to advantage.

Figure 7.7. Examination of cooling elements using a stiff borescope.

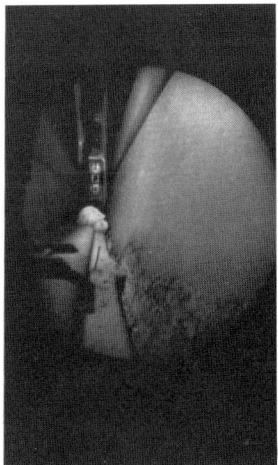

Figure 7.8. Examination of car door with a flexible fibrescope.

Training required/desired

An operator can be trained to use endoscopes relatively easily. However, the choice of objective/viewing direction, evaluation of small fields of view and the use of photographic equipment as well as TV and video equipment require some technical competence. The interpretation of defects, changes in colour shades, etc. in diverse materials also requires some knowledge of the materials under examination.

It is recommended that before commencing with independent endoscopy examinations the operator participate in a goal-oriented job training course including both theory and practical application of the endoscope.

Hardy Hansen

References

1) *Various technical information leaflets.* Classen & Co., Nekton Aps.

2) *Various technical information leaflets.* Olympus, Polack A/S.

3) *Various technical information leaflets.* Welch Allyn Video Probe 2000, Houlberg Aps.

4) *Various technical information leaflets.* Fibre Optics, Nordisk Materielkontrol A/S.

5) *Various technical information leaflets.* Volpi AG, Nordisk Optisk Compagni A/S.

Chapter 8

ER probe (electrical resistance probe)

NDE principle

The ER probe is based upon *the measurement of the loss of material* based upon changes in the electrical resistance.

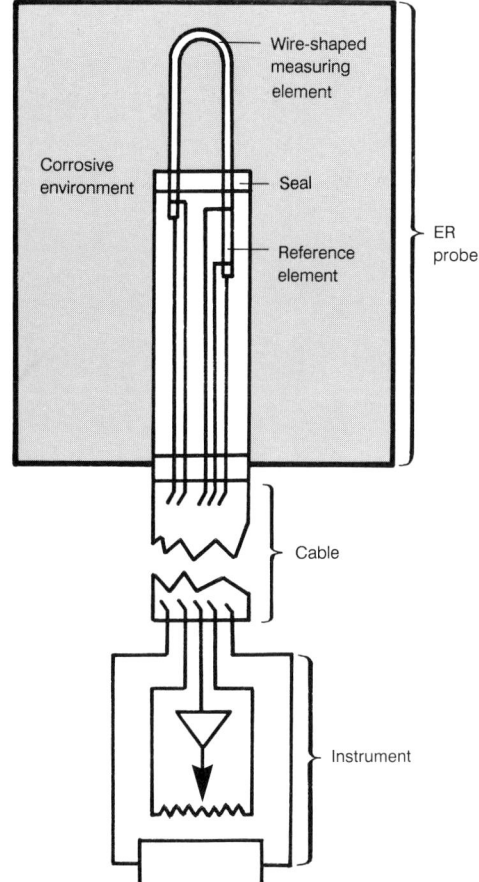

Figure 8.1. A sketch illustrating the principle behind the use of the ER probe with wire element and instrument.

The resistance along a conductor depends upon the length of the conductor and the cross sectional area. If the length remains constant, then the resistance will increase when the cross sectional area is reduced.

This principle can be utilized in a probe consisting of a wire as shown in Figure 8.1. If the wire is attacked by corrosion, the cross sectional area is reduced, and the resistance of the piece of wire increases.

Because the resistance is also temperature dependent, it is necessary to compensate for temperature changes. This is accomplished by placing a reference resistance which is not exposed to corrosion and which always has the same temperature as the measuring wire in the probe. Both of these resistances are part of a Kelvin bridge system, so that even very small changes in the resistance can be detected.

The probe is used in conjunction with an instrument which contains a power supply, balance resistor, amplifier and panel instrument for attaining balance in the bridge system.

Just as test coupons, ER probes measure, in principle, the average rate of corrosion. But because of the high sensitivity, one can operate with very short measuring periods, so that variations in the corrosion rate can be detected. Figure 8.2 shows typical data recorded with an ER probe.

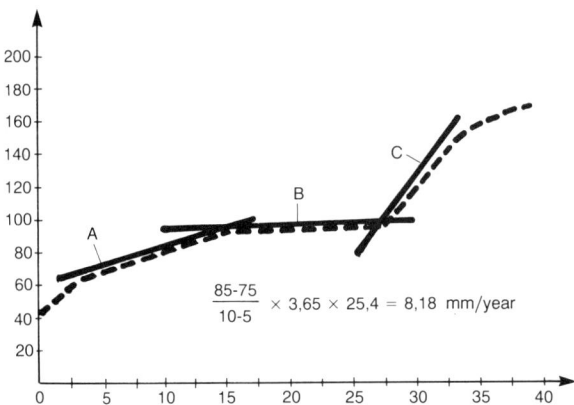

Figure 8.2. Plot of measurements performed with an ER probe.

It is apparent that the plot consists of a number of line segments, where the slope of each is related to the corrosion rate. Changes in the slope indicate changes in the corrosion rate.

Electrical resistance measurement, which is an indirect measurement of material loss, can be used to record both corrosion and erosion.

Development of the method

The method has been developed primarily for use in the oil industry. Originally a metal wire was used as the measuring element, and a portable instrument was employed for reading the probe. The wire-shaped measuring element has, however, a number of disadvantages. For one thing the metallurgy of a wire is quite different from that of plates and pipes which constitute the plant, and for another the element is sensitive to pitting. Just a single pit somewhere on the wire has a dramatic effect upon the resistance and thus upon the signal. As a consequence it would appear that corrosion increases very rapidly, which is not necessarily the case.

A third disadvantage is that the presence of conductive material within the arc of metal wire, whether liquid or solid deposits, will shunt the measuring element and thereby reduce the sensitivity of the system. Finally, the thin wire may not be strong enough to withstand high flow rates, e.g. where the flow is so great as to bend the wire. Therefore, tubular elements have been developed where the measuring object consists of a thin tube mounted around an insulating holder which also contains the reference resistance. A probe with tube element is shown in Figure 8.3. Using the tube element a more favourable ratio of surface area to cross section is attained and thereby a generally greater sensitivity with less influence due to localized attacks of corrosion.

The lifetime of a probe depends upon its thickness, but so does its sensitivity. The thinner the element, the lower the corrosion rates which can be measured, but at the same time the less corrosion it can tolerate before it is eaten away. As a general rule an element can be used until the thickness of the probe material has been reduced by about 50%.

One must thus select a probe type according to the application. If a long lifetime is desired or the environment is characterized by high corrosion rates, then wire elements with a thickness of 0.25-0.50 mm are used. If greater sensitivity is desired and the corrosion rate is low, then tube elements with thicknesses of 0.05-0.10 mm are used.

The method has also been developed with respect to instrumentation. From a simple portable, manual system (which is still available), equipment has been developed which can carry out continuous measurements on up to six probes simultaneously. Multiplexer modules are available for use in automatic systems. Each can read 15 probes, and they can be connected to form even larger systems.

Areas of application

Electrical resistance measurement with ER probes can be employed in conductive or non-conductive liquids as well as in gases. However, ER probes can not be used where conductive coatings are formed. In such cases a reduction in the metal cross section will not necessarily cause an increase in the electrical resistance. Quite the contrary, the resistance may fall steadily even though the element corrodes.

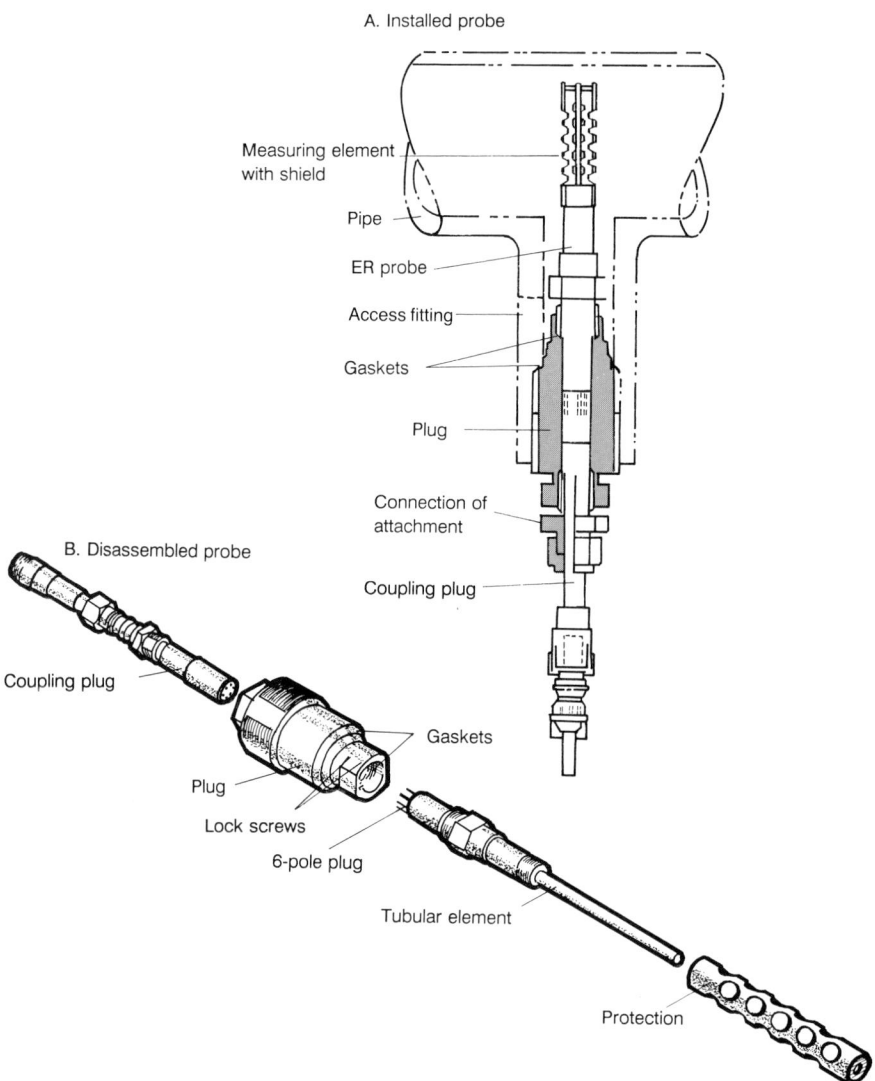

Figure 8.3. ER probe with tubular element mounted in an access fitting.

Because it is even possible to follow rather rapid variations in the rate of corrosion on the basis of recorded data, ER probes can be used for corrosion monitoring. This corrosion monitoring can be used to diagnose corrosion problems by investigating which changes in the process parameters cause corrosion to occur.

ER probes can also be used as part of a programme of preventive maintenance, where knowledge of the average corrosion rate and the state of the plant will determine the timing of the next inspection. In the intervals between pairs of inspections one can monitor the corrosion process and information from the ER probes can then determine how to adjust the date of the next inspection.

Finally, ER probes can be used to check that the corrosion situation is satisfactory. For example, if chlorine gas is to be distributed by means of ordinary steel pipes, then it is important that the gas is completely dry to avoid corrosion. Just a few ppm of water will cause the corrosion rate to increase dramatically. It is difficult to measure a moisture content of just a few ppm. The important point is, however, whether or not the gas is corrosive, and one can therefore monitor the corrosion rate with an ER probe and intervene at once if the gas proves to be corrosive, e.g. because there are problems with the drying unit.

Practical applications

As mentioned, corrosion measurement by the measurement of electrical resistance has been developed for use by the oil industry, and it is here that one finds examples of the use of this technique.

ER probes are used in refineries for corrosion monitoring, in some cases over 100 ER probes are in place in a single plant. These are read by the central computers of the refinery and the results are saved in a database which the computer can recall for subsequent reporting on plant condition and the planning of future shut-downs for maintenance. All these approaches contribute to a reduction in the number of corrosion accidents, fires and unpredictable shut-downs. Thus effective production time and efficiency (and thus earnings) of the plant are enhanced by these methods.

ER probes are also applied to corrosion monitoring tasks in the offshore industry. The purpose of the monitoring can be to acquire information for control and optimizing of the inhibitor dosage used to protect the pipes in the production wells and connecting pipes. A schematic overview of a system installed in the North Sea can be seen in Figure 8.4. Due to the distances between the production platforms and the main platform and land, a number of communications lines are necessary. All ER probes on an individual production platform are read by a local computer which transmits the data on to the central unit for the oilfield. From this point the data is sent via satellite for analysis on land, while tentative results are available to personnel on the platform.

Training required/desired

For measurements carried out manually, the NDE operator should be so well-trained in the measuring procedure that correct reading of the ER probes is assured. The operator should be familiar enough with the equipment and the fundamental principle that he or she can derive or compute corrosion rates on the basis of the data collected.

Ebbe Rislund

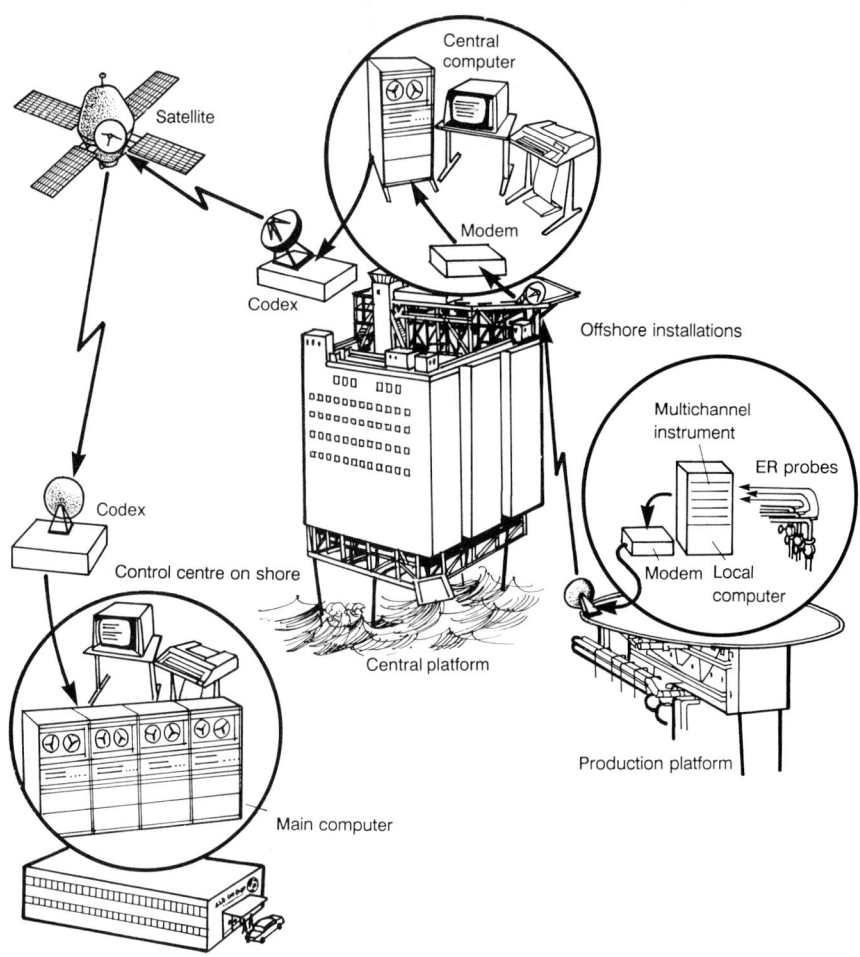

Figure 8.4. Schematic overview of automatic monitoring system on offshore installations.

References

1) *The Computer-based Data Gathering System for Internal Corrosion Monitoring at Greater Ekofisk*. C.J. Houghton, P.J. Nice & A.G. Rugveit.

2) *Procedings of the 2nd International Conference on Corrosion Monitoring in the Oil, Petroleum and Process Industries*. Oyez 1984.

3) *Corrosivity of water in absence of heat transfer (electrical methods)*. ASTM D 1776-79 Standard Test Methods.

Chapter 9

Ferrography

NDE principle

Ferrography is a technique which is based upon the systematic collection of oil samples from an oil-lubricated machine.

The method identifies, isolates and classifies wear particles from machine parts. A magnetic field is used to sort the wear particles in flowing oil. Registration of the quantity of "large" and "small" wear particles is used to monitor the development of the wear process between checks. Abnormal wear is revealed when there is a change in the distribution of particles, called the *wear index*, of the oil.

Two slightly different pieces of equipment are used for the evaluation of the S_A or the S_D wear indices of the oil.

Ferrography Analysis Apparatus for Preparation of Ferrograms (Figure 9.1)

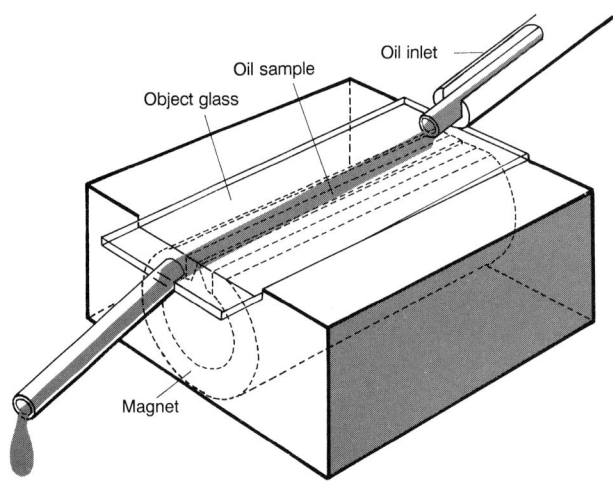

Figure 9.1. Ferrography analysis apparatus. About 2 cm³ of oil run across the object glass at a rate of 0.25 cm³/min, next the particles are rinsed and fixed on a substrate resting on the object glass called the ferrogram slide (Ref. 1).

Here the particles are separated on a treated object glass which due to its placement in a *special magnetic field* (with a very high field gradient) causes the particles to be sorted according to size. The largest particles are deposited first, while the smaller ones travel farther with the flowing oil (Figure 9.2).

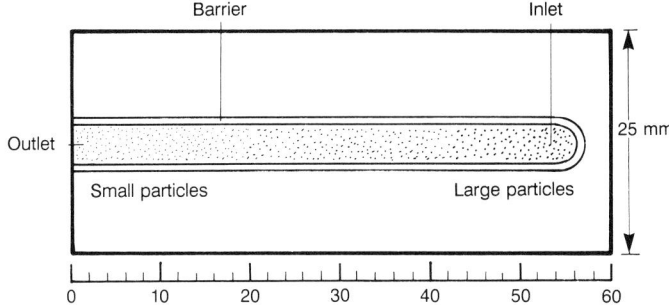

Figure 9.2. A ferrogram.

The density, i.e. the concentration of particles at a single location on the ferrogram, is measured with an optical densitometer by allowing light to pass through it.

The wear index $S_A = A_L^2 - A_S^2$

is obtained by comparison of the density A_L of large particles and the density A_L of small particles.

It takes about 30 minutes to produce a ferrogram. It is an essential prerequisite for the execution of analytical ferrography, i.e. the identification of the type and origin of the particles (see the section on practical applications).

DR (direct reading) ferrography (Figure 9.3)

This is a quick method for which direct reading of the wear index S_D can be achieved in about 5 minutes.

In this apparatus a controlled flow of oil passes through a calibrated glass tube which is mounted in a specially designed magnetic field. The separation process causes the particles to be sorted by size on the bottom of the tube.

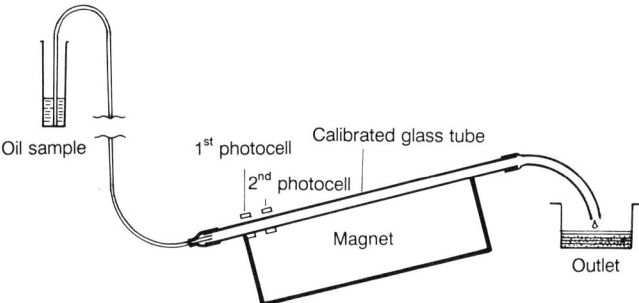

Figure 9.3. Schematic layout of a DR ferrograph (Ref. 7).

Ferrography

The apparatus uses photocells to convert the measured light intensities attained by passing light through the tube to electrical signals. The measuring region of the apparatus is from 0 to 190 DR units, where the maximum value 190 DR corresponds to the case where the bottom of the tube is completely covered with metal particles.

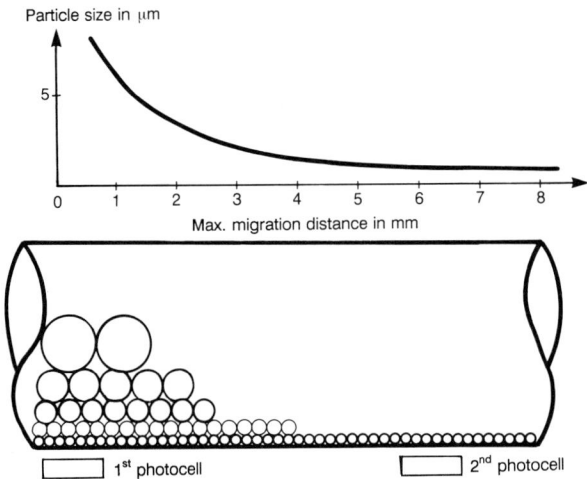

Figure 9.4. Particle arrangement in a DR tube (Ref. 7).

Densities at two fixed measuring points in the tube are used, corresponding to the densities of large and small particles, D_L and D_S respectively. The sum $D_L + D_S$ is termed the *total wear* and the difference $D_L - D_S$ is termed the *abnormality* of the wear (see Figure 9.15).

The wear index $\quad S_D = D_L^2 - D_S^2$

Development of the method

The ferrograph

The ferrograph is a relatively simple device. It was developed in the beginning of the 1970's because of the need for a simplified method for monitoring the wear of jet engines based on systematic oil sampling.

Most of the wearing parts in a machine are made of steel. Therefore, magnetic sorting methods were developed. A large magnetic field gradient sorts the wear particles by size (Ref. 1).

The DR ferrograph is a simple, portable instrument which on the basis of the DR measurements can determine which oil samples require more detailed analysis. Ferrograms are prepared from these samples for the purpose of performing analytical ferrography, i.e. an examination of the appearance and type of the individual particles.

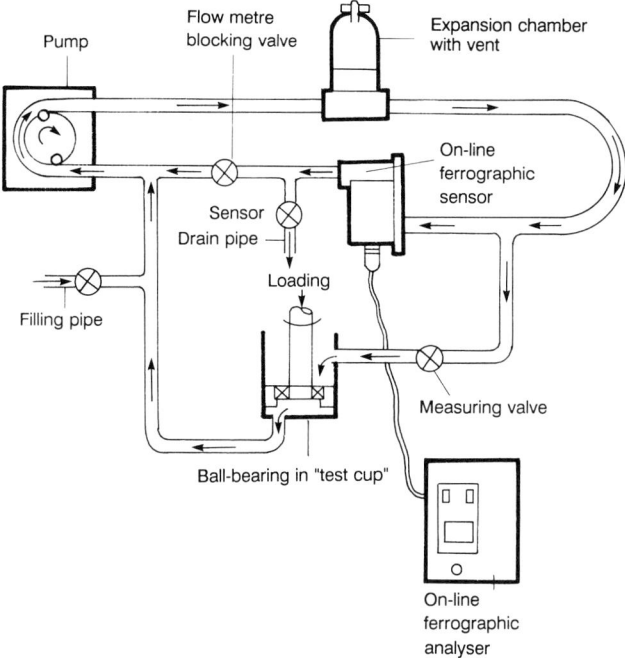

Figure 9.5. Schematic diagram of the lubricating system of a ball bearing test stand (Ref. 19).

Analytical ferrography

Microscopic examinations can be carried out with various types of illumination:

- Reflected white light
- Transmitted white light
- Polarized light
- Bichromatic light

These various forms of light will each cause some of the characteristics of the wear particles to be emphasized: form, colour, size and composition. The bichromatic microscope was developed in order to perform studies of particles on the ferrogram (Figure 9.6).

This microscope utilizes two sources of light: a green one and a red one. The green light is transmitted through the sample and the red light is reflected from the surface. It is then possible to distinguish between the various particles. Thus the pure metal particles appear bright red, while the oxide particles and the other metal compounds permit much more light to be transmitted. These particles will, therefore, depending upon their thickness, appear to be green, yellow or pink.

The use of tempera colours

Heating the ferrogram to various temperatures makes it possible to distinguish between particles of soft steel, cast iron, nickel and stainless steel due to the various tarnishing colours of the materials (Table 1, Ref. 3).

Temp. °C	Metals AISI 1090	AISI 52100	Cast iron	Nickel	304 Stainless
204	blue	partly blue	bronze	no change	no change
232	blue	blue	bronze	no change	no change
260	blue	blue	blue	no change	no change
287	blue-grey	blue-grey	blue	no change	no change
315	grey	grey	grey	no change	no change
398	grey	grey	grey	bronze	no change
420	grey	grey	grey	blue	bronze
471	grey	grey	grey	blue	mottled blue
510	grey	grey	grey	blue	mottled blue

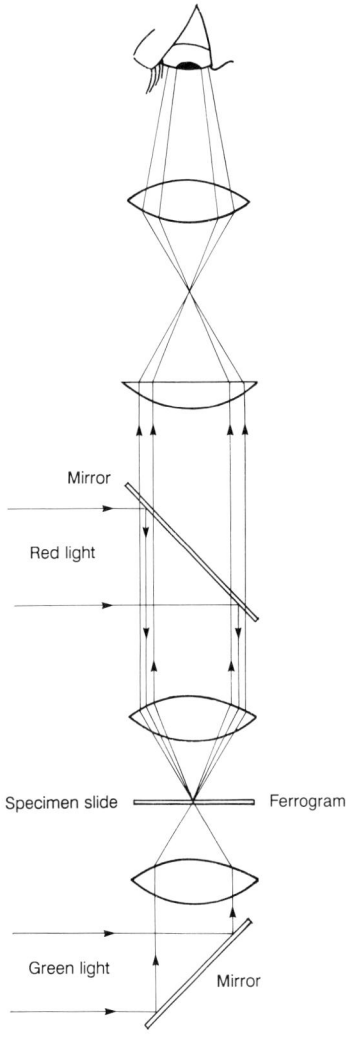

Figure 9.6. Overview diagram of the bichromatic microscope (Ref. 1).

Electron microscopy

Scanning electron microscopy with associated analyzing equipment (EDAX) is the best method for the determination of the form, type and origin of the sorted particles (Ref. 4).

Rotary particle depositor (RPD)

This is the alternative of the future to ferrographic analysis equipment (Figure 9.7). The particles are deposited here in concentric rings on a circular, rotating object glass which is placed over a magnet. The wear index is not measured directly, but the density of the various rings, which give an indication of the amount of particles of various sizes, can be measured (Figure 9.8).

The largest particles are deposited in the innermost circle, and the size diminishes further from the centre.

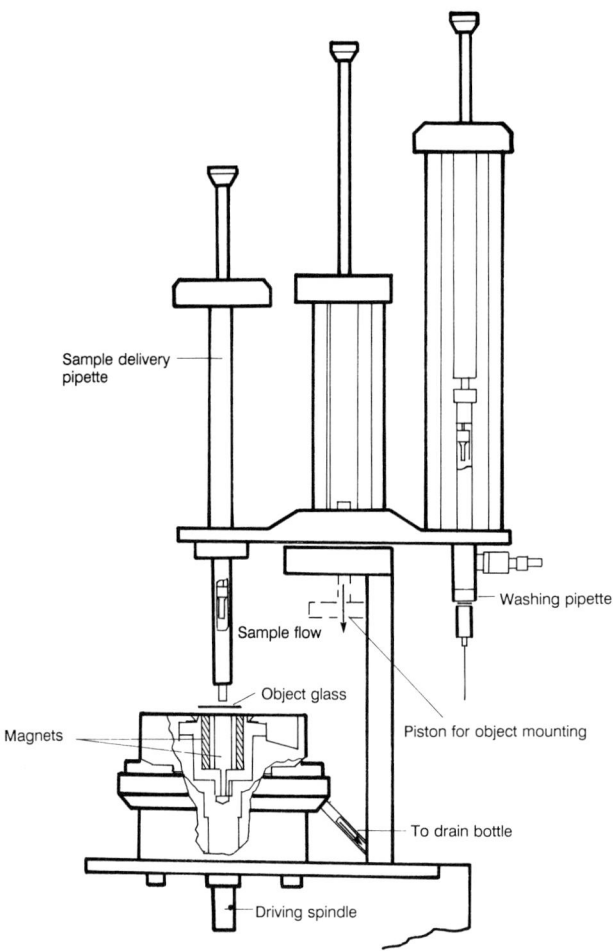

Figure 9.7. Sketch of RPD apparatus (Ref. 8).

The advantage of the new equipment is that it is less expensive to purchase and to operate, and the test can be carried out more quickly than the ferrogram. The preparation of an RPD sample takes five or ten minutes.

Dilution of the oil sample is not needed, for the particles are distributed over the entire periphery of a circle without clumping together. Dilution is often a problem in connection with the preparation of ordinary ferrograms. The RPD sample collects more large particles than the ferrogram.

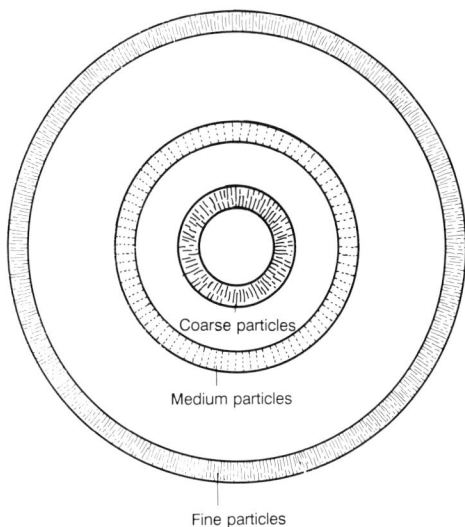

Figure 9.8. Wear particle precipitation on an RPD object glass (References 8 and 9).

Areas of application

Ordinary monitoring of oil-lubricated machinery has long been based on a spectrographic measurement of the metal content of the oil. The ferrographic DR measurement provides a warning of an incipient failure earlier than the standardized spectrographic method (see Figure 9.9).

Figure 9.9 shows measurements of oil samples from a jet motor where a spectroscopic analysis for iron has been performed and a ferrographic measurement of the optical density has been made. The ferrogram measurements show a continuous increase in the optical density from the 5-hour test, which is taken as a warning of an impending break-down. Only after the 20-hour measurement does the spectrometric test begin to show signs of increased iron content.

Particle identifications

The sorted particles on the ferrogram permit an identification of the characteristics of the individual particle types. Particles produced by specific wear mechanisms turn out to have characteristic shapes. It has thus been possible to prepare a particle atlas which can be used to make comparisons (Ref. 5).

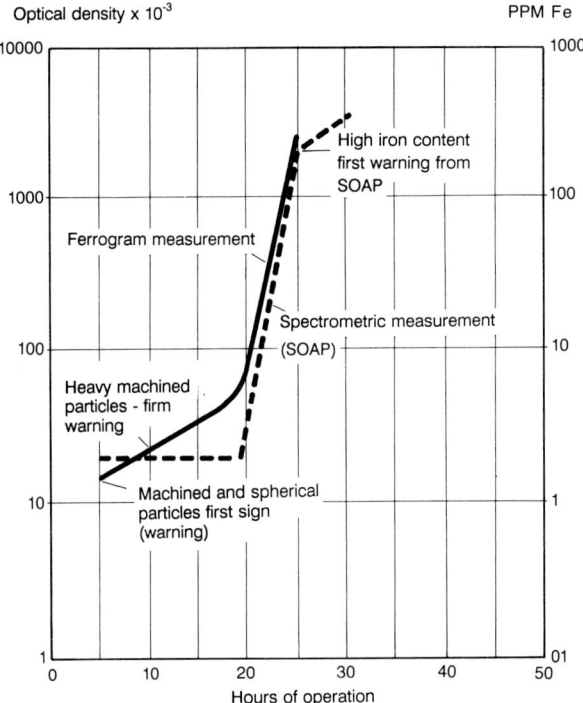

Figure 9.9. Comparison of spectroscopic measurements with ferrographic measurements on oil samples from a jet motor (Ref. 2).

Wear mechanisms and particles

Sliding adhesive wear particles are found in most lubricating oils. They are an indication of normal wear (see illus. 1 in Figure 9.10). They are produced in large numbers when one metal surface moves across another. The particles are seen as thin asymmetrical flakes of metal with highly polished surfaces.

Cutting, abrasive wear produces another particle type. The particles resemble most of all shavings from a metal shop, e.g. spirals, loops and threads (illus. 2, Figure 9.10).

The presence of a few of these particles is not significant, but if there are several hundred, it is an indication of serious cutting wear. A sudden dramatic increase in the quantity of cutting particles indicates that a break-down is imminent.

Surface fatigue: a consequence of periodic stresses with very high local tension in the surface which occurs with the meshing of gears. These wear mechanisms give the plate particles a rough surface and an irregular perimeter (illus. 4, Figure 9.10).

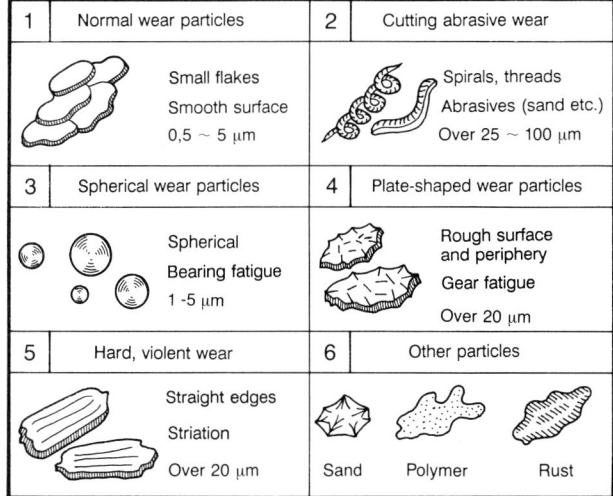

Figure 9.10. Typical types of wear particles (Ref. 6).

Small spherical particles often develop in connection with roller bearings (illus. 3, Figure 9.10).

Oil sample requirements

If a ferrographic examination is to provide a true picture of the condition of a machine, then the oil sample must contain a representative selection of the particles which are in circulation in the system. The following guidelines should be observed for taking representative samples:

- Consecutive control measurements taken at certain operational intervals from a machine must be taken from the same point in the lubrication system. Various places in the lubrication system can have different particle concentrations (e.g. before and after an oil filter).

- When samples are taken from a system in operation, the samples should be taken under the same specific operating conditions.

- If a sample is taken from a system which has been stopped, then the precipitation rate must be taken into account as well as the location of the sample.

- Oil samples from a system which has been out of operation for a period of time will not be representative of the particle content in the circulating oil when the system is operational.

Practical applications

Many articles have been published concerning the use of ferrography as a kind of offensive lubrication control which will permit a transition from planned maintenance to condition-based maintenance.

- Nippon Steel Co., Nagoya works (Ref. 6), introduced ferromagnetic monitoring of 600 machine units in 1980. In this case it has also been possible to optimize the selection of lubricating oils by means of ferrography.

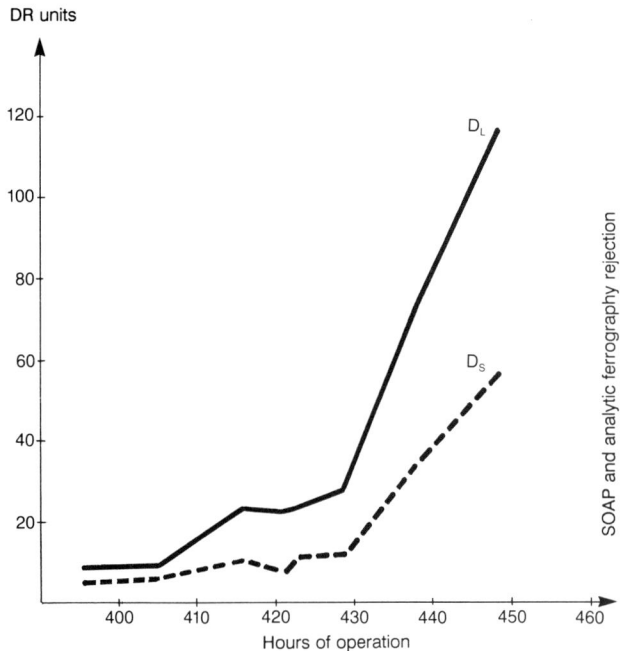

Figure 9.11. Napier Gazelle reduction gearbox No. 512. Defects: surface of a planet gear roller bearing is broken up due to fatigue. The other gear is badly scored (Ref. 7).

- The Naval Aircraft Materials Laboratory (Ref. 7) has used DR-ferrography for monitoring of the wear of gear boxes (see Figures 9.11 and 9.12).

- The Royal Danish Navy's Material Command introduced ferrography in 1977 (see Ref. 11).

    Figure 9.13 shows a continuous increase in the DR value for a gearbox in spite of oil changes. Disassembly of the gearbox revealed that the axis between the turbine and the gear had clear fatigue cracks. After replacement of the axle, the gearbox ran perfectly. Figure 9.14 shows the effect of replacing dirty oil after 196 hours of operation.

- In England ferrography is used for monitoring the condition of marine gas turbines (Ref. 12). A dramatic increase in particle production just before stopping for maintenance (Figure 9.15) is seen to yield a more normal wear particle production after this action.

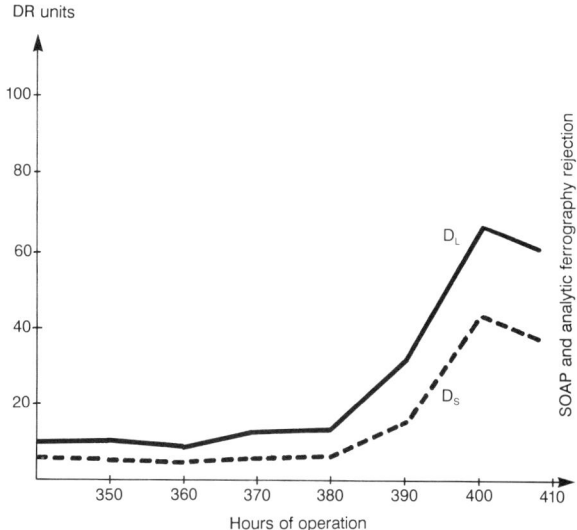

Figure 9.12. Napier Gazelle reduction gearbox No. 510. Defect: four teeth on the sun gear are badly pitted (Ref. 7).

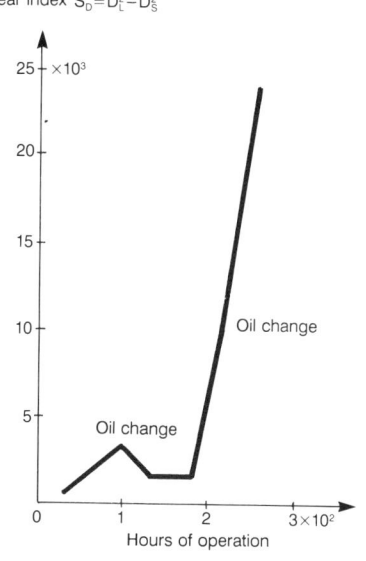

Figure 9.13. The result of a DR ferrograph on ship A, gearbox 2 (Ref. 11).

Ferrography

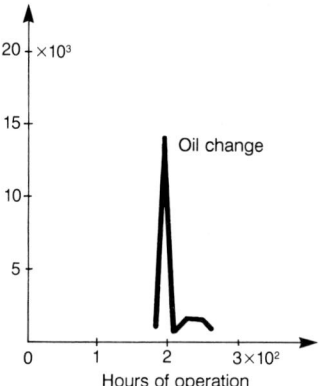

Figure 9.14. Changing impure oil in a gear after 196 hours of operation brings the particle production down to a normal level (Ref. 11).

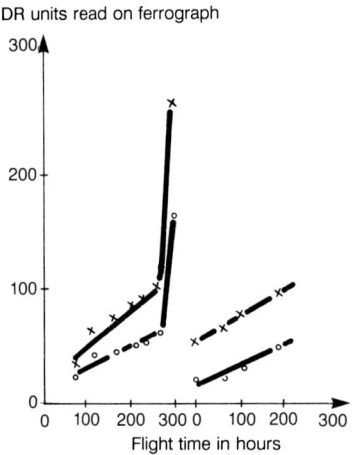

Figure 9.15. The particle production before and after a maintenance inspection. The data marked x: $(D_S + D_L)$ = total wear; o: $(D_L - D_S)$ = "seriousness" of the wear (Ref. 12).

Training required/desired

The DR ferrograph is a simple instrument which can easily be handled by a technical assistant after a brief training period. The preparation of ferrograms is also relatively simple, but to achieve a dependable interpretation of the ferrographic results, considerable experience is required.

Various types of machines are characterized by various break-down mechanisms which again produce differences in the ferrographic displays of the wear-down process.

When acquiring the ferrographic equipment, one should figure on several years of break-in time before a sufficient basis in experience has been acquired to perform completely dependable interpretations of the results.

Kirsten M. Dorph

References

1) *The Particles of Wear.* Douglass Scott, William W. Seifert, Vernon C. Westcott. **Scientific American,** May, 1974, pp 88-97.

2) *A Method for the Study of Wear Particles in Lubricating Oil.* W.W. Seifert, V.C. Westcott, **Wear, 21,** 1972, pp 27-42.

3) *The Use of Temper Colors in Ferrography.* F.T. Barwell, E.R. Bowen, V.C. Westcott, **Wear, 44,** 1977, pp 163-171.

4) *Debris Examination – A Prognostic Approach to Failure Prevention.* D. Scott. **Wear, 34,** 1975. pp 15-22.

5) *Wear Particle Atlas.* E.R. Bowen, V.C. Westcott, Foxboro/Trans.Sonics, Inc., Burlington, MA 01803 USA, July, 1976.

6) *Application of Ferrography to Iron and Steel Making Plant Maintenance.* Motofereni Kurahaski, Proceedings of International Conference on Condition Monitoring, University of Swansea, April, 1984, pp 537-540.

7) *The Role of Ferrography in the Monitoring of Helicopter Assemblies.* A.S. Yarrow, P.Gad, Proceedings of International Conference on Condition Monitoring, University og Swansea, April, 1984, pp 503-524.

8) *The Rotary Particle Depositor - a Response to Problems Experienced with Wear Particle Deposition.* A.L. Price, B.J. Roylance, Proceedings of International Conference on Condition Monitoring, University of Swansea, April, 1984, pp 596-607.

9) *The Evaluation of the Rotary Particle Depositor in the Monitoring of Gear Transmissions Used Underground.* A.L. Price, E.D. Yardley, **Condition Monitoring, 84,** Peneridge Press, pp 608-616.

10) *Continuous Wear Measurements by Outline Ferrography.* Wolfgang Holzhaner, S.F. Murray, **Wear, 90,** 1983, pp 11-19.

11) *The Use of Ferrography in the Danish Navy.* S.O.E. Schönthal, **Wear, 90,** 1983, pp 149-158.

12) *The Application of Ferrography to the Condition Monitoring of Gas Turbines.* D. Scott, **Wear, 90,** 1983, pp 21-29.

Chapter 10

Hardness testing

NDE principle

Hardness testing is a descriptive term for a number of methods for the measurement of the resistance of a surface to the action of a body which is forced into it under pressure or by means of an impact.

The values measured will depend upon the following characteristics of the object measured:

- The tensile strength and the elastic limit
- The elasticity modulus (Young's modulus)
- Dimensions
- The surface finish
- The homogeneity of the material

Furthermore, the geometry of the test body, the force (the loading) of the test body and the velocity during the application of pressure or impact as well as the loading time will affect the result.

The hardness values must, therefore, be accompanied by a unit which uniquely defines these parameters. For practical measurements a number of measurement techniques and associated parameters have been defined. The most of important of these are described briefly in the following sections.

Brinell

The test body is a spherical steel or tungsten carbide body with a diameter of 2.5, 5.0 or 10.0 mm. The loading can vary from about 1 kg to 3000 kg. The velocity of application of the load is less than about 0.5 mm/s.

The hardness value is determined by measurement of the resulting spherical impression, see Figure 10.1b, and substitution into the equation:

HB = constant x (loading / surface area of impression)

Example:

$$HB = 0{,}204 \cdot \frac{F}{\pi D(D - \sqrt{D^2 - d^2})}$$

or in a table of values. The unit designation has the format: HB (or possibly HBS or HBW) / test body diameter (mm) / loading (kg) / loading time (s) if it differs from 10-15 seconds.

Example: 136 HB/2.5/187.5/130

The measurement technique is standardized in ISO 6506.

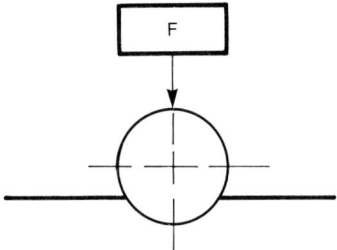

Figure 10.1a. Brinell measurement by the loading principle. The test body is a steel or hard metal sphere with a diameter of 2.5, 5 or 10 mm.

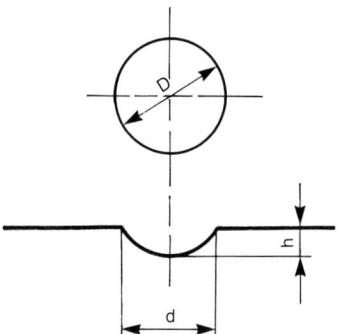

Figure 10.1b. Brinell measurement. Geometry.

Brinell with impact loading

The sphere diameter is most often 5 mm and the loading occurs by means of an impact. In one type of equipment the loading is controlled by means of a calibrated breaking pin which is designed to break under a given load (see Figure 10.2a). The diameter measurement is translated to a hardness value by means of a table which is valid for the pin in use. Other types have an undefined loading (only limited by the mass of the hammer and the enthusiasm of the operator). The results of the hardness measurement are found in this case by collecting the impressions both in the object of measurement and in a reference sample which is contained within the apparatus (Figure 10.2b). Brinell hardnesses can then be found in a table which requires the use of the impression diameters in both objects.

The most commonly used types of equipment are: Ernst type STE (breaking pin),

Zwick, Poldy hammer and Telebrineller (reference rod). See Figure 10.2.

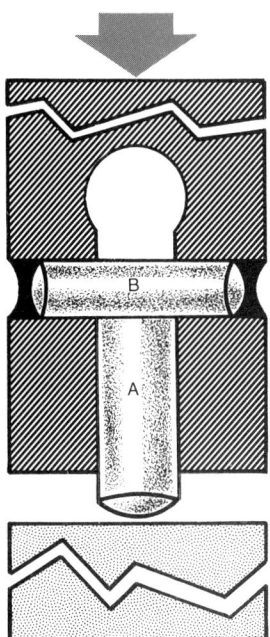

Figure 10.2a. Brinell with impact loading. The principle for the type with a calibrated breaking pin is illustrated. A = impression body, B = impact/breaking pin.

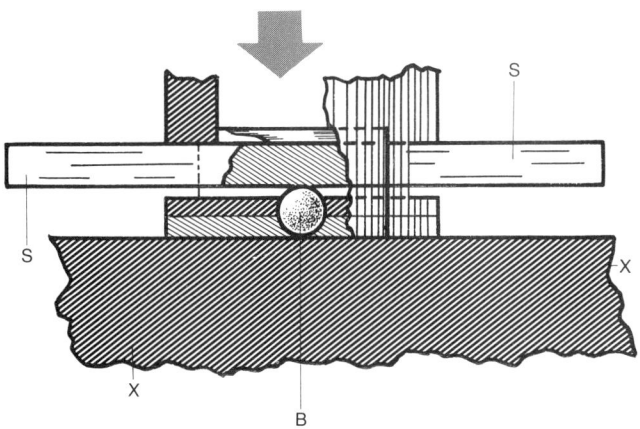

Figure 10.2b. Impact Brinell with reference rod, here Telebrineller. B = Impression sphere, S = Reference rod, X = Object of measurement. Note that an impact is applied from above.

Hardness testing

Vickers

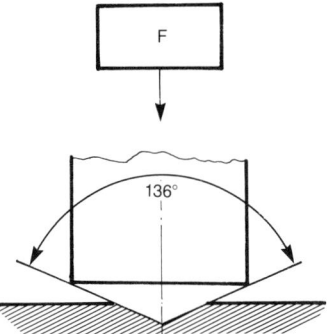

Figure 10.3a. Vickers measurement, loading principle. The impression body is a diamond pyramid.

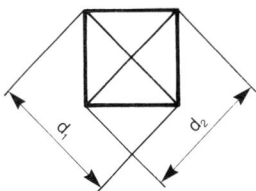

Figure 10.3b. Vickers measurement. Geometry.

The impressed object is a four-sided diamond pyramid with an apex angle of 136°. Loadings can vary from 0.01 to 100 kp, corresponding to 0.09807–980.7 N (1 kp = 1 kilopond = the gravitational force acting upon the standard kilogram in Paris). The loading time is normally 10-15 seconds. The measurement result is found by determining the mean diagonal length of the square impression (see Figure 10.3). These lengths are substituted into the formula:

HV = constant x [loading / surface area of impression]

Example (see Figure 10.3b):

Mean diagonal of impression d = $(d_1 + d_2)/2$

HV = 0.1891 · F/d_2

where F is the loading in newtons (N).

The HV value will, however, most often be found in a table which requires the use of the mean diagonal length and the current loading. The hardness unit is HV and the loading is specified in kp. The loading time is specified if it differs from 10-15 seconds.

Example: HV 10/30 s.

The measurements are standardized in ISO 6507/1 and 6507/2.

Rockwell C

The impressed body is a diamond cone with a vertex angle of 120° and a rounded tip. The loading is carried out in two steps: pre-loading 10 kp (98 N) followed by a total loading of 150 kp (1470 N). The velocity should be less than 0.5–1 mm/s. The loading time is 5–10 seconds.

The measurement result is found as follows: it is 100 minus the difference in the depth of the impression during pre-loading before and after total loading, measured in units of 0.002 mm (see Figure 10.4). The unit is termed HRC. The measurement is standardized in ISO 6508.

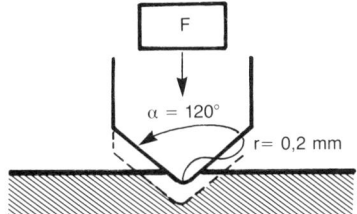

Figure 10.4a. The Rockwell C method. Loading principle.

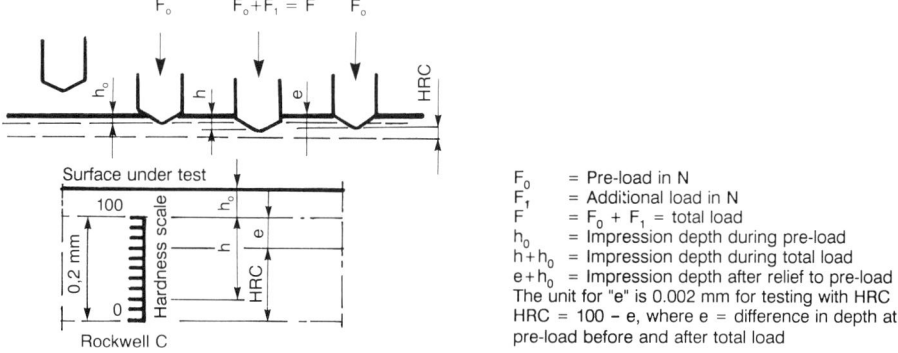

Figure 10.4b. Rockwell C. Geometry and definition.

F_0 = Pre-load in N
F_1 = Additional load in N
F = $F_0 + F_1$ = total load
h_0 = Impression depth during pre-load
$h + h_0$ = Impression depth during total load
$e + h_0$ = Impression depth after relief to pre-load
The unit for "e" is 0.002 mm for testing with HRC
HRC = 100 − e, where e = difference in depth at pre-load before and after total load

Hardness testing

Portable Rockwell superficial (low load)

The measurement principle is the same as for the Rockwell test but using low loading values. The measurement is carried out by pressing the apparatus by hand against the surface of the object under measurement. In this manner a spring-loaded cone is pressed into the surface. The apparatus is calibrated by means of a reference sample with known hardness, and the impression depth is converted, either mechanically to the calibrated hardness scale or electronically to a display. The equipment is available with Rockwell C, Brinell and Vickers scales.

The loading is typically 0.5 kp (4.9 N) pre-load and 5 kp (49 N) total load. Nearly identical apparatus of this type is available from Newage Inc. *USA*, Instrumatic *England*, Ernst (STP Eumetron) *Switzerland* and Brevetti Affri (DUROMETRI) *Italy*.

Equipment with an analogue readout scale is shown in Figure 10.5.

Figure 10.5. Measurement with low load Rockwell on crankshaft bearing surface.

Rockwell B and N

In addition to the Rockwell C test method, there is also a Rockwell B test where the impressed body is a 1/16" diameter sphere and the total loading is 100 kp (980.7 N). Furthermore, there is a Rockwell N (Superficial) test with a smaller pre-load and a lesser total loading. The pre-load is often 3 kp (28.4 N) in this case and the total loading is 15, 30 or 45 kp (147, 294 or 441 N, respectively).

Meyer

The hardness unit Meyer (HM) can be measured using Brinell equipment, where the projected area of the spherical impression is found by measurement of the mean diameter. The method yields results which to a close approximation are equal to HB.

Knoop

The Knoop method (HK) where the impressed body is a rhombic diamond pyramid is used particularly in the US for the measurement of very thin surface layers. Before any measuring technique is employed it is important to be sure that the surface has an appropriate finish and cleanliness. The lower the loading the better the surface finish should be.

Shore, Scleroscope, Equotip

The *Shore method* is based upon the dynamic measurement of the energy which is absorbed in the object of the measurement when a body with a spherical tip strikes it.

The *Scleroscope* measures the height of the bounce when a body falls onto a surface from a predetermined height.

The *Equotip* performs measurements of the initial speed and the bounce speed at a given height above the surface of the object measured by means of an induction principle. The impact body is fired by means of a spring. The Equotip is available with impact energies of 3, 11 and 90 N mm.

The equipment is illustrated in Figure 10.6.

Figure 10.6. Equotip measurement.

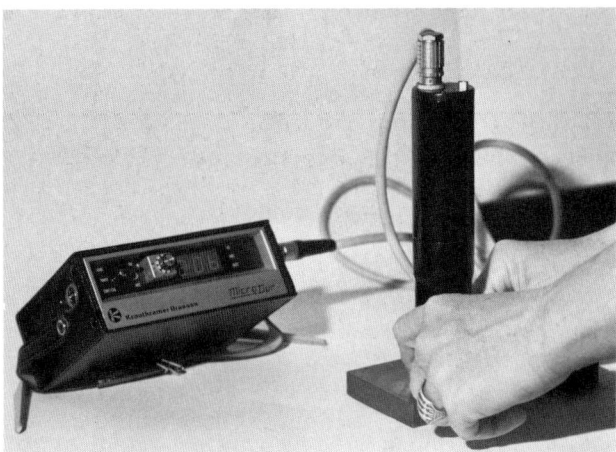

Figure 10.7. Measurement by the UCI principle with Krautkrämer Microdur.

UCI hardness measurement

The measurement is carried out by pressing a vibrating rod with a pyramid-shaped diamond tip down into the surface. The European equipment, *Krautkrämer Microdur*, operates with loadings of 0.2 kp (0.196 N) and 0.86 kp (8.4 N). In Japan Kawatetzu Instruments markets an apparatus called *Sonohard* with a number of loading options up to about 5 kg. Hardness values are a function of the contact area between the diamond and the object of the measurement. Microdur shows this value on the display in Vickers units. The equipment is also available calibrated for Rockwell C under the designation DHR 10. The accuracy of the Microdur is ±5 HV in soft, homogeneous materials and ±15 HV in hard, homogeneous materials. The measuring time can be adjusted from 4 to 60 seconds. Due to the dynamic character of the measurements, items which are thinner than 2 mm cannot be measured without some support behind the point of measurement. The 2 kp model of the Sonohard requires support for thicknesses of 7 mm and less. The Microdur apparatus is shown in Figure 10.7.

Development of the method

Hardness measurements began with the development of the Brinell and Rockwell principles designed, among other things, for checking canon and rifle barrels. Since that time the other methods have evolved to meet the need to measure using other levels of force and other materials. Some of the methods, particularly the most recent UCI (Krautkrämer Microdur) and Equotip have been developed to meet the requirement for rapid measurements and also to enable the equipment to be transported easily to the place of measurement.

Hardness measurements are widely used today to check the strength characteristics of objects either in direct comparison with a hardness specification for the given item or with the hardness to be expected for a material of the type in question. The check can be carried out during production, before the material is used or in connection with the examination of damage which has occurred.

For ideally hardened and tempered steel there is a direct connection between the hardness according to Meyer (HM) units and the tensile strength. The difference between HM and Brinell (HB) is as mentioned quite small, and the connection is therefore often used directly for data obtained by the Brinell method. The Brinell measured value divided by 0.3 gives a value which is close to the tensile strength of the steel. The tensile strength corresponding to various hardness measuring schemes are available in tables. Please refer to the references at the end of this chapter. In addition a conversion table is shown in Figure 10.8.

The relationship can be used with a somewhat greater uncertainty on normalized steel. On the other hand it cannot be used on cold-deformed/deformation hardened items. Metals without or with very limited linear elasticity working curves will not show this relationship. This is true of austenitic steel, aluminium and copper, for example.

Certain damage mechanisms have turned out to be directly dependent upon the strength of a material. Most prominent in recent years are cold cracks: hydrogen cracks, e.g. in connection with welding, and stress corrosion cracking due to hydrogen sulphide in acidic oil and gas environments. Experiments have formed the basis of the specification of maximum permissible hardnesses. The hardness limits are often about 325 or 350 HV where the risk of hydrogen cracks exists and 22 HRC or converted 248-250 HV in the case of hydrogen sulphide induced stress corrosion cracking.

Developments in the field of hardness measuring equipment are moving towards automatic conversion of the measurement results to hardness values, digital displays, automatic treatment of results (statistics, etc.) and printout of results. Work is also being done on semiautomatic measurement, e.g. with a digital ruler, where the operator adjusts the marker and records the measurement simply by pushing a button. Complete automation, e.g. by means of digital image recognition, is also being worked on. Experiments are underway with ultrasonic measurements of the area of the spherical depression which results from a Brinell measurement. The results appear promising, but they are not (yet) applied in mass-produced measuring instruments. A number of devices are available with automatic operation e.g. for use in the continuous control of a production line.

Considerable research has been performed concerning the determination of the hardness of a material based upon an eddy current signal, and equipment is presently available which can record metallurgical variations, among them hardness. A prerequisite for the signal recorded characterizing a change in hardness is, however, that all other material parameters such as grain size, microstructure, etc. remain unchanged.

Figure 10.8. Conversion table for hardness measurements.

Vickers HV (F - 98N)	*Brinell HB	Rockwell HRB	Rockwell HRC	Shore	Tensile strength N/mm²
63	60				200
65	62				210
69	66				220
70	67				225
72	68				230
75	71				240
79	75				250
80	76				255
82	78				260
85	81	41			270
88	84	45			280
90	86	48			285
91	87	49			290
34	89	51			300
95	90	52			305
97	92	54			310
100	95	56			320
103	98	58			330
105	100	59			
107	102	60			340
110	105	62			350
113	107	63,5			360
115	109	64,5			370
119	113	66			380
120	114	67			385
122	116	67,5			390
125	119	69			400
128	122	70			410
130	124	71			415
132	125	72			420
135	128	73			430
138	131	74			440
140	133	75			450
143	136	76,5			460
145	138	77			465
147	140	77,5			470
150	143	78,5			480
153	145	79,5			490
155	147	80			495
157	149	81			500
160	152	81,5			510
163	155	82,5			520
165	157	83			530
168	160	84,5			540
170	162	85			545
172	163	85,5			550
175	166	86			560
178	169	86,5			570
180	171	87			575
181	172				580
184	175	88			590
185	176				595
187	178	89			600
190	181	89,5			610
193	184	90			620
195	185				625
197	187	91			630
200	190	91,5			640
203	193	92			650
205	195	92,5			660
208	198	93			670
210	199	93,5			675
212	201				680
215	204	94			690
219	208				700
220	209	95			705
222	211	95,5			710
225	214	96			720
228	216				730
230	219	96,5			740
233	221	97			750
235	223				755
237	225	97,5			760
240	228	98			770
243	231		21	36	780
245	233				785
247	235	99			790
250	238	99,5	22	38	800
253	240				810
255	242		23	39	820
258	245				830
260	247		24	39	835
262	249				840
265	252			40	850
268	255		25		860
270	257			41	865
272	258		26		870
275	261			41	880
278	264				890
280	266		27		900
283	269				910
285	271			43	915
287	273		28		920
290	276			43	930
293	278		29		940
295	280			44	950
299	284				960
300	285			45	965
302	287		30		970
305	290				980
308	293				990
310	295		31	46	995
311	296				1000
314	299				1010
317	301		32		1020
320	304			47	1030
323	307				1040
327	311		33		1050
330	314			48	1060
333	316				1070
336	319		34		1080
339	322				1090
340	323			49	1095
342	325				1100
345	328		35		1110
349	332				1120
350	333			50	1125
352	334				1130
355	337		36		1140
358	340				1150
360	342			52	1155
361	343				1160
364	346		37		1170
367	349				1180
370	352			53	1190
373	354		38		1200
376	357				1210
380	361			55	1220
382	363		39		1230
385	366				1240
388	369				1250
390	371			57	1255

Vickers HV (F - 98N)	*Brinell HB	Rockwell HRB	Rockwell HRC	Shore	Tensile strength N/mm²
392	372		40		1260
394	374				1270
397	377				1280
400	380			57	1290
403	383		41		1300
407	387		41		1310
410	390			58	1320
413	393		42		1330
417	396				1340
420	399			59	1350
423	402		43		1360
426	405				1370
429	408				1380
430	409			60	1385
431	410				1390
434	413		44		1400
437	415				1410
440	418			62	1420
443	421				1430
446	424		45		1440
449	427				1450
450	428			64	1455
452	429				1460
455	432				1470
458	435		46		1480
460	437			64	1485
461	438				1490
464	441				1500
467	444				1510
470	447			66	1520
473	449		47		1530
476	452				1540
479	455				1550
480	(456)			67	1555
481	(457)				1560
484	(460)		48		1570
486	(462)				1580
489	(465)				1590
490	(466)			68	1595
491	(467)				1600
494	(470)				1610
497	(472)		49		1620
500	(475)			69	1630
503	(478)				1640
506	(481)				1650
509	(483)				1660
510	(485)			70	1665
511	(486)				1670
514	(488)		50		1680
517	(491)				1690
520	(494)			71	1700
522	(496)				1710
525	(499)				1720
527	(501)		51		1730
530	(504)			72	1740
533	(506)				1750
536	(509)				1760
539	(512)				1770
540	(513)			73	1775
541	(514)				1780
544	(517)		52		1790
547	(520)				1800
550	(523)			75	1810
553	(525)				1820
556	(528)				1830
559	(531)				1840
560	(532)		53	76	1845
561	(533)				1850
564	(536)				1860
567	(539)				1870
570	(542)			77	1880
572	(543)				1890
575	(546)				1900
578	(549)		54		1910
580	(551)			78	1920
583	(554)				1930
586	(557)				1940
589	(560)				1950
590	(561)			79	1955
591	(562)				1960
594	(564)				1970
596	(567)		55		1980
599	(569)				1990
600	(570)			80	1995
602	(572)				2000
605	(575)				2010
607	(577)				2020
610	(580)			81	2030
613	(582)				2040
615	(584)		56		2050
618	(587)				2060
620	(589)			83	2070
623	(592)				2080
626	(595)				2090
629	(598)				2100
630	(599)			84	2105
631	(600)				2110
634	(602)				2120
636	(604)				2130
639	(607)		57		2140
640	(608)			85	2145
641	(609)				2150
644	(612)				2150
647	(615)				2170
650	(618)			86	2180
653	(620)				2190
655	(622)		58		2200
675			59		2200
698			60		
720			62		
745			62		
773			64		
800			64		
829			65		
864			66		
900			67		
940			68		

* Computed as 0.95 x HV.

Hardness testing

Areas of application

The measurement of hardness is employed in connection with the monitoring of objects for which hardness/strength is a significant quality, e.g. gears, bearing races on axles, hardened axles, welds, etc. The hardness measurement is also a simple control technique in connection with deliveries, for sorting or for checking heat-treated items. Hardness measurement is often used for checking the state of a material after an unintentional (over-)heating, e.g. in connection with break-downs or fire. In addition to measuring equipment designed for metals, hardness measurement equipment is available for plastic and rubber, textiles, painting film, etc.

Equipment can be classified into two types, namely: *stationary apparatus* which only can measure objects of limited size or cutout samples; and the *transportable apparatus* which can perform measurements directly on the surfaces of large objects.

As a rule stationary equipment is the most accurate, but especially for the measurement of Vickers and low load Rockwell C, transportable equipment is available which can provide good accuracy, e.g. ±3-5 HV.

A transportable apparatus for performing Vickers measurements is shown in Figure 10.9.

In cases where the measuring apparatus operates on one principle, and results are desired in a different unit, and where the measured value is converted by means of an empirical table (e.g. Vickers measurement with Equotip), greater uncertainty will be associated with the results than is the case with direct methods. The magnitude of the maximum standard deviation will as a rule always be available in the tables. For example Equotip measurements which are converted to HV can vary within a standard deviation of up to ±13 HV in the interval 200-300 HV.

A number of standards are available for conversions between various hardness units for measurements on unalloyed and low-alloy steel.

For example there are: DIN 50150, ISO 4964 and ASTM E 140-84. In Figure 10.5 a comparison of several conversion tables used by steel wholesalers is provided.

Figure 10.9. A transportable apparatus for Vickers type measurements, type VEB, Werkstoffprüfmaschinen Leipzig, HMO 10.

Important limitations on transportable equipment

The *Vickers* measuring equipment provides as mentioned earlier a very accurate measurement. However such equipment is often so large that it is not in practice possible to perform measurements near joints or other irregular geometries. The measurement of the hardness of welding joints near flanges, bends and jumps in size can, therefore, present problems.

Portable Rockwell superficial (e.g. from the firms Ernst and Instrumatic) require a very finely polished surface and a very homogeneous material, and a reference material in the hardness range in question.

The most commonly used transportable Brinell equipment is loaded by means of an impact. The loading is often difficult to exert perpendicularly, which can be one of the reasons for the larger standard deviation experienced with such equipment. For the methods with reference rods the measurement uncertainty is of course doubled. In conclusion, the Brinell measurement with impact loading provides a hardness measurement with a somewhat larger standard deviation than a corresponding measurement performed with stationary apparatus. For the measurement of smaller items, the Ernst measuring device can be supplied with a clamp, whereby the loading velocity can be reduced and the measuring uncertainty thereby diminished.

Equotip provides a measurement which in practice has a standard deviation under 10 units after conversion for homogeneous items with more than 3-5 kg concentrated behind the point of measurement. Thinner items require support and coupling paste. For measurements on pipes a wall thickness greater than 10% of the pipe diameter is necessary, and pipes with wall thicknesses less than 5 mm should normally not be measured unless the pipe diameter is large enough for the measurement to be performed from the inside. The measurement point should be supported and coupled to the support by means of paste. The equipment provides very limited options for prior selection of a precise measuring point.

The low loading of *Krautkrämer Microdur* equipment means that particularly high demands are placed on the quality of the surface finish. Furthermore, the material must be homogeneous, otherwise the hardness can only be specified after computation of the mean of a number of randomly selected measuring points. Accurate measurement without a stand requires a relatively large supporting foot, and the measuring point can, therefore, not be selected precisely.

Practical examples

In European norms hardness requirements for welding joints are most often based upon Vickers with 49 or 98 N, HV 5 or HV 10. American requirements can, however, be formulated in Rockwell HRC. In the case of heated zones (HAZ), the conversion of the requirements between the units of measurement by means of the aforementioned tables will necessarily be uncertain. HRC is measured with a larger loading and the impression has a larger diameter. Rockwell measurements will thus yield lower values in the HAZ if the hardness maximum is in a zone which is narrower than the diameter of the HRC impression.

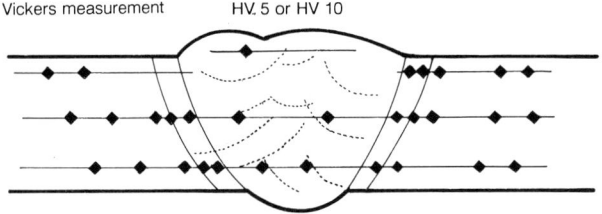

Figure 10.10. The measurement of the hardness distribution in a welding joint. Vickers measurements HV 5 or HV 10. If the risk of hydrogen cracks is present, then the maximum permissible hardness is 325 or 350 HV, depending on the particular specification. When the risk of sulphide stress corrosion cracking is present, the maximum permissible hardness is 248 or 250 HV, depending again upon the specification.

Monitoring of hardnesses in procedure and production test samples is performed on cross sectional samples in stationary Vickers equipment, see Figure 10.10. Monitoring of welding joints can be specified, for example, when doubt or disagreement arises concerning whether or not a particular construction has been welded according to the approved procedure.

The measurements in the field are carried out primarily with Vickers equipment. If space or geometry does not permit this, a different method must be used. However, the Brinell impact-type apparatus should probably be avoided, for it will yield an average hardness in a zone which can be up to 3-4 mm wide, and one can thus not achieve the desired check of maximum hardness.

Because the large supporting foot of the Krautkrämer Microdur hides the point at which the measurement is made, it is difficult to do measurements on narrow zones and similar regions. Due to both the dynamic measuring principle and the very low loading this equipment is difficult to use for this type of job. The equipment can be used without the supporting foot, but according to the manufacturer it requires considerable skill to achieve dependable results in this manner. Equotip also gives – due to the size of the ball, the dynamic measuring method and sensitivity to differences in the elastic modulus – limited correlation with Vickers measurements. Correctly performed, the results will nevertheless provide an indication of serious deviations from expected hardness levels. A measurement should be performed, after appropriate polishing of the surface, as shown in Figure 10.11.

Training required/desired

At present there are no requirements for certification of operators who perform hardness measurements. It is nevertheless always advisable to provide a written standard operating procedure and to ensure that the operator has acquired sufficient on the job training and experience. In the long run it is to be expected that, in view of the current rapid evolution in the field of quality control, that some form of certification will be required for operators of hardness testing equipment.

Training in connection with hardness measurements should, in addition to providing insight and familiarity with the equipment, also give a basis of understanding of the technical properties of materials. Only on this basis can an operator evaluate possibilities and limitations of the various methods and critically evaluate the results of measurements.

J. Vagn Hansen

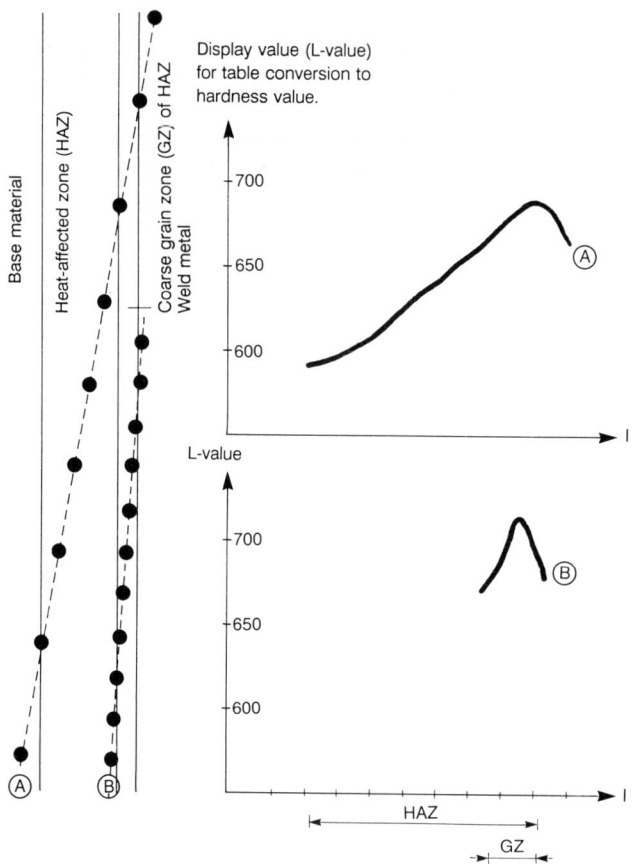

Figure 10.11. Hardness measurement of heated zone by means of Equotip equipment.

References

Standards

1) ISO 6506-1981 *Metallic Materials – Hardness Test – Brinell Test.*

2) ISO 6507/1, /2-1982 *Metallic Materials – Hardness Test – Vickers Test – Part 1:* HV 5 to HV 100, *Part 2:* HV 0.2 to less than HV 5.

3) ISO 6508 *Metallic Materials – Hardness Test – Rockwell Test* (scales A, B, C, D, E, F, G, H, K).

4) ISO 4964-1984 *Steel – Hardness Conversions.*

5) DIN 50150 *Umwertungstabelle für Vickershärte, Brinellehärte, Rockwellhärte und Zugfestigkeit (Conversion tables for Vickers, Brinell and Rockwell Hardnesses and Tensile Strength)*.

6) DIN 50163 *Prüfung Metallischer Werkstoffe, Härteprüfung an Schweissungen, Querschliff an Verbindungsschweissungen (Testing of metallic materials, Hardness testing of welds, Cross sectioning of connecting welds)*.

7) ASTM E 140-84, *Hardness Conversion Tables for Metals*.

8) *Härteprüfung in Theorie und Praxis (Hardness Testing in Theory and Practice)*. VDI-Berichte Nr. 583, 1986.

Books

9) *Hardness Theory and Practice*. Louis Small, Service Diamond Tool Co., Ferndale, Michigan, USA, 1960.

10) *Härteprüfung an Metallen und Kunstoffen (Hardness testing of Metals and Plastics)*. W. Weiler Band 155 Kontakt & Studium, Werkstoffe, Expert Verlag.

Articles

11) *Neues dynamisches Messverfahren zur Härteprüfung metallischer Werkstoffe (New dynamic measurement techniques for hardness testing of metallic materials)*. Dipl. Ing. D. Leeb, Zürich Schweiz VDI-Berichte Nr. 308, 1978.

12) *Quicker, Simpler Hardness Testing using Ultrasonics*, J. Szilaj, **Ultrasonics**, July 1984.

13) *Sammenlignende hårdhedsmålinger i HAZ (Comparative hardness measurements in the Heat Affected Zone)*. J. Vagn Hansen, DMS, Winter Meeting 1983.

14) *Usikkerhed ved hårdhedsmålinger (Uncertainties in hardness measurements)*, KC Seminar, Esbjerg, 1983.03.07.

15) *Ikke destruktiv metallurgi (Non-destructive metallurgy)*. J. Vagn Hansen, DMS, Winter Meeting 1986.

Chapter 11

Hydrogen cell

NDE principle

In many corrosion processes generation of hydrogen occurs to a greater or lesser extent. The hydrogen can be liberated and measured in the environment, e.g. in the steam from a power station boiler. It can also permeate the metal (hydrogen permeation) and cause damage, but this can also provide a means for measuring the generation of hydrogen non-destructively on the external surface of the container wall. A newly developed hydrogen cell can carry out these measurements with great sensitivity and perhaps lead the way to new applications of this technology.

Hydrogen generation

A corrosion process, e.g. iron which rusts, is an oxidation process, where the metallic iron is oxidized while something nearby is reduced. If we ignore corrosion in particularly aggressive oxidizing agents such as chlorine, iron chloride or potassium nitrate, then the corrosion will involve a reaction with either:

1) oxygen (no hydrogen generation),

2) water (the water is "broken down" to form oxygen and hydrogen, i.e. hydrogen generation occurs),

3) acid (the hydrogen ions of the acid are changed to free hydrogen).

For completeness it should be mentioned that hydrogen generation and hydrogen uptake can occur during the process of cathodic protection, where corrosion does not occur at the same time, and hydrogen generation can occur due to reactions between the first corrosion products formed and water.

Process (1) is dominant in non-acidic environments where the oxygen in the air has access, but during the corrosion of iron some of the corrosion, less than 1%, occurs with hydrogen generation by process (2). If iron is pitted, there will be an acidic environment at the bottom of the pits which can cause hydrogen generation by means of process (3). During the corrosion of less noble metals, zinc, aluminium and particularly magnesium, a greater proportion of the corrosion can be due to process (2).

During hydrogen generating corrosion of steel, particularly corrosion in acids, there can be great differences in how great a proportion of the hydrogen is absorbed in the steel, and there are many examples of materials which impede the rate of corrosion in acids, but simultaneously increase the quantity of hydrogen absorbed. The amount of hydrogen which is absorbed in the steel can vary from a few to almost 100% of the quantity of hydrogen generated.

When using the evolution of hydrogen as a measure of corrosion rate it is of course a source of uncertainty if one does not know how great a proportion of the corrosion occurs with the generation of hydrogen, or if one does not know whether the hydrogen is primarily generated freely or if it is absorbed by the metal. On the other hand hydrogen absorption in the metal provides new possibilities for non-destructive testing, and often it is the absorption of hydrogen in the metal in particular, which one wishes to study, because it is precisely this effect which makes corrosion dangerous.

Development of the method

A known technique with power plant boilers is the measurement of the hydrogen content of the steam. Hydrogen is generated especially in new or newly chemically cleaned boilers, and the generation of hydrogen falls off quickly when a protective layer of magnetite has been formed in the pipes. Continued monitoring of the quantity of hydrogen can reveal abnormal new corrosion.

In the oil and gas industry hydrogen damage in steel is a big problem in connection with the handling of sour gas (gas with hydrogen sulphides), and measurement of the hydrogen generation and hydrogen permeation is often employed. Often a very simple and rough measurement technique is used, namely a manometer which records a pressure increase in the hydrogen generated. Insertion probes may also be used, often at the end of a closed pipe which is inserted into the corrosive medium and which is expected to corrode in the same manner as the materials used in the construction. Patch probes are also used. They measure the hydrogen which has diffused through the steel wall and which is collected in a narrow space between the pipe and an affixed plate.

The manometric measurement techniques just mentioned are quite inaccurate, but nevertheless often used in practice, perhaps because the alternative electrochemical measurement methods have been too expensive and inconvenient to use until now. Ref. 1 provides an overview of the status of the technique a few years ago.

The electrochemical methods make use of the electrochemical conversion of hydrogen to hydrogen ions:

$$H \rightarrow H^+ + e^-$$

The measurement can be carried out potentiometrically, i.e. the activity of the hydrogen or its partial pressure is converted to an electrical potential which is measured by means of a high-impedance voltmeter. The sensor does not consume hydrogen during the measurement. The amperometric sensor consumes the hydrogen received and converts it to a current which is proportional to the hydrogen flux.

The hydrogen cell is an example of an amperometric hydrogen detector which has been developed by the Corrosion Institute (Refs. 2-3). It has the shape of a little "button cell" and contains an oxidizing cathode with an appropriately high oxidation potential, an alkaline electrolyte and a membrane of either iron or palladium which constitutes the flat side of the cell. Both iron and palladium are permeable for hydrogen, but the iron membrane absorbs in practice only part of the hydrogen which is generated due to a cathodic reaction on the surface of the iron, e.g. in a galvanic process. Palladium is the only metal which easily absorbs gaseous (molecular) hydrogen, and the palladium cell can therefore measure hydrogen generated elsewhere.

In both cases the hydrogen is measured by means of the oxidation of hydrogen ions on the inner surface of the membrane, whereby the cell produces a current which is proportional to the hydrogen flux. One might say that the diffusing hydrogen acts like the fuel in a fuel cell.

During measurement the cell must therefore be connected to a sensitive ammeter with low internal resistance. In practice the cell is supplied with a "short-circuit" resistance of 1000 ohms. This can be used as a current measurement shunt (1 μA corresponds to 1 mV output), and in addition it ensures that the lowest possible background current has been achieved at the outset of the measurement. Typical data for the hydrogen cell are as follows:

Background current: 0.01-0.05 μA at 20°C
Measuring range: 0-200 μA
Conversion factor: 1 μA corresponds to 1.25 x 10^{-7} ml H_2/s.
Lifetime: Shelf life over 12 months.
                      Capacity ca. 0.2 A·h

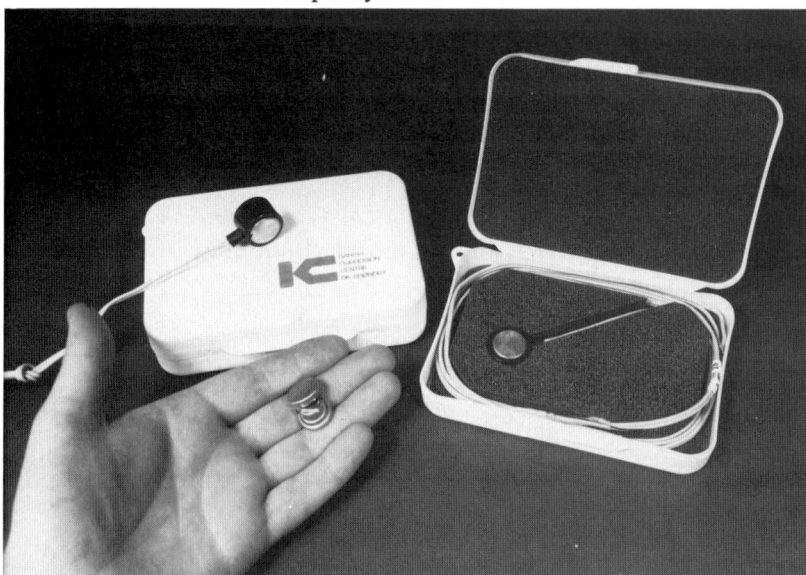

Figure 11.1. Hydrogen cell mounted in a plastic housing.

Areas of application

The cell with an iron membrane is designed to be inserted directly into the corrosive medium, either water based or non-water based liquids or moist gas. Figure 11.1 shows a cell mounted in a plastic housing so that only the iron membrane is exposed when the cell is subjected to corrosion. With this sort of mounting it is easy to dip the cell into a corrosive liquid or let the iron membrane of the cell be exposed to a galvanotechnical process.

Even the corrosive environment which is as mild as that present in a freshly poured glass of carbonated mineral water clearly produces hydrogen absorption and a current of 1-3 µA. Because hydrogen diffuses quite quickly in pure iron, it only takes two or three minutes before the cell begins to produce a current.

Among the more serious applications are that the cell (with iron membrane) can be used to measure the hydrogen absorption which steel can encounter in various cleaning or staining baths, phosphating processes, etc. It can also be subjected to galvanic processes and possibly follow the treated items through all stages in an automatic surface treatment process. The cell can thus be used to monitor all processes which entail a risk of hydrogen embrittlement for high strength steel.

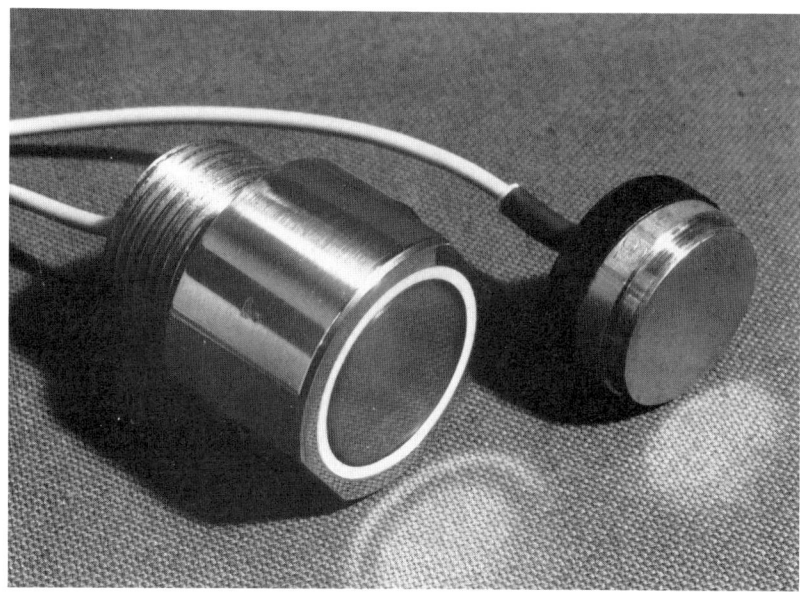

Figure 11.2. Hydrogen cells. Access fittings allow placement inside pressurized systems.

A version of the iron cell has also been developed where the membrane can be a little thicker and can be formed from any given type of steel. In this case the membrane is welded onto a housing of stainless steel, and the connection from the cathode is through a glass seal. The cell contains no plastic sealant and it can handle somewhat higher temperatures. This cell can be mounted in a stainless steel housing which fits commercial access fittings and can be placed into pressurized systems in this manner (Figure 11.2). Because the cell is 100% water-filled, it can be subjected to any isostatic pressure, and it can as mentioned be fabricated from the steel which is used as a construction material in the system of interest.

Cells of this type have been in use at the Corrosion Institute for the measurement of hydrogen in steel subject to cathodic protection in seawater and have operated completely submerged in seawater or in the seabed for considerably more than a year.

So far the most important application of hydrogen cells is the use of palladium cells for external measurement of the hydrogen which has permeated out through a steel pipe or container wall due to internal corrosion. In this case the palladium cell need only be placed in intimate contact with a flat, ground part of the surface, where the narrow space is filled with a little silicone grease, which allows the hydrogen to diffuse through and keeps air and contaminants out. The palladium surface of the cell must be cleaned by light abrasion with emery paper immediately before use, but the cell can otherwise operate for a long time when it has been positioned and secured by means of a hose clamp, magnet or in some other way.

Practical examples

Figure 11.3 shows how one can carry out an external measurement of diffused hydrogen. The internal corrosion produces hydrogen, some of which is absorbed by the steel and diffuses through it.

The palladium cell is placed in close contact with a cleaned area on the steel wall, and the narrow space is filled with a bit of silicone grease. The current from the cell is measured by measuring the voltage drop across a measuring resistance, e.g. 1000 ohms, so that a current of 1 μA yields an output of 1 mV.

Figure 11.4 shows a number of cells employed in an experiment with a steel pipe filled with a corrosive test solution under pressure. The hydrogen flow is measured at eight locations around the pipe (the cell can operate in any position), and the results are printed out via a datalogger.

It takes one or two days before the hydrogen permeates a 10-12 mm (0.4-0.5") thick steel wall.

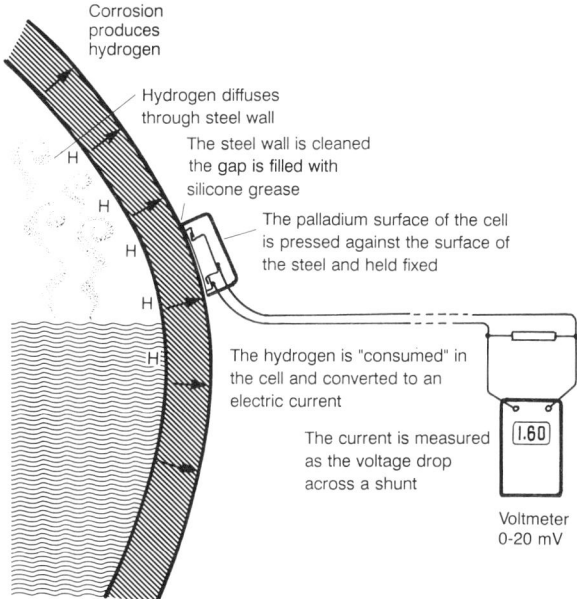

Figure 11.3. A palladium cell is mounted on the wall of a steel tank.

Hydrogen cell

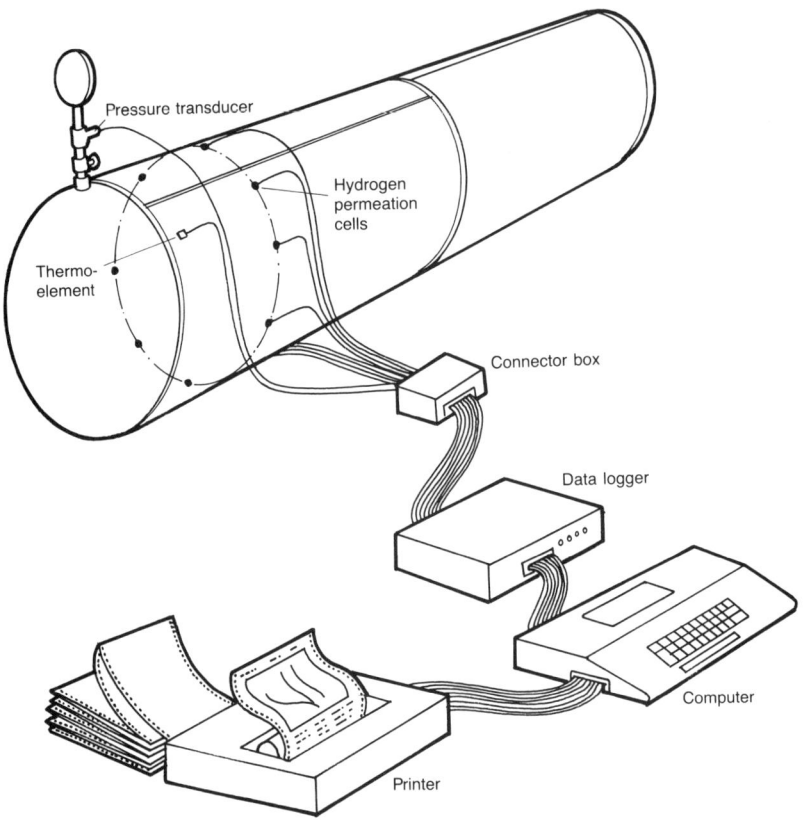

Figure 11.4. Examination of a steel pipe using the hydrogen cell technique.

Training required/desired

The method is simple to use and requires only a brief period of instruction.

Hans Arup

References

1) *Barnacle Electrode Measurement System for Hydrogen in Steels*. F. Mansfeld, S. Jeanjaquet, and D.K. Roe, **Materials Performance**, February, 1982.

2) *Measuring Hydrogen Activity with a Sealed Devanathan Cell*, Hans Arup, 9th Scandinavian Corrosion Congress, Copenhagen, 1983.

3) *Hydrogen Pick-Up in Steel Exposed to Corrosion or Cathodic Protection and Examples of Damage Caused by Hydrogen*, Hans Arup, 10th Scandinavian Corrosion Congress, Stockholm, 1986.

Chapter 12

Isotope techniques

NDE principle

Isotope techniques is a descriptive term for a wide range of applications of radioactive isotopes and measuring instruments to the measurement of radioactivity from the isotopes. The methods can be divided into two main categories: *tracer techniques* and *radiometry*.

Tracer techniques

The radiation emitted from a *tracer* is due to the radioactive decay of unstable atomic isotopes. The tracer technique takes advantage of the fact that various isotopes of the same element have identical chemical properties and follow the liquid, gas or solid material containing the radioactive isotopes in a system. Normally gamma-active isotopes are used (gamma radiation is electromagnetic radiation just as X-radiation), and measurements can, therefore, be carried out through pipe walls or even certain thicknesses of concrete. The tracer technique is used particularly for the detection of leaks from hidden pipe installations.

In general, the tracer technique can also be used with non-radioactive isotopes. In the following sections several examples of the use of radioactive tracers are provided, and a single example of a non-radioactive tracer is given.

Radiometry

Radiometry is a technique which corresponds in principle to X-ray photography, but instead of using a radiation-sensitive film to create an image of the distribution of radiation, a radiation detector and a measuring instrument are used. One does not obtain an image of the interior of the item under examination, but instead a quantitative measure of the absorption or scattering of the radiation in a region of the item is obtained. In this manner it is possible to carry out an accurate measurement of thickness, layer thickness and density or to localize defects such as inhomogeneities or cavities.

The absorption and scattering of the radiation depends upon the chemical composition of the object. When radiation particles (photons or electrons) collide with atoms, X-radiation may be reemitted with a wavelength which is characteristic for the element. Both transmitted radiation (absorption) as well as scattered and secondary radiation (X-ray fluorescence) can, therefore, be used for a number of types of non-destructive examination analyses.

Development of the method

Isotope techniques were developed primarily after World War II, when the construction of atomic reactors made a large selection of artificially prepared radioactive isotopes available. Both the tracer methods as well as the radiometric methods are most commonly used in medicine, and in many cases there is a strong analogy between the technical and medical applications. For example, the principle for the measurement of flow rates in industrial piping is the same basic principle upon which the measurement of the flow of blood to parts of the brain or other organs is based. In both cases the liquid flow is followed by means of the addition of radioactive tracers, and the movement of fluid is revealed by means of external detection of radioactivity.

Safety

It should be emphasized that the use of radioactive isotopes does not normally pose a serious danger due to radiation. All applications are performed according to standard operating procedures which limit the radiation as much as possible. When tracers are used, the measuring instruments are so sensitive that only a slight increase of the radiation level compared to the natural background radiation level occurs. When a radiation source is used in connection with radiometry, the source is always shielded in such a manner that only a very narrow beam of radiation is directed at the object of examination. Detection in this case is much more sensitive than when a film is used. The intensities of the radiation sources used in radiometry are, therefore, typically a factor 100 times less than corresponding radiography sources.

Measurement

The detection of radioactivity is usually a matter of detecting individual particles (alpha or beta radiation) or photons (X-rays or gamma rays). The traditional measuring instrument has been the Geiger counter. Present day detectors are far more sensitive than the Geiger counter, but in principle it is still individual electrical pulses which are counted. Isotope measuring techniques are, therefore, well-suited to digital data handling, and currently used measuring techniques are closely linked to the use of microprocessor technology and electronic data processing.

Areas of application

Here is an overview of the most important areas in which isotope techniques are used.

Tracers

- *Leak detection*
  Hidden pipes, tanks, heat exchangers.

- *Flow measurement, closed pipes*
  liquids: ISO Standard 2975
  gases: ISO Standard 4053
- *Residence time measurement*
  chemical processing systems
  water purification facilities
- *Mixing time measurements*
  continuous mixing
  batch mixing
- *Air exchange measurement:* Nordtest BUILD 256
- *Wear measurement*
- *Ball test*

Radiometry

- *Checking joints, etc*: Nordtest BUILD 123
- *Moisture and compression*
- *Pressure tanks*, checking filling of CO_2 fire extinguishers
- *Coating thickness*
- *Smoke detection*
- *Process monitoring*
- *Collimated photon scattering*

Other applications

- Steel qualities (metal alloys) with portable equipment based upon X-ray fluorescence (see introduction)
- Worn parts in lubricating oil (X-ray fluorescence)
- Silver nitrate in fixer liquid (absorption, gamma)
- Sulphur in oil (absorption, gamma)
- Ash in coal (reflection, gamma)
- Area weight of rugs, textiles, plastics (reflection, absorption)
- Dust emission (absorption measurements of filter samples)

Practical examples

Leak detection in *hidden piping installations*, particularly pipes in single-storey homes, snow melting systems and other locations where pipes are hidden inside cement floor constructions are by far the most widely used application of the tracer technique. The basic idea behind this routinely used method is illustrated in Figure 12.1. Leaks as small as a few drops per day can be located to within a radius of c. 10 cm. A prerequisite for being able to localize leaks in this manner is that the pipes are no more than about 50 cm below the surface. If they are deeper, e.g. buried in the ground, the method can sometimes be used by sealing off a section of pipe and pumping a charge of tracer into the pipe. The charge will move forward towards the leak and spread out there. The movement of the tracer can be followed by measuring at a few holes dug or bored along the pipe.

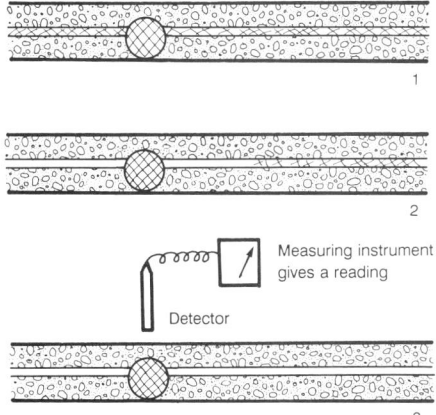

Figure 12.1. Leak detection in pipes cast in cement, e.g. floor heating installations.
1. The radioactive tracer is added to the system and circulated under pressure.
2. The tracer is flushed out, and the pipe is flushed with pure water.
3. The leak is localized with the aid of a sensitive measuring instrument.

Tanks and watertight membranes. Freely standing, vertical, cylindrical tanks can be examined by means of a technique which corresponds to the detection of leaks in hidden pipes. The tank is filled to a certain level with a tracer solution. After a time the tank is emptied and the inside surface is flushed clean. Measurements inside the tank can then reveal those places where the tracer has penetrated. A similar technique can be used for water-tight membranes provided that the membrane is at the top of the construction, and provided it is possible to dam up a region containing tracer solution. In many cases one can apply the neutron method which is described in the section of this chapter on radiometry.

It is often of interest to determine whether leaking water or other liquid stems from one of several possible tanks before a more detailed search is initiated. In this case a tracer can be mixed in one or more tanks, then samples of the leaking fluid can be collected and analyzed for their tracer content.

In these cases a so-called *activatable tracer* can be used. Such a tracer is not radioactive at the time of the examination but it can be detected with great sensitivity (10^{-11} to 10^{-13} g) by measurement of the radioactivity in the sample after it has been irradiated by a neutron source, e.g. in a reactor.

Flow measurement

Flow measurement with radioactive tracers is used particularly for the monitoring and calibration of flow meters which are permanently installed in piping systems in industry and in oil and gas processing systems on off-shore installations or installations on land. The principle is illustrated in Figure 12.2. The linear flow rate can often be measured with an accuracy better than 0.5%. In most cases one can find an existing valve or other part into which the tracer can be injected, and the measurement can thus be carried out without disrupting the system.

In open channels a dilution technique is used as shown in Figure 12.3. Depending upon the dosing equipment available, accuracy will be between 1 and 5% of the volume flow rate.

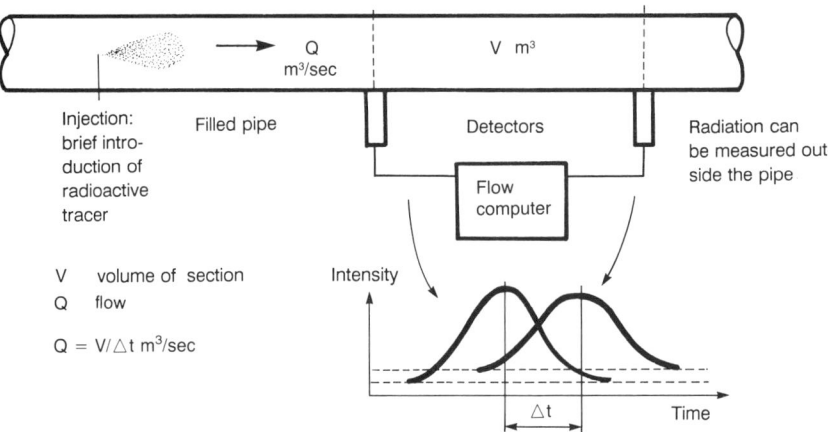

Figure 12.2. Flow measurement in filled pipe. Measurement of volume flow rate.

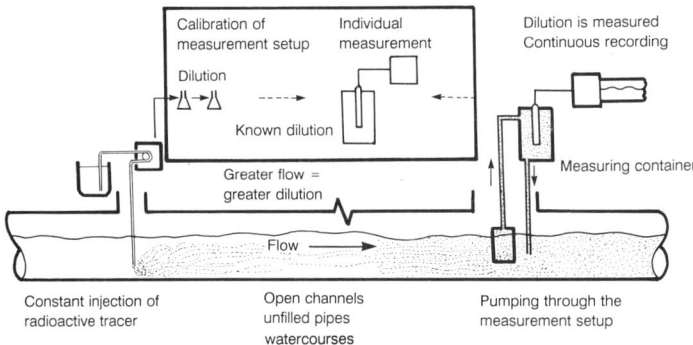

Figure 12.3. Flow measurement in open channels and unfilled pipes. The dilution method.

Residence-time measurement.

In systems where chemical processes, heat treatment, sedimentation, etc. occur, both the mean residence-time and the standard deviation of the residence-time are extremely important factors affecting the efficiency of the system. The residence-time and its distribution are measured simply by performing a brief injection of a radioactive tracer at the inlet to the processing unit, and measuring the tracer concentration at the outlet. If the system is closed a radioactivity detector is placed at a convenient place on the outside surface of a pipe or tank wall. The method is shown in Figure 12.4. Short circuit flow is revealed quite clearly, and on the basis of the measurements it is normally possible to develop a mathematical model for the system. If the volume of the system and the flow are known with sufficient accuracy, the measurement will reveal the size of any "dead" volumes through which flow does not occur effectively.

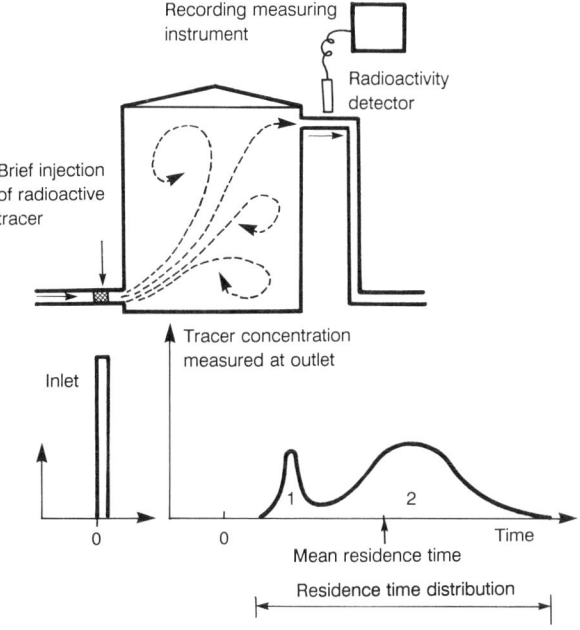

Figure 12.4. Measurement of the residence time (residence time distribution) in a container by means of a radioactive tracer. The signal recorded at the outlet shows: (1) tendencies to short circuit currents, (2) the normal deviation of the residence time of the material.

Mixing time measurements

Continuous mixing processes can be examined as described above. Using the recorded residence time distribution one can compute the evening-out effects of input variations of various magnitude and duration. Charge (batch) mixers such as cement mixers and similar systems can be studied by adding a portion of the material marked with a radioactive isotope. The mixing process can then be followed by placing one or more detectors on the outside surface of the mixing machine.

Air renewal measurements

Radioactive tracer gases can be used in the same manner as many other tracer gases for the measurement of air renewal in *closed rooms*. The tracer gas is dosed and kept mixed in the volume of the room. The concentration is measured and recorded. Air renewal is computed by measuring the rate of decay of the concentration of the tracer material. The advantage of using a radioactive tracer gas is that it can be measured continuously with great sensitivity, and a detector installed in the room will be affected by radiation from a volume on the order of one cubic metre in contrast to other measuring techniques where a flow of air is forced through an analyzing instrument, or where air samples are collected for later analysis. Typical air exchange rates determined by this method will be between 0.1 and 10 air exchanges per hour.

In very *open, freely ventilated rooms* this method can not be used, for it is impossible to hold the tracer gas mixed in such rooms. In such cases a technique is applied in which the tracer gas is forced into the entire inlet cross section an appropriate number of times while the passage of the material through the outlet cross section is recorded. In this case it is very important that the isotope method allows the registration of transient tracer concentration (down to 0.01 s) and that the detector receives signals from a one square metre cross section of the air flow. By means of an appropriate number of measurements the total air exchange for the room is determined. But it is important that it is also possible to ascertain the minimum air exchange which is found in a "dead" region in the room, even if this region is not identified physically. In open off-shore processing modules this method has been used to measure air renewal rates of up to 900 times per hour.

Wear measurements

Wear measurements with radioactive labelling is used in connection with the development of new or altered machine parts and in connection with the examination of various corrosion phenomena. Using a special labelling technique, *thin layer activation* (TLA), it is possible to create radioactive isotopes of a number of the elements which are found in the steel or metal alloy of the object under examination. Activation is typically undertaken to depths of 0.1 mm on the wearing surface and can be applied to areas from about 10 mm^2 and up. The measuring method is illustrated in Figures 12.5 and 12.6. TLA is presently carried out routinely on metal and plastic parts. The method is often used to examine the cylinders of ship motors and other objects with a chemically prepared nickel finish.

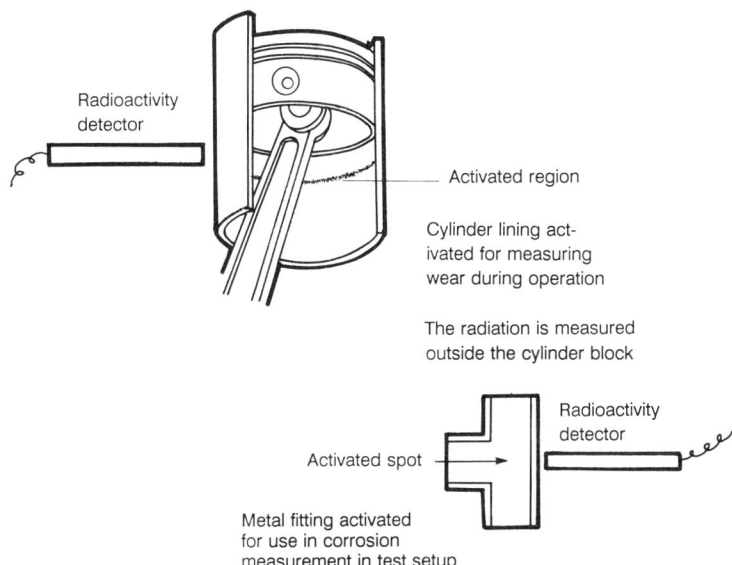

Figure 12.5. Wear and corrosion measurements by means of thin layer activation.

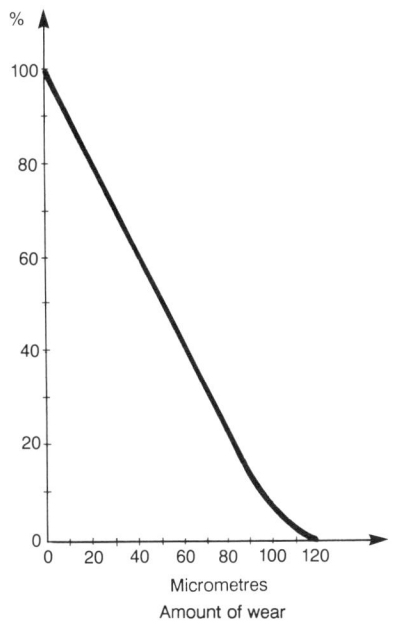

Figure 12.6. Calibration curve for wear measurements.

Ball test

Pipelines and other large piping systems can be cleaned by pumping various types of cleaning brushes, also called "pigs" or "spheres" through them. Particularly in connection with the cleaning of new systems there is the risk that the cleaning pig will get stuck. An appropriate radioactive source can be mounted on the brush so that it is easier to find if it does get stuck.

Joint monitoring

When modular buildings are being assembled the elements are not placed directly on the underlying deck. In order to even out variations in the height of the concrete elements, they are placed 2-4 cm above the deck, supported by a couple of leveling bolts. After mounting, the gap below the elements is filled up with a cement mortar, and it is of course very important for the carrying capacity of the element that the joint is filled in compactly and uniformly.

A radiometric measurement can be used to check the mean density of the cement mortar. A radioactive source is placed on one side of the joint. It sends a narrow beam of gamma rays through the joint. On the other side of the joint a detector is centred in the radiation beam (Figure 12.7). The measured radiation intensity can then be converted to the density of the material in the joint at the point of measurement (Figure 12.8).

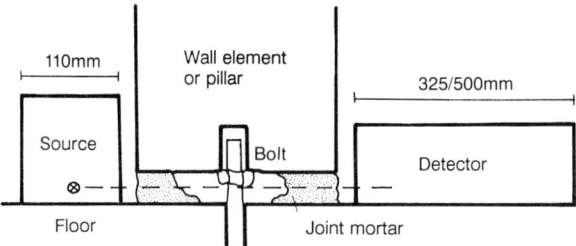

Figure 12.7. Source and detector setup. In the case illustrated the surface is all right, but the middle of the joint is not filled in. The mean density of the mortar in the joint, i.e. the mass per unit volume, will be too small. The space required on either side of the joint for standard equipment is indicated in millimetres.

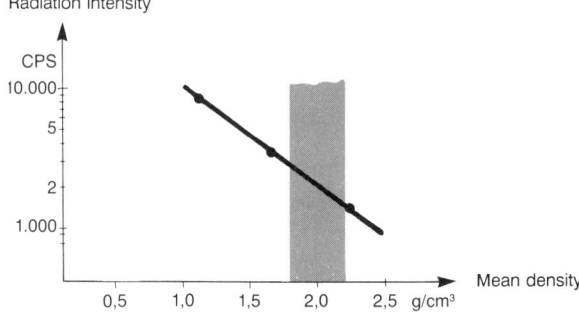

Figure 12.8. The calibration curve converts the measured intensity to mean density. The region within which the density normally is required to lie is shown as a grey band in the figure.

The method is routinely applied to make spot checks of the quality of the work and is described in detail in Nordtest Build 123. Radiometric measurement is not limited to the checking of joints. It is also used in various forms as an alternative to X-ray or radiographic control for example for checking the concrete casting around and the position (and presence!) of iron reinforcement in concrete. The method is also used to check the density of concrete elements and to study the water movements (change in water content) in various construction materials.

Moisture and compression

The density (compression) and the moisture content on the surface as well as in deeper layers of earth can be determined today by means of commercially available isotope measurement equipment.

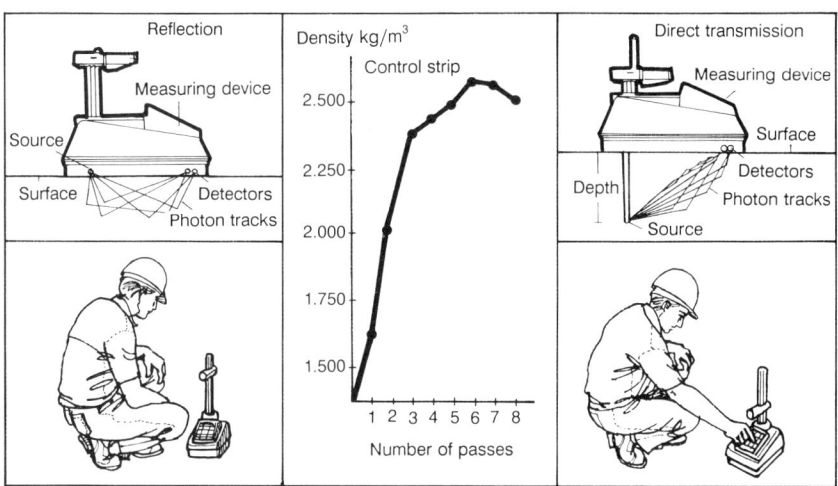

Figure 12.9. Compression monitoring. Surface – deeper levels.

Isotope techniques

The measuring probe (Figure 12.9) contains a gamma and/or a neutron radiation source, one or more radiation detectors and a signal processing unit. When measurements are performed in surface layers (at depths of 5 to 10 cm), then both radiation sources are placed in the measuring probe. The reflected gamma radiation is a measure of the density (see Figure 12.9, compression increases with the number of passes with a roller), while the reflected neutron radiation is a measure of the hydrogen content (and thus the water content). To perform density measurements at greater depths the gamma source can be forced down to a depth of 30 cm, then a measurement of the transmitted radiation can be converted to density.

The method has been used for about twenty years (particularly in the US). The method has become more and more important in Europe in recent years for checking the quality of extensive repair work in streets and highways due to the distribution of natural gas, district heating and telecommunications lines.

Among other uses for the technique one might mention the checking of moisture penetration in roof constructions and tank walls as well as the measurement of the bitumen content (hydrogen content) of asphalt.

Pressure tanks

The filling of CO_2 fire extinguisher tanks (and pressure cylinders containing liquids) can be checked quickly and simply by means of a small fork-shaped instrument containing a weak gamma source and a Geiger counter. The significant difference in the absorption of radiation in the liquid and gas phase provides a reliable level indication (see also joint control).

Coating thickness

The coating thickness, particularly of noble metals on plated dinnerware, circuit boards, connectors, etc. can be checked by means of measuring equipment based upon the reflection of beta radiation from radioactive isotopes. Measuring equipment with various radiation sources typically covers the region from a few micrometres to several hundred micrometres (see Figure 12.10).

A variation on this equipment can be used to determine the proportion of tin to lead in soldering materials.

With more advanced (and expensive) equipment which utilizes X-ray fluorescence, it is possible to perform analyses of the chemical composition of coatings and substrates.

The areas of application of these techniques are the electronics and noble metal industries.

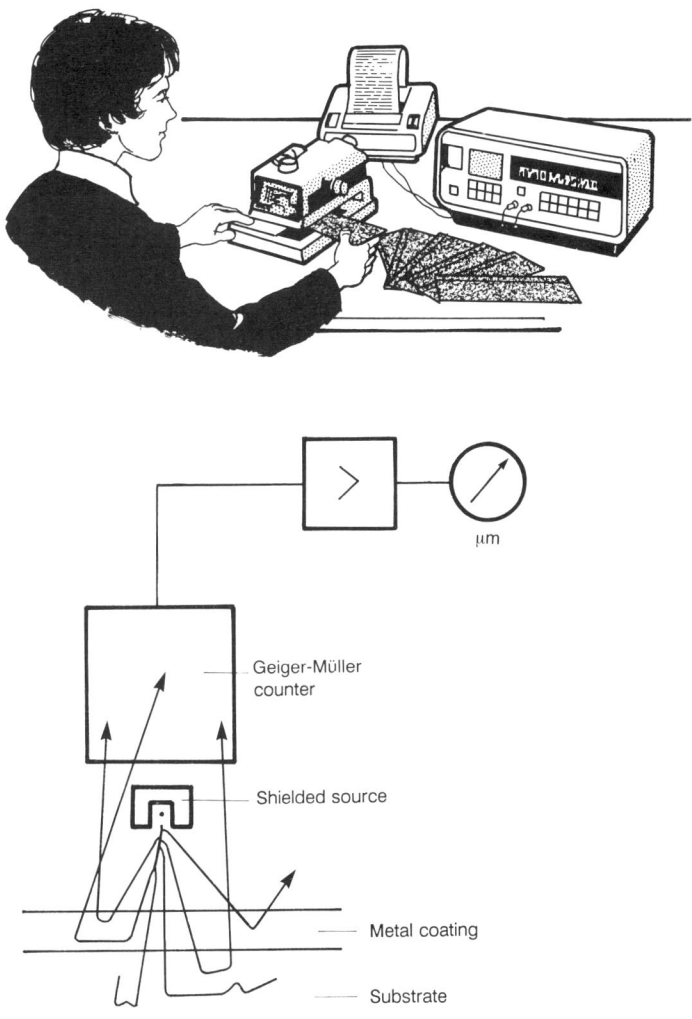

Figure 12.10. Coating thickness measurement. Beta reflection.

Smoke detection

Smoke detectors in automatic fire alarm systems are often based upon small radioactive sources. A radiation source which emits heavy, charged particles (alpha particles) ionizes the air between two electrodes and causes a small current (ion current) is measured. When smoke particles enter the device and come between the electrodes, the ion current is reduced and the alarm is set off.

Isotope techniques

Process monitoring

All of the measuring techniques mentioned are also used extensively in production industries for process monitoring (level measurement, filling control, thickness measurement, area weight, density and chemical analysis), and every measurement of a process parameter provides in principle a check of the state of a production line (see the references).

Collimated photon scattering

A technique which is termed *collimated photon scattering* (the photons being gamma quanta) appears to possess important potential as a non-destructive examination technique.

As suggested by the term, a narrowly collimated beam of gamma radiation is sent towards the object of the measurement, and one or more collimated detectors measure the gamma radiation which is scattered from a small, well-defined region in the object (Figure 12.11).

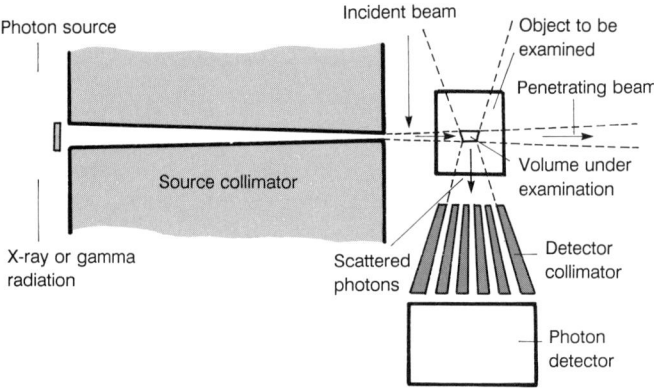

Figure 12.11. Collimated photon scattering.

The method is well suited e.g. to the measurement of density or to the localization of defects in light materials which are encapsulated in denser, heavier materials, where the contrast with gamma radiography or radiometry in general is poor (neutron radiography can, however, be an alternative). American studies point out, among many possibilities, the use of this method to check ammunition (both during production and for examining the cause of misfires), examination of cooling channels in the turbine blades of jet engines and the checking of complex castings, see Reference 3.

The method is also well suited to the examination of materials or constructions which are not suited to inspection by means of ultrasonics, e.g. inhomogeneous or air-filled structures with cellular plastic.

In connection with the latter application the Danish Isotope Centre has developed measuring equipment for checking the foam density (insulation) in prefabricated district heating pipes (Polymeter, see Figure 12.12). The equipment, which is used routinely by several pipe producers, can also be used to check another quality parameter in district heating pipes: the eccentricity between the inner pipe (the medium pipe) and the outer pipe after foam filling. A new version is under development for use with foam fill checks in freezers, refrigerators and various other types of sandwich constructions.

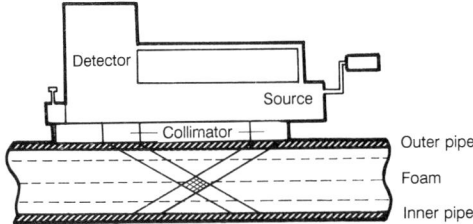

Figure 12.12. Polymeter. Foam density.

Development trends

In the border discipline between *radiometry*, with real time dynamic measurements with high sensitivity detectors but poor geometric resolution, and *radiography* with fine geometric resolution but relatively long (stationary) exposure times, important developments are predicted due to the extensive research which has been carried out in the field of medical imaging.

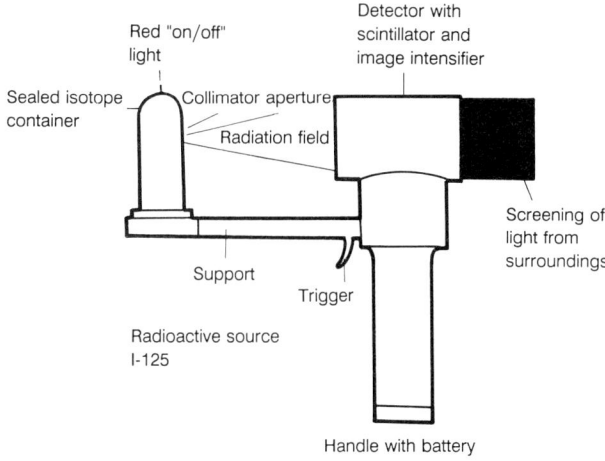

Figure 12.13. Lixiscope.

Isotope techniques

A scintillation screen (where gamma or X-radiation causes the emission of light) attached to a high sensitivity image intensifier system is the immediate successor to the radiography film (e.g. baggage checking in airports). More sensitive systems which do not require the use of X-ray equipment or intense radiographic sources are presently available commercially, e.g. see the Lixiscope (Low Intensity X-ray Image system) in Figure 12.13. It is used for the quality control of multi-layer circuit boards in the electronics industry.

Corresponding equipment with more penetrating radiation (Cobalt-60, Caesium-137) for the examination of larger metal items is under development, among other places in West Germany.

By complete or partial rotation of a radiation source and detector system around an object (e.g. a patient) the medical CAT-scanning technique (Computer Assisted Tomography) can provide a complete cross sectional image. CAT-scanners have great potential as a non-destructive examination tool, although the current prices of £500,000 to £1,000,000 will place some constraints on its dissemination. For medical purposes only X-ray tubes are used as radiation sources, but for industial purposes the first experiments using radioactive isotope sources have begun. See Reference 4.

Reference 5 provides an excellent introduction to the measuring principle and demonstrates the use of the technique for the location of defects (rot and knots) in timber.

Training required/desired

The use of radioactive tracers and radioactive sources requires a permit from the appropriate government supervisory agency. A prerequisite for such a permit is that the individual responsible for their use has adequate practical and theoretical knowledge of radiation safety practices and the correct procedures for handling radioactive materials.

The use of radioactive tracers requires extensive education and training, and those uses which have been described can often be requisitioned from a speciality centre with trained and experienced personnel. Leak detection in hidden piping systems is, however, only carried out by certain specialized firms.

The use of closed radiation sources requires training which can vary a great deal depending upon the type of equipment. Often the supplier of the equipment can provide the necessary training programme. In other cases a government institution or agency may offer special courses of instruction in this area.

Erik Mørch
Steen Teller

References

Tracer Techniques

1) *Guidebook on Radioisotope Tracers in Industry*, International Atomic Energy Agency (IAEA), Vienna, to be published in 1989.

2) *Safe Use of Radioactive Tracers in Industrial Processes*, IAEA safety No. 40, 1974.

Radiometry
Research results
3) *Some New Applications of Collimated Photon Scattering for Non-Destructive Examination*. J.A. Stokes, et al., **Nuclear Instruments and Methods, 193**, 1982, 261-267.

4) *A Tomographic Gamma-ray Scanner for Industrial Applications*, W.B. Gilboy et al., **Nuclear Instruments and Methods, 193**, 1982, 209-214.

5) *Detection of Defects in Logs Using Computer Assisted Tomography (CAT) Scanning*, P.D. Tonner, L.R. Lupton, Atomic Energy of Canada Ltd, AECL-7742, 1985.

Handbook
6) *Radioisotope Instruments*, J.F. Cameron and C.G. Clayton, Pergamon Press, 1971.

Overviews
7) *Industrial and Analytical Applications of Radioisotope Radiation Sources*, E.A. Lorch, **Jour. of Radioanal. Chem, 48**, 1979, 209-212.

8) *Industry, product and information data sheets from the manufacturer and government agencies.*

Safety training programs

9) *Report on the Applicability of International Radiation Protection Recommendations in the Nordic Countries*, Liber tryk, Stockholm, 1976.

10) *Safe Handling of Radionuclides*, IAEA safety series no. 1, Vienna, 1973.

Chapter 13

Laser distance measurement

NDE principle

The optocator is an opto-electronic instrument which employs a triangulation principle to measure the distance to an object without physical contact.

The measuring equipment can have a variety of configurations depending on the measuring setup. The system consists of a laser, a camera and normally a computer which collects and handles the measurement data. A typical distance measurement arrangement with a laser and a camera is shown in Figures 1a and 1b.

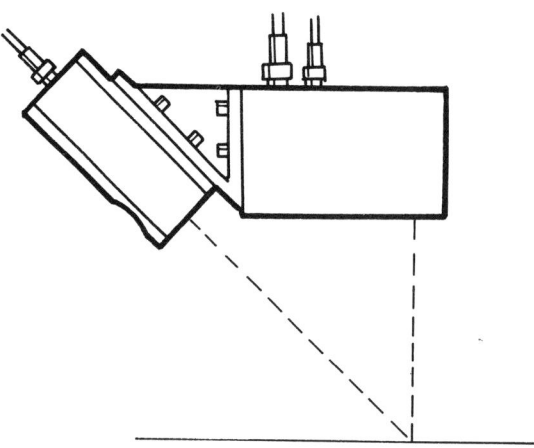

Figure 13.1a. Distance measurement.

This discussion refers to Figure 13.1b on the following page. A beam of light is transmitted from a laser L down to the surface of the object to be measured (O_1). The beam will strike the surface of the object at point A. The reflected light is focused via a lens at point A'.

If the distance from the measuring instrument to the object changes by an amount X, then the beam of light will strike the surface (O_2) at a new point B. The image of the point B in the detector B' is then changed from A' with the distance X'.

The connection between the change in distance X and the translation X' of the spot of light is dependent upon the geometry of the measuring apparatus. The connection is non-linear. Because the light source, the lens and the detector are always mechanically fixed, the connection between X and X' is known and can be used to linearize the data.

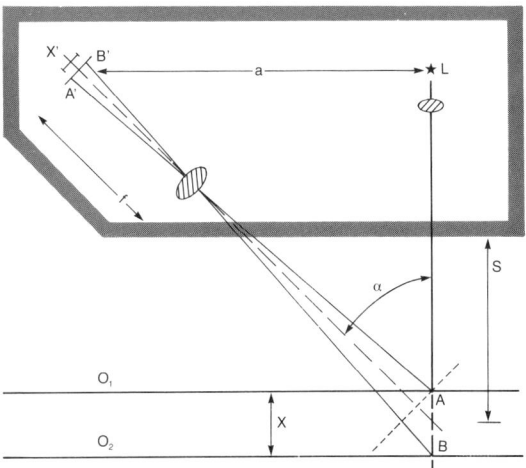

Figure 13.1b. Width measurements.

Region of measurement

Both the length ($O_2 - O_1$) of the measuring region and the stand-off distance S are determined by:

- the focal length (f) of the lens in front of the detector,
- the distance between the light source and the detector (a),
- the angle α between the beam and the optical axis of the detector.

Measuring accuracy

Various "ready-to-use" systems are available on the market today, complete with respect to both hardware and software. The systems are characterized by their high accuracy, their measurement frequency and the fact that they contain no moving parts.

The measurement accuracy is in the order of 1/4000th of the length of the measuring region with up to 30,000 measurements per second. The stand-off distance can vary from 5 mm to 1 metre.

Laser distance measurement

Areas of application and practical examples

Almost all types of materials can be measured by means of this method. The only requirement which must be fulfilled is that it must be able to reflect infrared radiation.

Wood, plastic and metals

Many applications exist in connection with the fabrication of wire and plates. Measurements are carried out directly on the production line where profiles and thicknesses of the objects of interest can be measured.

In the metal processing area laser measurements are used, e.g. in connection with determinations of the thicknesses, widths, flatness and waviness of both glowing hot and cold items. Furthermore laser devices are used to check windings and to check the symmetry of rods and for measuring the profiles of the blades of a turbine in operation.

Robot inspection

Industrial robots equipped with optocators can examine welds, castings and surfaces (e.g. automobile bodies) with a high degree of reproducibility. The principle is illustrated in Figure 13.2.

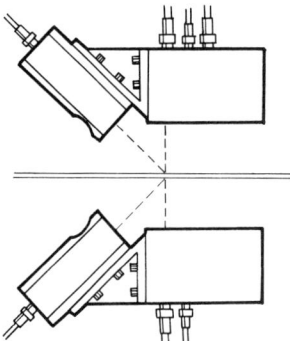

Figure 13.2. Triangulation.

Road surfaces

Laser measurements can also be used to record the condition of the surfaces of ordinary roads. With a row of optocators mounted on the front of a car, it is possible to chart the state of a road surface at speed of up to 90 km/h (55 mph).

Train wheels

With equipment attached to a large computer system it is possible to measure the flanges and running surfaces of train wheels with an accuracy of about ±1 mm. The computer can then calculate how much the individual train wheel has been worn since the previous passage. The measurements can be carried out on all wheels of the train by using four optocators, and the train can pass the measuring point at speeds of 80 km/h (50 mph) or more. Figure 13.3 illustrates the principle behind this type of measurement.

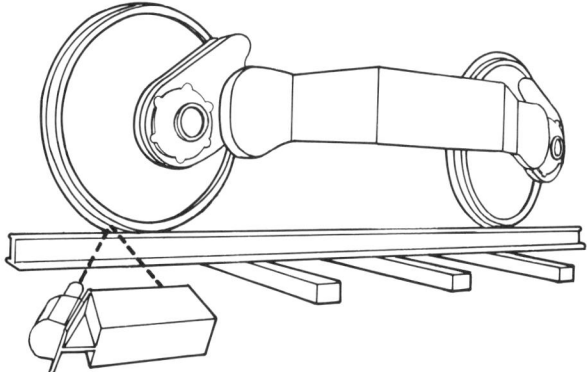

Figure 13.3. Laser measurements on the wheels of a passing train.

Training required/desired

The are no special training requirements in this area, but the use of laser measuring techniques requires thorough knowledge of the equipment, and safety regulations must also be observed. For example, if the power output is greater than 2.5 mW, then total shielding must be established in order to prevent the laser beam from striking the eye.

Søren Poulsen

References

1) *Technical Manual, Optocator and Application Reports.* Selcom Electronic C.O. AB, Box 550, S-43325 Partille 1, Göteborg, Sweden.

Chapter 14

Leak testing

NDE-principle

A *leak* can be considered to be a channel through the wall of a pipe or container. Due to irregularity of form and most often small dimensions, it is rarely possible to measure or characterize such a leak directly. It must be evaluated via its consequences which also correspond to its practical significance.

Leak testing includes, as the name suggests, a qualitative and sometimes also quantitative demonstration of the existence of leaks, and perhaps localization of the leaks in the (container) walls with respect to various media.

Media

The media in question can be:

Powder material or agglomerates

It is generally quite easy to achieve tight seals with these materials for they must often actually be "shaken" to pass through leaks due to the high friction coefficient of these materials at rest. *Litres/second* or *kilograms/second* can be used as units of measure for the size of a leak with respect to a particular material.

Liquids with various viscosities

The flow through a leak will often obey Poiseuille's law, and in such cases the rate of flow will depend – in addition to characteristics of the imperfection causing the leak – upon the pressure difference between the two sides of the leaking wall (directly proportional) and the viscosity of the liquid (inversely proportional). The unit *litres/(second·bar)* can be used as a measure of the size of the leak with respect to a particular liquid.

Gases

It is generally true of all containers and systems that perfect air-tightness does not exist. In practical situations a system or container will be considered to be air-tight provided the sum of the leaks is less than a specified limit. The concept *leak rate* has been introduced to permit quantitative descriptions of leaks.

Leak rate

In vacuum technology the leak rate L is defined as the quantity of air (or gas) which passes from the environment (under standard temperature and pressure conditions: 1013 mbar and 20°C) through leaks into the low pressure system of volume V and causes a pressure increase per unit time equal to: $\Delta p/\Delta t$.

$L = \Delta p/\Delta t \, V$ units: *mbar ·litre/second*

Instead of the above units, the units *Torr ·litre/second* are often used, where:

1 *Torr ·litre/second* = 1,333 *mbar ·litre/second*

For practical reasons it can be of interest to find the location as well as the size of the individual leaks in a system or container. In such cases the leak rate equals the quantity of air (or gas) which passes a leak per unit time (at standard temperature) due to a pressure difference between the sides of the leak. As above the units:

mbar ·litre/second

can be used. In the above case it is of course also necessary to specify the relevant gas. When comparing different gases, one can expect (all other things being equal) that

- the flow rate in the laminar flow region (which is the case for larger leaks) is inversely proportional to the viscosity, while
- the flow rate in the molecular region (true for the smallest leaks) depends upon the molecular weight of the gas.

In refrigeration technology one sometimes observes that the outward flow rate from a cooling system is inversely proportional to the square root of the molecular weight of the coolant.

Development of the method

The need to determine that leaks exist and to localize them has led to the development of a wide range of methods, some of which will be described in the following. In connection with this description the development of each method will be touched upon briefly.

Areas of application and practical examples

The accessibility of surfaces which must be examined for leaks can vary a great deal. For example, the walls of a large, detached storage tank will, as a rule, be accessible from both sides, while the bottom may only be accessible from one side. A cooling spiral welded into a tank or a double walled tank will not be accessible from any side. A pressure vessel can be tested by increasing or decreasing the pressure, while a storage tank may not be designed to withstand this.

A container can itself be a part of a piece of equipment (e.g. watch casings, transistor housings, radioactive source capsules, light sources, X-ray tubes, automobile wheels, tennis balls, air-filled shock absorbers, gas-insulated electrical switches or high voltage transformers); or containers which can be used for the storage, transportation or production of solid, liquid or gaseous products.

Choice of method

It is apparent that with such a wide variety of (container) sizes and shapes, applications and air-tightness requirements, it is difficult to generate a systematic overview of recommendations and guidance regarding specific methods of examination. Therefore, considerable all-round knowledge of leak testing principles is required in order to choose the most appropriate examination method in each individual case.

Because the options are innumerable, the description provided here must be limited to the most commonly used methods which are applied in connection with condition monitoring and maintenance.

In order to achieve some systematization these will be classified in the subsequent treatment into the following groups:

- methods based upon checking the quantity of a material.

- methods based upon pressure differences and capillary action, and

- methods based upon the detection of easily identified tracers or tracer materials.

Where comparisons are possible at all, the methods have been arranged in order of increasing sensitivity.

Finally, it should be noted that in the following sections exact procedures cannot be given, and with respect to the most appropriate use of the rather advanced equipment which is described, the user is referred to the detailed instructions provided by the manufacturers and to the relevant regulations and standards.

Quantity control

As an illustration of this method one should mention that unexpectedly large quantities of liquid have been known to disappear from storage tanks (via leaks in the bottom) and from buried tanks (for gasoline and heating oil). An up to date record of addition and withdrawal of liquid and of the current quantity will quickly reveal any anomalies which may occur.

Monitoring of pressure difference

When approving a pressurized system it should, as a rule, be liquid pressure tested for a period of at least 15 minutes (at a test pressure of 1.3 times the design pressure for products made from rolled stock, and at a pressure of 1.5 times the design pressure for cast items). The surfaces must not be insulated and they must not be covered with corrosion-protective coatings or paint. To avoid the deposition of lime in possible cracks it is often required that items be cleaned with demineralized water. In addition a wetting agent may have been added to the water. Pressure testing provides of course a simultaneous total impression of the integrity of the system, for leaks cause the pressure to decrease.

With respect to localizing leaks the system should be examined carefully after the hydrostatic pressure has been applied for a certain minimum required period. It can be an advantage to add dyes or fluorescent materials to the test liquid, making it easier to find small drops on surfaces.

It should be noted that in connection with possible later testing for leaks with air, foams or tracers that water remnants which are left behind can "block" narrow cracks. Such water remnants can be difficult to remove; air under (the highest allowable) pressure can be used for this purpose. See also the section on pressure testing.

The dye penetrant method

The dye penetrant method – previously referred to as the capillary liquid method – is based upon the principle that a penetrant enters or passes through a leak. The penetrant can be a dyed or fluorescent hydrocarbon with very low surface tension. Due to the pressure which arises due to the capillary effect and the low viscosity of the penetrant, it readily enters very small cracks. The method is described in detail in the chapter dealing with this subject: *Dye Penetrant Examination*.

The penetrant is applied to the one side of the wall under examination by spraying, painting with a brush or using an oil can; possibly repeatedly with a view to extending the penetration time (from 20 minutes up to 12 hours). After most of the penetration period has passed, a thin layer of penetrant developer (consisting of a strongly absorbent powder suspended in a volatile liquid) is applied on the opposite side of the item under examination. Any leaks will appear as very clear (red or fluorescent) markings. The penetrant type is selected with or without emulsifier content according to the application of the container. A penetrant with an emulsifier can be rinsed away with water immediately after the examination, while penetrants without it must be removed with an organic solvent or be scrubbed off with soap and water. Penetrants must of course always be used within the specified temperature interval (as a rule 10-30°C), and other instructions from the manufacturer should also be adhered to.

The penetrant itself will as a rule consist of low viscosity hydrocarbons with a high boiling point and with a flash point in excess of 90°C. Effective dyes and materials which reduce the surface tension will have been added. Due to the low volatility of the penetrant, it does not present a particularly high poisoning or fire risk. The developer can be a suspension of a strongly absorbent powder in a vehicle based upon 1.1.1. trichlorethylene, which is permitted even in spray cans. Good ventilation and the use of a face mask are recommended when working with both the penetrant and the penetrant developer.

For use with exotic metals sulphur- and halogen-free chemicals are required. The penetrant itself will probably fulfil these requirements, while the developer selected should be based upon ethyl acetate instead of chlorinated solvents.

The penetrant method is quite an effective leak testing technique, and has proved to be superior to the low vacuum test method (described later) in certain cases. See also the section about dye penetrant examination.

Ultrasonics

The leakage of a gas through small openings produces, even for small pressure differences, noise in the ultrasonic region. Apparatus equipped with a highly directionally sensitive microphone with emphasized sensitivity in the relevant frequency range (around 40 kHz) has a favourable "signal-to-noise ratio" and it can consequently detect leaks of a reasonable size many metres away. The method is described in greater detail in the chapter of this book entitled *Ultrasonic Leak Detection*.

Checking pressure drops

A general impression of the gas tightness of a pressurized system can be achieved by injecting air at a suitable pressure (but not greater than the allowed working pressure), sealing the system and checking the ability of the system to maintain the pressure. The leak rate will then be proportional to the measured pressure drop per unit time and to the volume of the system.

Foam generators

By applying foam generators to the surface of the system under examination, the pressure differential method described above can be extended to include the localization of leaks. When performing leak localization with foam generators, all surfaces must be free of paint and temporary corrosion protection. To avoid "blocking" all traces of water should be removed from possible leaks before performing the test. This can be accomplished, for example, by pumping the system up to the maximum permissible pressure for 10 minutes or more. Then the pressure is reduced somewhat, and foam generators (leak detection spray or 5-10% aqueous solution of Nekal BX possibly with thickener and inhibitor added according to the manufacturer's (BASF's) recommendations) are applied. In practice only 1-2% is used followed by rinsing and possibly also the application of corrosion protection. Leaks are clearly revealed by visible bubble formation.

Checking by pressure increase

In exactly the same manner as described above a general impression of the gas tightness of a vacuum container can be gained by evacuating the container, sealing it and then checking the pressure increase per unit time.

Vacuum and foam generators

The combination of vacuum and foam generators is an extremely useful method for the examination of the bottoms of storage tanks and their connected piping. Instead of applying pressure or vacuum to the entire tank, which could entail the risk of explosion or collapse, the relevant areas can be examined section by section by means of so-called "vacuum boxes". These devices, which consist of transparent acrylic glass and are provided with foam rubber seals and a manometer, are attached to a vacuum pump. During the examination the vacuum box is pressed down onto the section of interest after the section has been brushed with a "foamer" such as a 5% solution of Nekal BX, possibly with thickener added, according to the manufacturer's (BASF's) instructions; in practice only 1-2% need be used, followed by rinsing and perhaps also corrosion protection. The vacuum box adheres tightly to the surface; the applied liquid has a double purpose, for it contributes to sealing the vacuum as well as serving as an indicator for leaks. Normally an 80% vacuum is used.

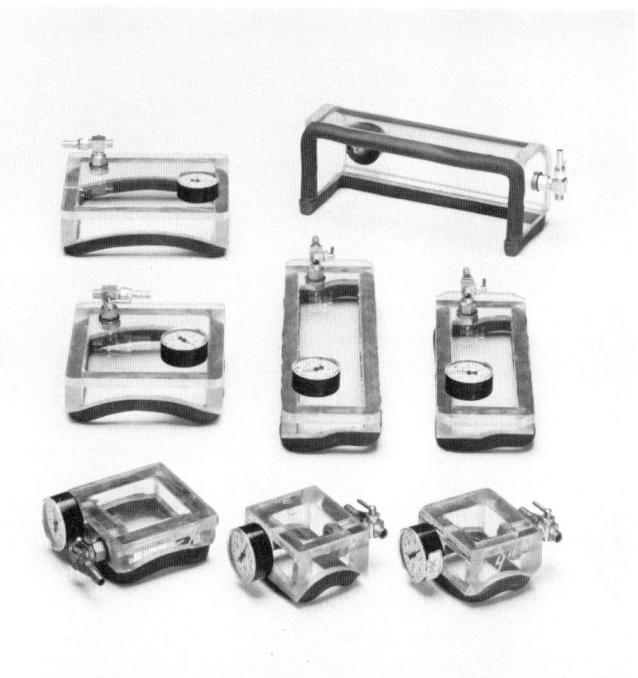

Figure 14.1. An assortment of NMK vacuum boxes for vacuum testing.
Middle row, middle and right: Standard boxes for air tightness testing of tank bottoms. Note the special profile which allows dependable testing of welded lap joints.
Upper row, right: Corner box for corner joints between tank bottom and lowest tank strake.
Left and bottom row: Examples of boxes for checking circumferential welds on pipes etc. Those shown are from 4" (100 mm) to 16" (400 mm).

Leak testing

Because each individual type of welded joint requires a special vacuum box, many modifications have been developed, e.g. for lap joints between offset surfaces, for corner welds, for welds in corners and on bends and for welds around pipe stubs (aquarium-shaped boxes). A variety of vacuum boxes have been produced for use in inaccessible places. A selection of vacuum boxes is shown in Figure 14.1. The vacuum box principle is very useful in many cases where the wall to be examined is only accessible from one side.

Tracers

These methods are quite varied in character. In the following some examples will be provided starting with the examination of hidden pipes using radioactive isotopes and then describing some techniques which utilize gaseous tracers. Those methods using gaseous tracers are particularly attractive when the tracer material is the material which is normally in use in the container; e.g. refrigerators with fluorinated chlorine hydrocarbons or transformers with sulphur hexafluoride insulation.

Radioactive isotope techniques

This method is used a great deal to locate leaks in hidden or buried pipes, central heating systems, etc. The basic principle behind such an examination is that a radioactive isotope, e.g. Br-82 in the form of an ammonium bromide solution (half-life 36 hours), is injected into the circulation fluid. After the fluid with the tracer has circulated for a period of time under pressure, some of the weakly radioactive circulation fluid will have escaped from the system through leaks and accumulate outside the system in the immediate vicinity of the leaks. The radioactive circulation fluid is now removed from the piping system, the system is flushed thoroughly, and the locations of the leaks can be determined using a geiger counter. The radioactivity disappears quickly because of the short half-life, and the excavation work needed to locate and fix the pipes can be limited to regions around the leaks. This method is described in somewhat greater detail in the chapter of this book on *Isotope techniques*.

Sulphur sticks or ozalide paper

In refrigeration systems which use ammonia as the refrigerant, the refrigerant itself serves as the tracer gas, and leak detection in such systems has – in the past – been carried out by burning "sulphur sticks" (a pipe cleaner impregnated by dipping it in melted sulphur) to produce sulphur dioxide. When ammonia gas interacts with sulphur dioxide, ammonium sulphite is formed, and it can be observed as white smoke in the neighbourhood of the leaks. Unfortunately, it turns out that the mixture of ammonia and air in certain proportions is explosive. For this reason the use of moist phenolphthalein paper is preferred.

Halogen leak detection

The detection of leaks in refrigeration systems which use fluorinated chlorine hydrocarbons (F-gas) is easy because the refrigerant itself is an excellent tracer material, and the technical evolution of detection equipment for this purpose is so advanced that even systems which do not normally use F-gas can be leak tested by filling them with F-gas under appropriate concentrations and pressures.

When filling a system with F-gas, it should be filled from above because of the high density of the gas.

The halogen torch

For detection of the F-gas which leaks out the halogen torch was previously used quite extensively. This consists of a burner, generally for bottled gas. The special properties of this burner are that: (1) copper is heated to a high temperature, and (2) the entire air intake to the burner takes place via a tube. This tube is passed close by the suspected leaks, and where leakage of F-gas occurs, the F-gas which has been drawn into the burner will be broken down due to the high temperature, and the liberated halogen will react with the copper of the lamp to a volatile compound which causes the flame to become bright green.

Electronic halogen leak detection

The electronic halogen leak detector is one of the most widely used pieces of equipment for leak testing, and it is based upon the following principle. An appropriately activated platinum filament heated to 900°C emits positive alkali ions in the presence of halogen gases and can thus be used as the anode in an ionization chamber. By means of a small fan air is drawn through this ionization chamber, and the magnitude of the anode current, as shown on an indicator instrument, becomes a measure of the halogen gas content of the air drawn through it. Most equipment can be adjusted to provide an audible alarm when the F-gas concentration of the input air exceeds a certain threshold.

The ionization chamber can be designed in the form of a measuring probe or "sniffer" by providing it with a short intake tube. This intake tube must be as short as is practically possible, and the length/volume ratio is optimized in such a manner that the time delay between passage of the leak and the alarm signal is as small as possible.

For example, the "sniffer" can be moved along a welded joint at the speed of about ½-1 metre per minute. The sensitivity, which is quite high to begin with, can be further enhanced by an order of magnitude by collecting the leaking gas under tape or a hood (the accumulation method).

A disadvantage of the halogen leak detector is that the anode filament can be "poisoned" by excessive exposure to high concentrations of halogens. It is, however, often possible to regenerate the filament by prolonged heating in halogen-free air.

It is very important to avoid spilling F-gas when filling the system to be tested. Gas which is lost due to leaks in the filling tubes and fittings contaminates the surrounding air and makes searching for leaks more difficult. Most equipment designed for halogen leak detection has an "automatic" adjustment which allows the presence of a certain background concentration of F-gas which may have escaped due to leaks (but not from filling the system!). When using this option only sudden changes in the concentration are recorded. Obviously, an examination performed under these conditions is carried out with reduced sensitivity.

When commencing and during a test with a halogen leak detector it is desirable to be able to check the order of magnitude of the sensitivity of the instrument. This can be done by means of a leak standard (usually with a shelf life of about 5 years) with a known leak rate (e.g. 0.3 g CCl_3F per year at 22°C). The regular use of such a standard is a fine way to check the reliability of the examination. The standard can also be used to evaluate the size of the individual leaks.

Leak detection tubes, leak detection probes

By means of leak detection probes (low vacuum region) or leak detection tubes (high vacuum region) the halogen leak detector can be used for the detection of leaks in vacuum systems. Tubes or probes are mounted on the connecting pipe between the container under test and the pumping system. The test gas fluorinated chlorine hydrocarbon (F-gas) can either be added to a plastic hood which is pulled over the vacuum container (hood test), or it can be "sprayed" onto the surface. After the absorption of the F-gas via leaks and during its subsequent movement past the leak detector on its way to the vacuum pump, even the smallest leaks in the vacuum container can be detected. In contrast to the "hood method" the spray-on method allows leaks to be localized.

Sulphur hexafluoride detector

SF_6 is used (primarily under pressure) instead of oil for electrical insulation in high voltage transformers and similar installations. The SF_6 molecule is extremely stable and is difficult to detect with a halogen leak detector.

Because SF_6 is used extensively in electrical technology, advanced equipment has been developed to detect it. In the same manner as described above the gas to be examined is drawn in by means of a "sniffer" via a cylindrical cell. Slow electrons emitted from a tritium source applied to the inner surface of the cylinder wall move to a centrally positioned anode, causing a weak current to flow through the cell. The isotope Ni-63 is also used in newer equipment. Pure nitrogen which passes through the cell has no influence on the electron current. On the other hand SF_6 is a strong absorber of electrons, whereby the electron current, even with a very small content of SF_6, will drop dramatically. This phenomenon can be displayed on a milliammeter or cause an acoustic signal when the concentration exceeds a specified threshold. Unfortunately the oxygen present in the surrounding air which is drawn into the tube is also electron absorbing, but fortunately much less so than SF_6, so by mixing in a considerable quantity of pure nitrogen the leak detector can detect very small concentrations of SF_6. In a manner similar to that described in connection with the halogen leak detector the escaping gas can be accumulated under tape or hoods (the accumulation method), whereby the sensitivity is enhanced substantially.

Helium testing

Helium is in many respects the ideal tracer gas. In addition to being very inactive chemically, the helium atom is the second smallest (after the hydrogen molecule), and it can thus pass through very small leaks.

The helium leak detector operates on the principle of the mass spectrometer. This principle will be described briefly here.

The gas which is to be investigated for its helium content is ionized by electron bombardment under low pressure, and the helium ions created are accelerated to very low energies and then passed into a sorting chamber in which they are deflected by means of a strong magnetic field to follow circular paths (with heavier atoms having paths of larger radius and lighter atoms of smaller radius). Via the output of the sorting chamber those atoms with the proper energy/mass ratio are allowed to pass through an aperture and to strike an ion trap. The ion current, which is measured by means of an electrometer amplifier, is thus a measure of the concentration of helium which is present. Present day advanced mass spectrometers are of course equipped with many refinements to achieve high stability and particularly efficient selective sorting.

In a manner similar to the method described under the halogen leak detector the helium leak detector can be used in connection with evacuated systems as well as systems under pressure.

In connection with evacuated equipment even the smallest concentrations of helium (and thus the smallest leaks) can be detected. The equipment under test can be covered with a plastic bag which is filled with helium (hood test, integrity method), or helium can be "sprayed" onto the surface with a pistol. The former method is the most sensitive, while on the other hand the latter technique permits the localization of leaks.

For systems under pressure and to which helium has been added, the "sniffer" method can be used. A limit for the sensitivity of this method is the helium content of atmospheric air.

Training required/desired

Many of the methods described are quite easy to use. However, experience has shown that – perhaps for this very reason – errors in interpretation often do occur in connection with these tests.

It is therefore recommended that the NDE operator is trained to use these methods. In many countries brief instructional courses are provided as well as longer courses (of about a week's duration) which lead to certification of the participant.

O.L. Høppermann

References

1) British Standard 3636:1963: *Methods for proving the gas tightness of vacuum or pressurized plant.*

2) *Handbuch der zerstörungsfreien Materialprüfung (Handbook of non-destructive testing of materials).*

3) ASTM Specification E 432-71. *Standard recommended guide for the selection of a leak testing method.* More complete descriptions of the methods and apparatus will as a rule be available in the instruction manuals supplied by manufacturers.

4) Suppliers of most equipment can be located in the current editions of professional buyers' guides.

Chapter 15
LPR probe (linear polarization resistance)

NDE principle

The measurement of the *instantaneous corrosion rate by means of polarization resistance measurement* is an electrochemical measuring technique based upon knowledge of the electrochemical nature of corrosion. The measurement is carried out by means of a linear polarization resistance (LPR) probe (also called a PAIR probe). PAIR is an acronym for Polarization Admittance Instantaneous Rate.

Electrochemical techniques are commonly used for studying the nature and the rate of corrosion processes. It is usually carried out by means of three electrodes placed in the corrosive environment. The three electrodes are: a *working electrode* composed of the material to be examined, a *reference electrode* with which one can measure the electrochemical potential, and an *auxiliary electrode*. A simple schematic diagram is shown in Figure 15.1.

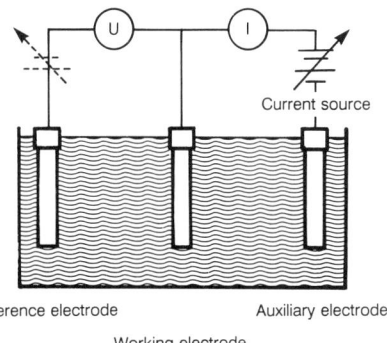

Figure 15.1. Schematic of setup for electrochemical corrosion measurement.

When performing electrochemical measurements a current is passed through the working electrode and the corresponding change in the potential of the electrode is observed. The result is the so-called current/potential curve from which one can derive information about the nature and the rate of the corrosion processes. A current/potential curve for iron in water is shown in Figure 15.2. It is apparent from this figure that the current/potential curve has an inflection point precisely where the current is zero. The slope of this tangent ($\Delta E / \Delta I$) is called the polarization resistance, as it has the units of resistance (volt/ampere). This polarization resistance can be shown to be inversely proportional to the instantaneous corrosion rate under conditions of free corrosion. When it is possible to determine the polarization resistance, then a quantity is available which varies with the instantaneous corrosion rate and which thus can be used as a measurement value.

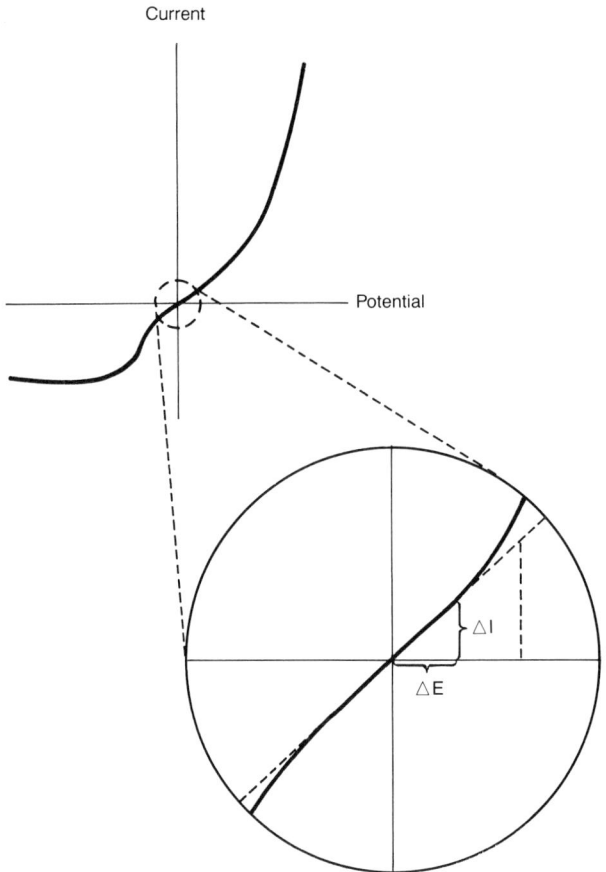

Figure 15.2. The derivation of the polarization resistance from a current/potential curve.

From the above figure it is also apparent that for minor variations in the corrosion potential the current/potential curve lies quite close to the tangent. The slope of the tangent can thus be determined by performing measurements as described above, but where the current is adjusted so that the potential is only changed by a few millivolts, typically 5-10 mV. The measured values of current and potential displacement (polarization) forms the basis for the calculation of the polarization resistance R_p, which is again dependent upon the instantaneous corrosion rate. The mathematical relation is as follows:

$$S = B(1/R_p), \text{ where}$$

S is the instantaneous corrosion rate
R_p is the polarization resistance
B is a constant

The constant B is known from experience, or it must be determined by calibration. For example it is possible to measure the polarization resistance of a working electrode over a period of time. From the data the average polarization resistance R_p during the period is calculated. From weight loss measurement on the same working electrode the average corrosion rate S for the period is determined. By substituting these values into the equation mentioned above, B can be calculated.

Development of the method

The greatest difficulty with the measurement of the polarization resistance over a prolonged period is to provide a maintenance-free and stable reference electrode. This problem can be solved if one keeps in mind the fact that it is the displacement of potential and not the actual value of the potential which corresponds to a change in the current which is of interest. Therefore, a reference electrode consisting of the same material as the working electrode is used, for the two electrodes can be assumed to have the same potential under free corrosion conditions. In this manner the measured potential between the two electrodes directly indicates the potential displacement to be used for the calculation of the polarization resistance. The three necessary electrodes thus normally consist of the same material and are all built into a probe (LPR probe) in a configuration appropriate for insertion into the system.

Figure 15.3 shows some examples of such probes. Notice the construction with concentric electrodes which can be used e.g. in the wall of a pipe and which can use a thin film of liquid as an electrolyte. The design with threaded electrodes permits convenient weight loss measurements to be made in conjunction with the calibration of the system.

Figure 15.3. Examples of the construction of LPR probes.

By making certain assumptions – among others that no significant voltage drops occur between the working and the reference electrode due to the conductivity of the liquid – the polarization measurement can be carried out with just two electrodes. Thus systems for electrode measurements may be offered for sale either as two-electrode or three-electrode probe types. Two-electrode systems thus require higher minimum conductivity in the environment to provide usable results.

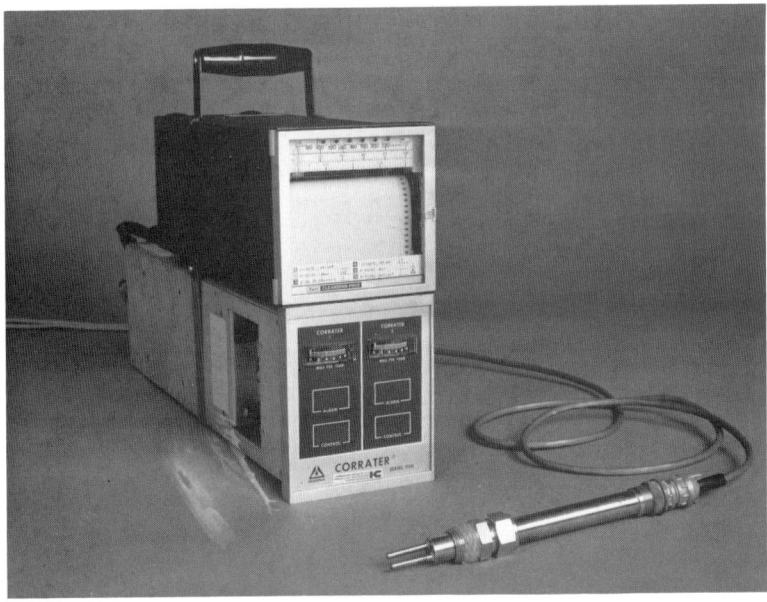

Figure 15.4. LPR corrosion monitor with two-electrode probe.

With respect to instrumentation both portable, *manual instruments* containing adjustable power supply and display for direct readout of corrosion rate as well as *automatic multi-channel instruments* which can continuously read several probes are available. These are often supplied with alarms for high corrosion rate as well as control outputs which can be used to regulate operating conditions on the basis of information about the corrosion rate. Figure 15.4 shows a two-channel LPR corrosion monitor with a chart recorder and a two-electrode probe (*Corrater, Rohrback Instruments*).

Both portable as well as automatic instruments are, in addition to the polarization resistance measurements, able to perform measurements of a possible current between the working and the auxiliary electrodes due to electrode short circuiting. Since both electrodes are made of the same material, a current between the two electrodes is a qualitative expression for the occurrence of localized corrosion (pitting) in the system. There will, therefore, be a switch on the instrument for displaying corrosion, pitting and possibly also the sum of the two signals. This provides greater flexibility, because e.g. in water systems localized corrosion will often be most significant and therefore that factor which must be monitored closely.

Areas of application

Because LPR probes measure corrosion kinetic parameters, these can only monitor *corrosion* and not *erosion*. It is also a condition that an electrolytic conductive connection must exist between the electrodes in use. LPR probes can, therefore, not be used in non-aqueous media such as oil or gas. These are two important limitations as compared to ER probes (see the chapter devoted to this subject).

The polarization resistance measurement is the corrosion measurement method with the fastest response. It is, therefore, well suited to the monitoring of systems where the corrosion rate can change quickly, e.g. in cyclic processes where the corrosive environment can change quickly during the transition from one phase of a process to another. However, it is important to keep in mind that there must be sufficient conductivity in the various phases.

Practical applications

Corrosion monitoring of large tower cooling systems is sometimes carried out using LPR probes connected to automatic instruments. With the rapid response of these instruments the control output can be used as an alarm and also to control the dosing of inhibitor to protect the coolers, piping system and cooling tower. Such a system is shown schematically in Figure 15.5.

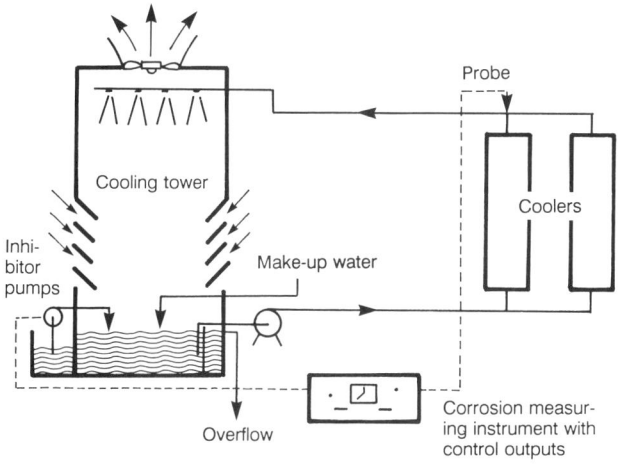

Figure 15.5. Automatic corrosion protection with LPR monitor used as a regulator.

Similarly, LPR systems are also used to monitor corrosion conditions in overhead systems of crude oil distillation in refineries. It is also possible in these cases to allow the monitor to control the addition of inhibitor to protect condensers and storage tanks.

Training required/desired

The use of the LPR system requires substantial knowledge of the electrochemical properties of the systems under investigation, and in many cases it is necessary to analyse the system electrochemically in advance of the establishment of the monitoring system. This is essential to make sure that the monitor does not react to other electrochemical parameters than those determined by corrosion. It is therefore necessary for the operator, in addition to being familiar with the instrument, to have contact with engineers with a thorough knowledge of corrosion and electrochemistry.

Ebbe Rislund

References

1) *Corrosion Monitoring using Polarization Resistance Measurements. I. Techniques and Correlations.* L.M. Callow, J.A. Richardson & J.L. Dawson, **British Corrosion Journal, 11**, 3, p. 123, 1976.

2) *Corrosion Monitoring using Polarization Resistance Measurements. II. Sources of Error.* L.M. Callow, J.A. Richardson & J.L. Dawson, **British Corrosion Journal, 11**, 3, p. 132, 1976.

3) *Technique and Instrumentation for Polarization Resistance Measurements.* P.J. Moreland & J.C. Rowlands, **British Corrosion Journal, 12**, 2, p. 72, 1977.

4) *A Flush Mounted Probe for Instantaneous Corrosion Measurements in Gas Transmission Lines.* E.C. French & P.E. Eaton, **Materials Performance, 17**, 7, p. 13, 1978.

5) *Corrosivity of water in absence of heat transfer (electrical methods).* ASTM D 2776-79 Standard Test Methods.

Chapter 16

Magnetic plugs

NDE principle

Magnetic plugs are used in oil-lubricated machines. The monitoring equipment is mounted directly in the lubricating system of the machine, see Figure 16.1.

The principle for the magnetic plug system is that ferromagnetic particles in the oil are attracted by the magnetic plug.

Because magnetic plugs are an "on-line" control, they are not – in contrast to SOAP and ferrography – dependent upon the sampling procedure.

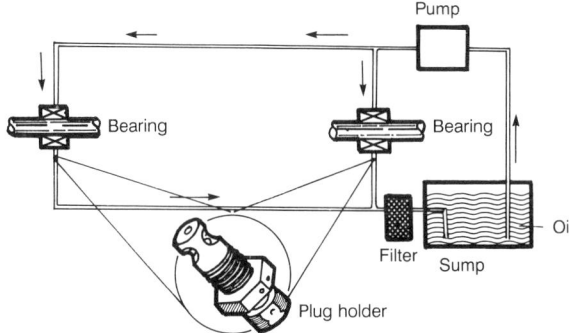

Figure 16.1. Placement of magnetic plugs in a lubricating system (Ref. 1).

The quantity of particles collected depends upon the path of the oil flow and the placement of the plugs in relation to the critical parts which are to be monitored. The magnetic plugs are therefore placed so that they provide a maximum amount of information about wear (particle production) of the critical parts.

Regular examination and evaluation of the coating of the plug, Figure 16.2, allows the process of particle production to be followed.

Development of the method

The use of magnetic plugs is a well known technique. They were originally used to capture the largest ferromagnetic particles to keep them from causing damage. Thus their use was similar to the use of an oil filter.

Later, magnetic plugs were used for condition monitoring. Regular registration of the coating of the plug, i.e. the quantity and size of particles, often follows a typical "bathtub" curve, see Figure 16.2.

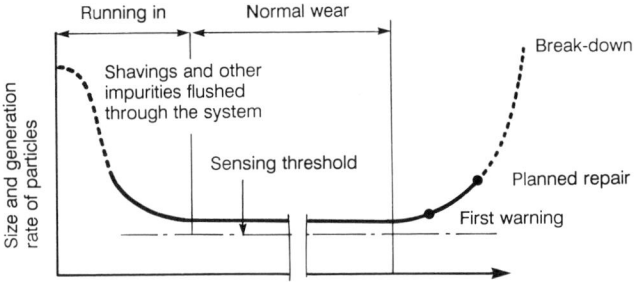

Figure 16.2. The size and the production rate of particles as a function of time.

By means of such a graph it is possible to identify appropriate times for the performance of preventive maintenance.

As experience with the use of magnetic plugs increases, the classification of the individual particles in a stereoscopic microscope is possible. If the particles must be classified as abnormal, then they can be examined more closely in modern analysis equipment such as an SEM (scanning electron microscope). At present such equipment permits an almost complete classification of particles and their origin.

Automatic magnetic plugs

Developments have been moving in the direction of automatic magnetic plugs, i.e. an MPD (metal particle detector), which can supply an electrical signal when the wear particle content of the lubricating oil exceeds a certain limit.

A new wear particle transducer of this type is illustrated in Figures 16.3a and b.

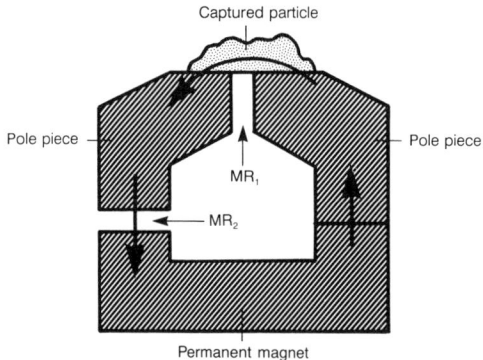

Figure 16.3a. Wear particle transducer.

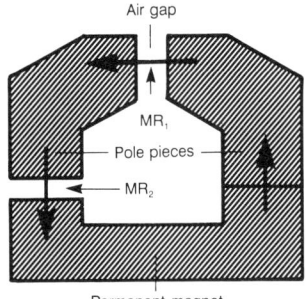

Figure 16.3b. Wear particle transducer (see Reference 3).

It consists of a permanent magnet with a pair of finished poles and two magnetic resistance sensors. The capture of particles will alter the magnetic circuit, and this change is recorded by the resistance sensors.

Oil cleaning

The original purpose of magnetic plugs, the collection of ferromagnetic particles in the oil flow, has also been pursued and developed as, for example, high gradient magnetic separator (HGMS) equipment; see Reference 4. In this manner it has become possible to effectively separate magnetic particles in the μm range. The problem of removing the particles from the HGMS for closer examination can be partly solved by means of ultrasonic cleaning.

Areas of application

The magnetic plug technique supplements the two other oil monitoring methods (SOAP and ferrography). Figure 16.4 illustrates how the sensitivity of each method depends upon the particle size. The magnetic plug captures particles from about 100 μm and up, i.e. a large number of particles are detected which would not normally be recorded by means of ferrography.

The magnetic plug is thus in a position to capture the larger flakes which are formed due to the breakdown of surfaces by fatigue. The magnetic plug is, therefore, used particularly in connection with the monitoring of gear boxes and bearings.

An acceptable coating of the magnetic plug corresponding to normal wear is illustrated in Figure 16.5a. A scattering of black particle fragments ("whiskers") is seen.

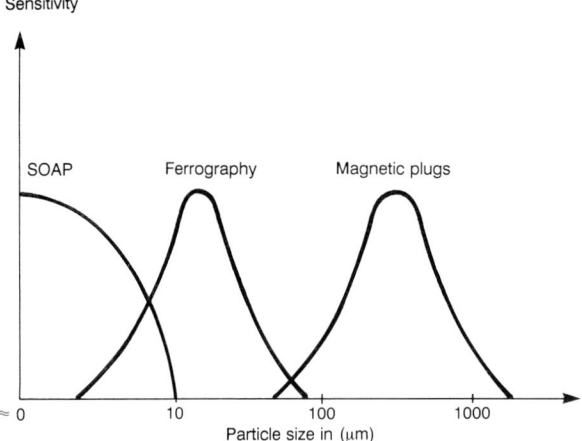

Figure 16.4. Comparison of three different wear monitoring techniques: SOAP, ferrography and magnetic plugs (Ref. 5).

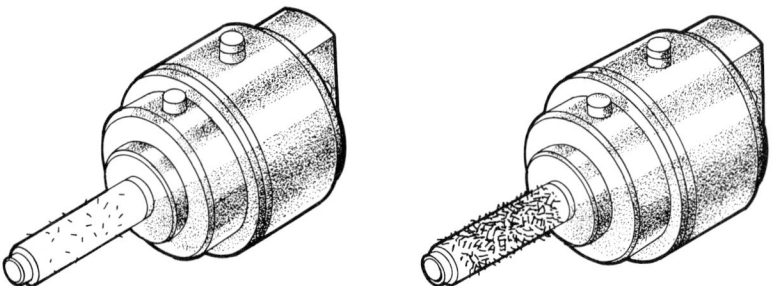

Figure 16.5. Magnetic plugs with acceptable coating (a) and unacceptable coating (b) ("whiskers").

An *unacceptable* coating is visible in Figure 16.5b. This indicates abnormal wear. An unacceptable coating can be characterized by the following conditions:

- Large, individual fragments
- Pieces which can be identified as flakes from a bearing
- Flat fragments
- A large number of whiskers (magnetic particle fragments)
- Particularly long whiskers.

Pictures of such "typical" abnormal particle coatings can be included as a standard in the NDE manuals for each individual machine.

Practical applications

Magnetic plugs are used in modern aircraft engines, where particle sizes in the order of 0.2-1 mm are found.

The Corrosion Centre (*Korrosionscentralen*) in Denmark has performed quantitative analysis using a SEM 435 (scanning electron microscope) to provide guidance in this connection. The results of the analysis of wear particles could e.g. in one case cause attention to be directed to one of 14 different alloys which were indicated in the test manual for the motor. In other cases the identification of the origin of the particles has been more difficult, but the analyses almost always provide important information about the motors and are, therefore, a contributing factor in optimizing the service lifetimes of the motors.

Training required/desired

Removal and insertion of magnetic plugs is relatively easy to perform by simply following standard operating procedures. This part of the monitoring process can be performed by a technical assistant, although it does require a certain degree of care on insertion and particularly on withdrawal to avoid the loss of any particles.

The visual examination of the coating and the classification of the abnormal particles under a magnifying glass or stereoscopic microscope does require some training and experience and access to a certain amount of reference material. The final identification of the larger particles requires access to the more complex SEM and EDAX (energy dispersive analysis with X-rays) equipment which can only be operated by specially trained personnel.

Kirsten M. Dorph

References

1) *Mechanical fault diagnosis and condition monitoring.* R.A. Collacott, Chapmann & Hall, 1976, pp. 260-263.

2) *The development, proving and application of an in-line metal particle detector (MPD).* R.A. Mason, **Condition Monitoring 84**, Pineridge Press, 1984, pp. 637-659.

3) *An improved magnetic plug for continuous monitoring of wear debris.* Robert W. Bogue, **Condition Monitoring 84**, Pineridge Press, 1984, pp. 628-635.

4) *Advances in quantitative analytical ferrography and the evaluation of a high gradient magnetic separator for the study of diesel engine wear.* David N. Andersen, Christopher J. Hubert, John H. Johnson. **Wear 90**, pp. 297-333, 1983.

5) *Wear debris analysis for machinery health monitoring.* G. Pocock, **Annual Conference**, Birmingham, Sept. 1978.

Chapter 17

Magnetic particle examination

NDE principle

The *magnetic particle examination* technique is a non-destructive examination technique which is well suited to the detection of surface defects such as cracks, lack of fusion and laminations, etc. in ferromagnetic materials.

The use of this method exploits the fact that a surface defect in a magnetized ferromagnetic item will disturb the magnetic field in the object of the test. The defect will cause some of the lines of magnetic force to depart from the surface and thus to form a magnetic leak field. This leak field can be found by placing fine iron particles on the surface. The leakage field will hold the magnetic particles in a ridge on top of the crack (see Figure 17.1).

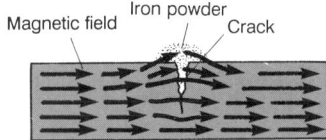

Figure 17.1. The leakage field holds magnetic particles in a ridge on top of a crack and a clear indication of the location of the crack is obtained.

By taking advantage of this effect the magnetic particles will form a ridge which is many times wider than the crack itself, thus making an otherwise invisible crack visible.

The flux lines are disturbed most by the crack if the crack is orthogonal to (i.e. perpendicular to) the lines of magnetic force. In this situation the leakage field is strongest and provides the clearest indication of a defect. Figure 17.2 illustrates the importance of the orientation of a crack in relation to the direction of the magnetic field. In this example a cylindrical object has been used. It was circumferentially magnetized by means of an electric current.

A good magnetic particle examination should, therefore, always be carried out with the magnetic field in two different directions in succession, preferably in two orthogonal directions.

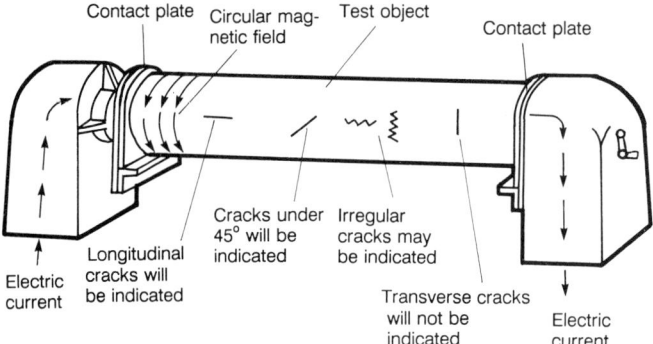

Figure 17.2. Stationary bench unit. The importance of the orientation of a defect in relation to the magnetic field lines is shown.

Defects which do not reach the surface but which lie below it will cause a weak leakage field (see Figure 17.3).

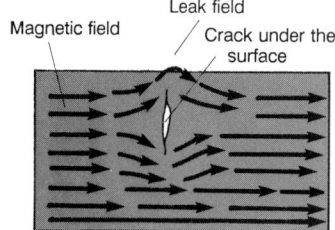

Figure 17.3. The leakage field due to a defect just below the surface is weaker.

In order to obtain even a weak indication of a defect, it must not be more than a few millimetres below the surface.

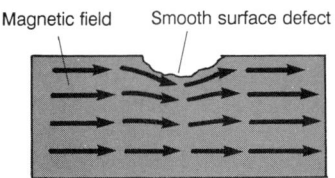

Figure 17.4. In this case no leakage field appears near a smooth surface defect.

If a surface defect is not too deep and it is uniform with a smooth transition to the surrounding material, then no leakage field will be formed. The defect will, therefore, not cause any indications during a magnetic particle examination (see Figure 17.4).

Advantages of the method
The magnetic particle method has many valuable advantages in connection with the testing of ferromagnetic materials. Here are some of them:

- It is one of the most dependable and sensitive methods for finding surface defects.

- The method is fast, simple and inexpensive.

- The indications are directly visible on the surface of the subject.

- The method is unaffected by possible deposits in the cracks, as e.g. oil, grease or other metals.

- The method can be used even on objects with thin layers of paint.

- Special preparation of the surface is not required before the measurement can be performed.

- The examination can be readily documented by means of a photograph or tape impression, etc., for the indication is an actual mapping of the defect.

- Simple and durable equipment means low investment and maintenance costs.

Limitations of the method
As is the case with all other NDE methods the magnetic particle method has certain limitations.

It is important that one be aware of these limitations before making use of the magnetic particle examination method.

- The method can only be applied to ferromagnetic materials.

- Defects below the surface will not always be indicated.

- The direction of the magnetic field has an important bearing upon the results of the examination.

- Certain objects must be demagnetized before and after the examination.

- The item examined may be subject to burn scars in connection with the current magnetization process.

- Experience and knowledge of the material is necessary in order to perform a correct interpretation of the results.

Testing procedure

Cleaning
In order for defects to be revealed it is important that the surface be free of loose rust, flakes due to heating, grease and oil which can hinder the free movement of the magnetic particles on the surface and thereby hinder many of the particles from being captured in a leakage field, if present. The sensitivity of the method will thus be significantly reduced and defects may be overlooked.

Cleaning can be carried out with solvents, rotating steel brushes, sand blasting or grinding.

NB! Excessive mechanical cleaning can mask possible defects.

Demagnetization
If the object has previously been magnetized or has been magnetized during use, e.g. in magnetic fields, then it may be necessary to demagnetize the item to eliminate poles which otherwise would disturb the examination (see under *Demagnetization* later in this chapter).

Contrast dyes
To make the indications which the magnetic particles form even clearer, it is advantageous to paint the surface of the object with a thin layer of paint (max. 50-75 μm) in a colour which will provide a good contrast to the colour of the magnetic particles. For example if black magnetic particle is used, it is a good idea to use white paint (available as spray paint).

As an alternative to white paint one can use fluorescent magnetic powder which emits visible light when exposed to ultraviolet light, providing a fine contrast to dark surroundings and thus high sensitivity.

A contrast paint is not generally required for machined surfaces.

Magnetizing the object
The object to be examined must be magnetized in such a fashion that the defects which one seeks are revealed in the best possible manner.

One distinguishes normally between two different forms of magnetization: *longitudinal magnetization* (pole magnetization) and *circumferential magnetization* (current magnetization).

Longitudinal magnetization of an object can be achieved by means of a permanent magnet, an electromagnet or a coil.

Figures 17.5a, b and c show examples of longitudinal magnetization.

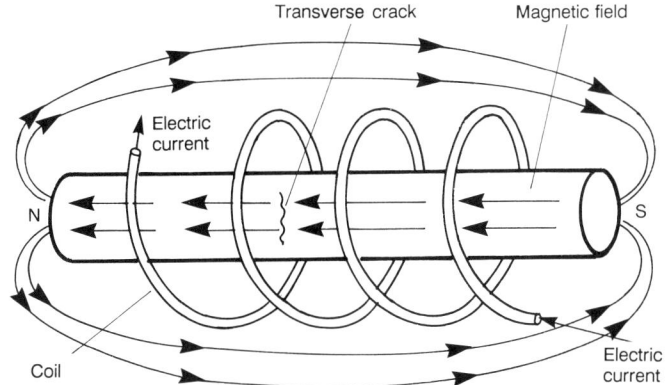

Figure 17.5a. The longitudinal magnetization of an axle achieved by means of a coil.

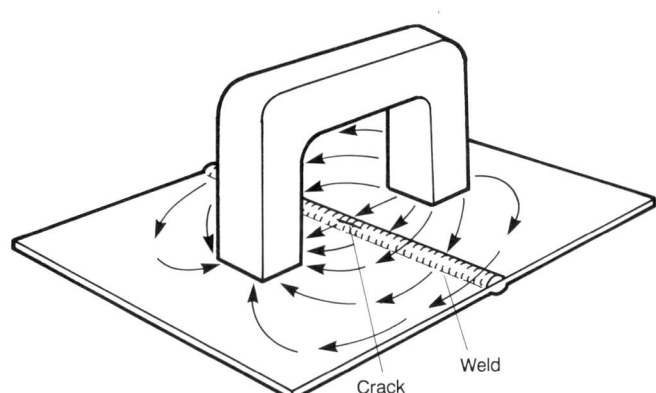

Figure 17.5b. Magnetization using a permanent magnet.

Magnetic particle examination

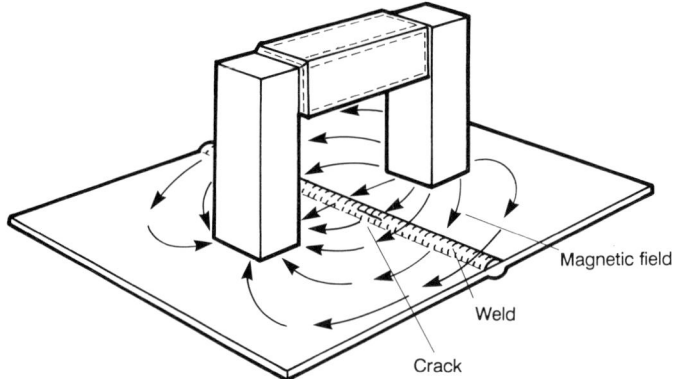

Figure 17.5c. Magnetization using an electromagnet.

Circumferential magnetization can be achieved by sending an electric current through the object to be examined or, in the case of objects with holes, by passing a current through a conductor passed through holes in the object.

Figures 17.6a and b show examples of circumferential magnetization.

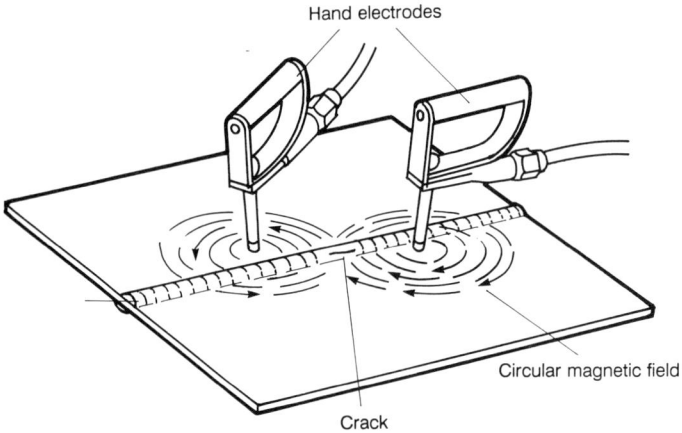

Figure 17.6a. Magnetization using hand held electrodes.

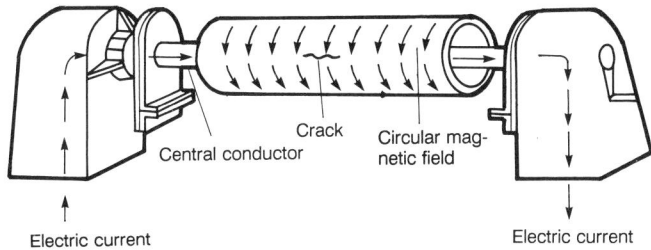

Figure 17.6b. Circumferential magnetization of a pipe by means of a central conductor.

The most commonly used currents employed in connection with magnetic particle examination is alternating current (AC), pulsed direct current (HVDC) and direct current (DC). Due to the displacement of the current (the skin effect), which is a characteristic property of alternating current, the magnetic field is strongest at the surface of the material. Alternating current is, therefore, used mainly when very high sensitivity to surface defects such as lack of fusion, fatigue cracks and the like is desired. Pulsed direct current can be produced by simply rectifying single phase alternating current. This type of current causes only a minor current displacement, i.e. the current flows over the entire cross section of the object. The object will thus be magnetized in depth and the method is well suited to the location of defects below the surface.

A constant direct current can in principle be obtained from storage batteries, but usually fully rectified three-phase alternating current is used with the resulting current source only slightly pulsing. Direct current has a good penetrating effect and is used for large, stationary tests.

Choice of current strength
There are no precise formulas or rules which indicate the correct current to apply. The following rule of thumb can, however, be applied:

Direct current magnetization of cylindrical objects:

$$I = 20 D$$

where I = the necessary current in amperes
D = external diameter of the cylindrical object in millimetres

Local electrode magnetization of larger surfaces:

About 40 amperes per centimetre electrode spacing

Coil magnetization of long cylindrical objects:

$$I = (X/n)(D/L)$$

Magnetic particle examination

where I = the necessary current in amperes
X = a constant depending upon the current type
for AC: $X = 22000$
for DC: $X = 32000$
D = the diameter of the object in millimetres
L = the length of the object in millimetres
n = the number of windings in the coil

Notice that the ratio L/D should be greater than 2 and less than 15 when the above formula is used.

The addition of magnetic particles
While the object is under the influence of the magnetic field magnetic particles should be applied in such a manner that the particles can move freely over the entire object and become attached by any leakage fields caused by defects in the surface.

The magnetic particles to be used must be so fine that they must be obtained in powder form, the so-called *magnetic powder*.

Pulverized iron oxide (Fe_3O_4) or carbonyl iron powder (chemically pure iron) can be used as the magnetic powder.

Often coloured magnetic powder is used thereby making the indications easier to see. Magnetic powder is available in many colours, e.g. black, red, yellow, green and fluorescent.

Powder suspended in liquid
The magnetic powder can be used suspended in an appropriate liquid, such as petroleum or water. The magnetic liquid is sprayed over the object, and the particles flow readily over the surface.

One must be sure to follow the manufacturer's recommendations or the specifications in the relevant testing standard with respect to the concentration of powder to be used in the liquid (see the section on *Checking the magnetic liquid* later in this chapter) and to keep the particles in suspension by periodically shaking or stirring the mixture.

Dry powder
When testing using dry powder the powder is generally distributed over the object by means of a special powder spray or a spray can which can provide an even distribution.

Dry powder testing is not in general as sensitive as testing with suspended powder, for the grains are – due to environmental concerns (danger due to dust) – considerably larger than the powder used in the suspension. The dry powder is used in connection with the testing of hot objects or for detection of sub-surface defects in connection with direct current magnetization.

Fluorescent magnetic powder

If the highest possible sensitivity is desired with magnetic particle examination, then one should use a fluorescent magnetic powder. It is actually just ordinary magnetic powder where a thin layer of fluorescent material has been applied to the individual grains.

When the object is observed under ultraviolet illumination in the dark or in an environment with subdued lighting, then defects will be visible as bright indications (see Figure 17.7).

Figure 17.7. The test object is viewed under ultraviolet illumination.

Illumination during inspection

If coloured powders are used it is important to ensure that there is adequate illumination in the area to be examined. Ordinary daylight is best, but it is possible to manage with artificial light, e.g. a strong hand lamp which can be moved around by hand to illuminate the object.

When fluorescent powder is used a sufficiently strong ultraviolet light must be available (see the section on *Checking the ultraviolet light* later in this chapter). To achieve maximum sensitivity as much visible light as possible must be excluded by darkening the room.

Interpretation

On the basis of the build-up of indications and their final appearance it is the task of the examiner to evaluate the cause of the indications. All indications need not necessarily be due to actual defects. Such indications are termed "false indications".

They can be due to the geometrical shape of the object, changes in the composition and structure of the material, cold deformation of the material or local, unintentional magnetization of the object.

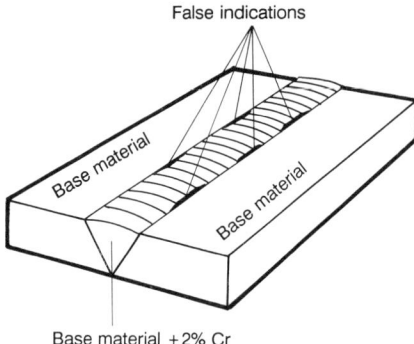

Figure 17.8. False indications caused by differences in the composition of the material.

Crack indications
The most serious surface defect which one can encounter is a crack. A crack which is open to the surface is easy to detect with magnetic particle examination, for it yields a very pronounced local disturbance in the magnetic field (see Figure 17.9).

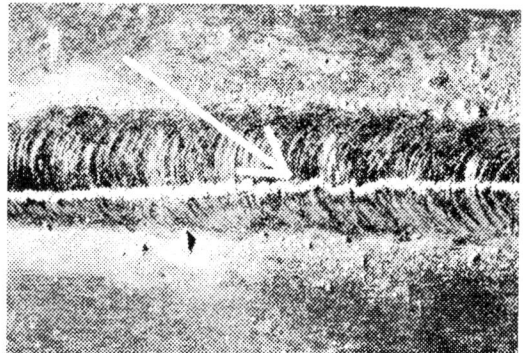

Figure 17.9. Crack indications in a weld joint.

In questionable situations a strong magnifying glass can be helpful to determine the existence and type of defects present. In other cases light polishing may be required in order to determine whether an indication reveals a crack or whether the indication is false.

Documentation / reporting

A permanent record of the appearance of defect indications can be made by means of photographs or by making tape impressions. When making a tape impression the indications are transferred to a report sheet by means of tape which is pressed down on top of each indication. The tape impression is easiest to use in connection with dry powder. All test data should be included in connection with a report of the examination.

A magnetic particle examination is of no value without a written report.

Demagnetization

After a magnetic particle examination has been carried out there may be some residual magnetism (remanence) left in the object. It can, therefore, in some cases be necessary to demagnetize the object.

The most commonly used method of demagnetization of a object is to allow it to pass through the magnetic field of a coil. As the object is gradually removed from the magnetic field of the coil the residual magnetism will be gradually reduced. Complete demagnetization is often difficult or impossible due to the shape of an object.

Examples of components which should be demagnetized:

- Aircraft components where the residual magnetism can affect sensitive instruments.
- Machine parts such as ball bearings, gear wheels and crankshafts where the residual field can attract iron particles and thus cause abnormal wear in moving parts.

Development of the method

The magnetic particle method has - just as all of the other non-destructive examination methods - undergone a rapid evolution since the 1930's. Developments have primarily been in the area of improved test equipment and supplementary equipment designed to make this testing method more effective.

Test equipment
Today many different types of equipment are available.

Permanent magnets can be used where a power supply is not accessible or is dangerous (e.g. due to the risk of explosion). NOTE: In all other cases permanent magnets should be avoided, for the magnetic field is weak and even large defects may be overlooked.

Electromagnets are easy to handle, provide a strong magnetic field and can be used as all-round equipment, particularly for objects where burn marks must not occur. Welded joints in ordinary steel structures, storage tanks, containers, etc. are tested in so far as it is possible by using the electromagnet.

Transportable current-magnetization equipment consists as a rule of a transformer which via a pair of heavy-duty welding cables can provide a current of about 1,500-2,500 amperes at a voltage of a few volts (see Figure 17.10).

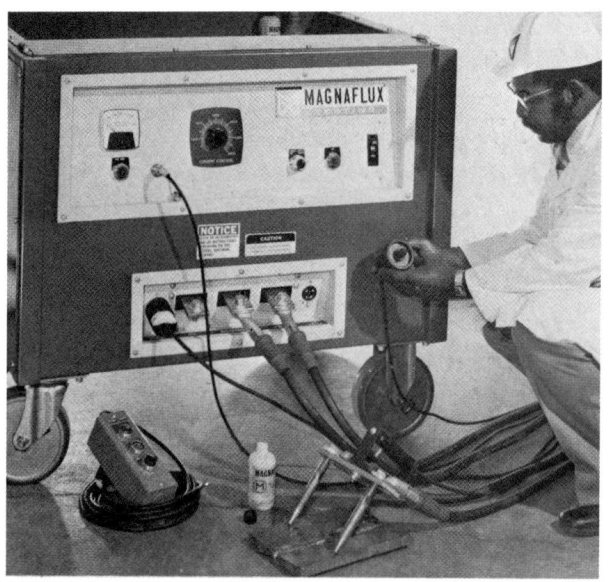

Figure 17.10. A transportable current magnetization apparatus with hand-held electrodes.

Figure 17.11. Magnetic test bench for longitudinal and circumferential magnetization.

In addition to hand-held electrodes coils can be attached to this equipment for use with the coil magnetization method. The use of current magnetization is only appropriate where burn marks are regarded as harmless, or where any burn marks can be ground away after the test. Objects composed of cast iron or forged steel can generally be examined using current magnetization with electrodes without significant damage.

Stationary equipment which can supply currents of up to 5000 amperes has been developed specifically for checking machined parts such as crankshafts, axles and similar parts. Figure 17.11 shows an example of such equipment, often referred to as a magnetic testing bench. The equipment is normally supplied with a transformer, rectifier, adjustable current plates for circumferential magnetization, coils for longitudinal magnetization, containers with pumps and stirring arrangements for suspended magnetic liquid and devices for current regulation, including demagnetization. In addition ultraviolet lamps can be attached for use with fluorescent magnetic particle examination.

The equipment mentioned above has the advantage that it is possible to examine the object to be examined for longitudinal as well as transverse defects during the same test.

Areas of application

Ordinary carbon steel, low allow steel, cast iron and steel castings are materials which are easy to magnetize. Therefore, in most situations it is permissible to use the rules of thumb which were mentioned earlier in this chapter in the section on *Choice of current strength*. Notice that before testing one should be sure that the material is in fact ferromagnetic. Use a small permanent magnet. If it is attracted to the object, then the test can be carried out. If it is not, then another testing method must be used.

Under ideal test conditions even small cracks of 2-3 millimetres in length can be revealed.

Quality control
During critical examinations of materials which are difficult to magnetize and in questionable cases the strength of the magnetic field (the field strength) should be checked with an appropriate instrument (e.g. a Hall effect probe), or at least an "artificial crack" should be checked for indications. Figure 17.12 shows an example of a field strength measurement during which the strength and direction of the field can be measured. The field strength is usually measured in "Ørsted" or kA/m (kiloampere per metre).

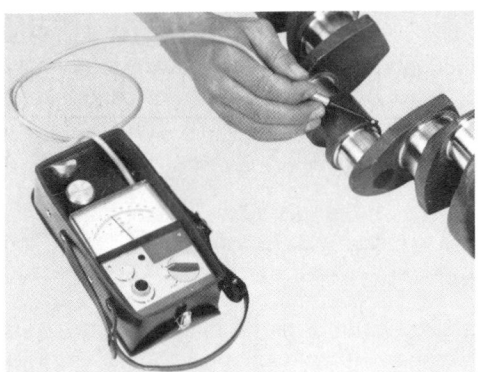

Figure 17.12. Field strength measurement by which the magnitude and the direction of the field can be determined.

Checking the magnetic liquid
In order to achieve the highest possible sensitivity using the magnetic particle method it is important to use the magnetic liquid in the proper concentration. This can be checked in the following manner.

After carefully stirring the magnetic liquid, pour 100 millilitres into a "Sutherland bottle" (see Figure 17.13). The bottle must normally be allowed to stand completely still for at least 30 minutes so that the powder can settle. After that the content of magnetic powder can be read from a scale on the narrow section at the bottom of the bottle and compared with the required amount.

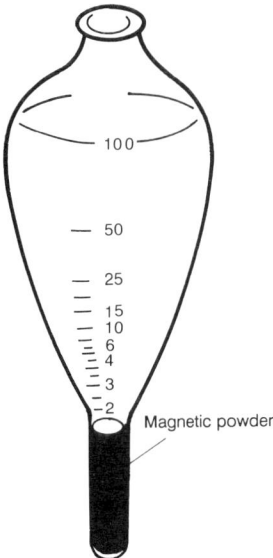

Figure 17.13. Sutherland bottle used for checking the magnetic liquid.

Checking the ultraviolet light
When one uses fluorescent magnetic liquid, the indications must be examined under ultraviolet illumination. It is very important to be sure that adequate UV illumination is available. Devices are available on the market to determine the intensity of ultraviolet lamps.

Environmental and safety requirements
The magnetic particle test method does not present substantial environmental hazards. However, it is important to observe the following precautions:

- The magnetic liquid can irritate the skin. Use, therefore, gloves or protective cream.

- Smoking or other open flames must not occur near oil-based magnetic powder mixtures.

- Use a protective mask when applying dry magnetic powder.

- When applying contrast colour in the form of spray paint, be sure to ventilate properly or use a fresh air mask when doing big jobs.

- The magnetic particle method must not be used in rooms or areas where flammable or explosive gases are present. However, testing using a permanent magnet is acceptable in certain situations.

- Use protective glasses when grinding areas with cracks.

- A magnetic field can affect pacemakers and watches.

- Magnetic particle examination equipment should be handled with the same precautions which are used for other heavy-duty electrical equipment.

Standards
A large number of standards have been developed relating to magnetic particle testing. Some of them are mentioned here:

Swedish Standard
SIS 114401:   Magnetic particle examination.
                Ferromagnetic material.

British Standards
BS M.35:   September 1970,
                Aerospace Series:
                Magnetic Particle Flaw Detection of Materials and Components.
BS 4397:   Methods for magnetic particle testing of welds, 1969.
BS 6072:   Magnetic particle flaw detection, 1981.

ASTM Standards
ASTM E 109-63: Dry powder magnetic particle inspection.
ASTM E 125-63: Standard reference photographs for magnetic particle indications on ferrous castings.
ASTM E 138-63: WET magnetic particle inspection.

Practical applications

There are many factors to be considered before initiating a magnetic particle examination. Here, briefly, are a few of them:

- Is the material ferromagnetic?

- What is the size and geometry of the object?

- Expected defect type – size, location, orientation.

- What test equipment and method should be used?

- What will the examination cost?

Figures 17.14, 15 and 16 show examples of practical solutions in connection with the choice of test equipment and test method.

Figure 17.14. Coil magnetization of a fork-lift truck arm.

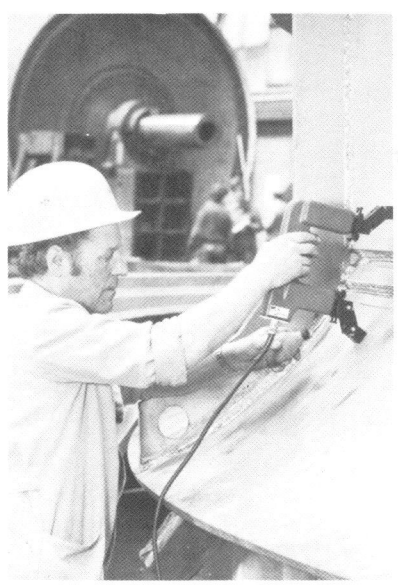

Figure 17.15. Magnetic particle examination of a welded joint performed with an electromagnet with flexible pole pieces.

Figure 17.16. Circumferential magnetization of a crankshaft performed using transportable current magnetization apparatus.

Magnetic particle examination

Training required/desired

The operator can relatively easily be trained to use the magnetic powder method so that he or she can reliably find possible defect indications. On the other hand it is difficult to train an operator to *interpret* the indications. This requires years of experience and knowledge of materials. In practice it often turns out that the easier the examination method is to carry out, the less careful people are when performing it – obviously an unfortunate tendency. The NDE operator should, therefore, as a minimum have participated in a course of instruction in the basics of magnetic particle examination including both the theoretical background as well as practical applications of the method. Such a course of instruction can be concluded with a qualifying examination according to NORDTEST or ASNT rules.

Hardy Hansen

References

1) *Non-destructive testing, magnetic particle*, 1977. Programmed Instruction Handbook, General Dynamics.

2) *Principles of magnetic particle testing*, Magnaflux Corporation, Carl E. Betz.

3) *Non-destructive testing handbook*, Society for Non-destructive Testing, Robert C. McMaster.

Chapter 18

Mechanical calibration

NDE principle

Mechanical calibration in this connection means measurements which can be carried out with relatively simple tools to record changes in the dimensions of an object due to breakdown or wear during use.

When such measurements are carried out in connection with condition monitoring it is normally done to follow a progressive breakdown of material and not to accept or reject a particular piece of equipment. The measuring equipment and thereby also the measuring accuracy must, therefore, be adapted to the particular measuring task. This requires, as with all other work with condition monitoring, knowledge of the breakdown mechanisms which are operative for the object of interest.

In the following some of the most commonly used tools for mechanical calibration are discussed, but it should be mentioned that there is a wide range of tools available for this purpose, e.g. height indicator, bastard compass, rod compass, thread gauge, etc.

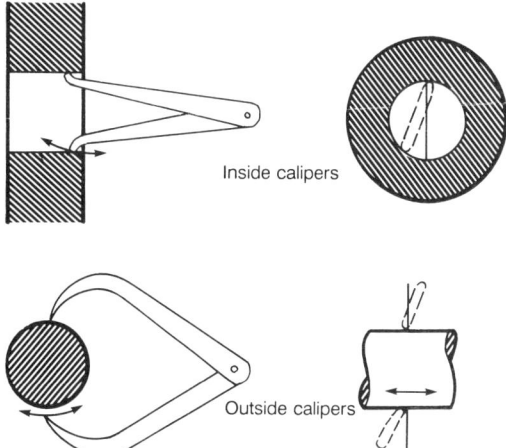

Figure 18.1. Tools used for mechanical measurement.

Outside calipers and inside calipers

Transfer calipers have been used since olden times to perform measurements. In connection with condition monitoring tasks – where it is a matter of comparing results between measurements – the measurements performed with these instruments must be converted to the metric system or some other system of units. There is some uncertainty associated with these measurements to begin with (Figure 18.1), and the results become more uncertain due to the indirect reading of the units (see Figure 18.2).

The advantage of using these instruments is their relative sturdiness and the fact that their accuracy is not affected by the onset of wear.

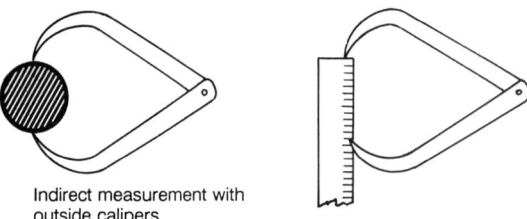

Figure 18.2. Indirect measurement with calipers.

When using these instruments it can be a problem to avoid changing their adjustment before reading the units. Both instruments are available in spring loaded versions (see Figure 18.3) so that this source of error can be avoided. Transfer calipers can also be obtained complete with a measuring scale (see Figure 18.4) or with a fixed measuring gauge mounted on them. This latter option is, however, generally not sturdy enough for use with inspections in processing plants. If these instruments are used, one must keep in mind that errors can occur during the measuring process itself, see Figure 18.1.

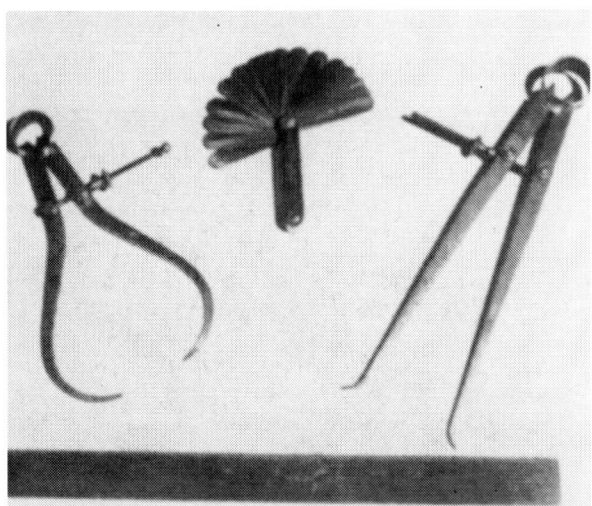

Figure 18.3. Some tools for mechanical measurement: spring calipers, thickness gauge and scale.

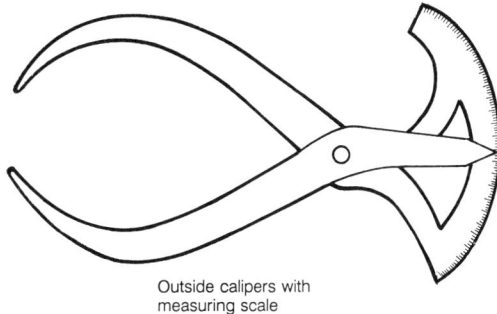

Outside calipers with measuring scale

Figure 18.4. Calipers with measuring scale.

Measuring scale

In connection with condition monitoring work the measuring scale - whether it be a foldable ruler, a steel measuring tape or a fixed steel measuring stick - is only used for rough field measurements.

Vernier gauge

Measurements performed with a vernier gauge can be a satisfactory solution to many condition monitoring problems (see Figure 18.5).

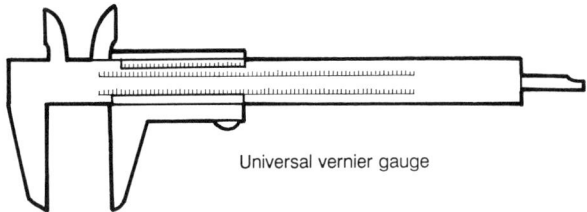

Universal vernier gauge

Figure 18.5. Universal vernier gauge.

By means of a vernier gauge measurements can be carried out with relatively high accuracy: from 1/10 to 1/50 mm depending upon the division of the vernier scale.

Present day vernier gauges are available with digital display of the measurement value. For the condition monitoring of heat exchanger tubes this can be a big time saver, for in this case it is usually necessary to perform a large number of measurements. Vernier gauges are manufactured in a wide range of various types depending upon the task which they are designed to solve. The jaws can be designed for measuring height or depth (see Figure 18.6).

Mechanical calibration

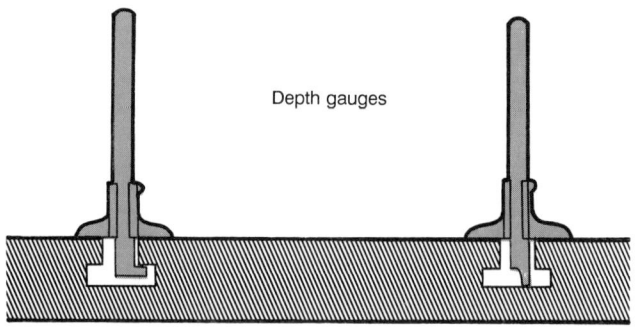

Figure 18.6. Depth gauges.

Micrometer tools

If even greater accuracy is desired when measuring the size of an object, then can use micrometer measuring tools. Tools of this type permit measurements to be performed with a precision to the nearest 1/100 mm. A measurement of such accuracy will in many cases be irrelevant in matters of condition monitoring. If it is a case of checking the internal or external corrosion of pipes, then a vernier gauge with an accuracy of 1/10 mm would be more appropriate.

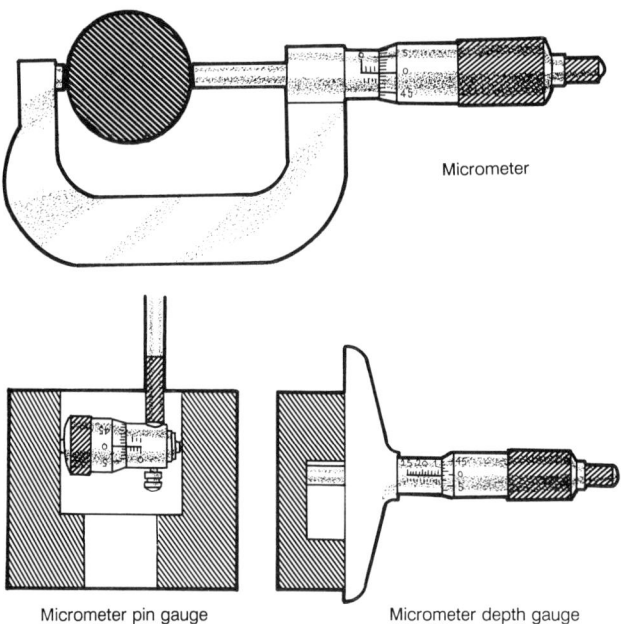

Figure 18.7. Various micrometer tools.

On the other hand if it is a matter of checking the wear of rotating parts or examining close fits, then a micrometer tool is relevant. The micrometer measuring tool is available in various designs depending upon the nature of the job to be done (see Figure 18.7).

Tolerance forks and dorns (plug gauges)

Tolerance forks and dorns can also be used for this type of measurements (see Figure 18.8). Tools of this type do not provide exact measurements but make it possible to accept or reject an item on the spot.

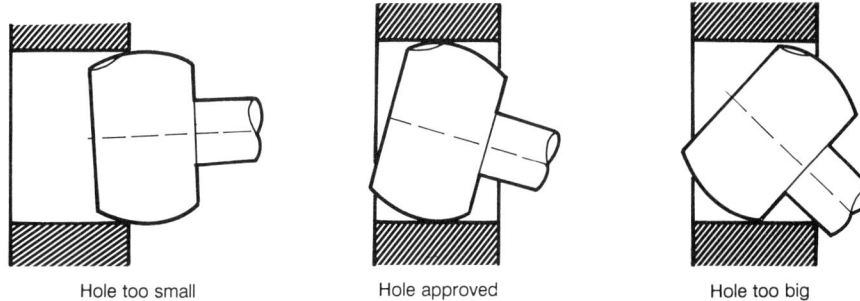

Hole too small   Hole approved   Hole too big

Figure 18.8. Measurements with a tebo-dorn (plug gauge).

Normal measuring tools

Among the simple but more specialized measuring tools the following can be mentioned: *feeler gauges* for measuring narrow gaps, valve clearances, electrode separation, etc.

Feeler gauges are classified along with tools such as *roundness gauges, nozzle gauges* and similar tools as normal measuring tools. Use of normal measuring tools places rather heavy demands on the experience of the user. It is important to be aware of the fact that wear on the tool can lead to incorrect evaluations of the condition of the equipment which is examined.

Measuring gauges

The measuring gauges which are familiar in connection with production control can of course also be used for condition monitoring. Figure 18.9 shows a measuring gauge mounted on a special measuring caliper designed to reach in around flanges, etc.

Handling of measurement results

In addition to appropriate selection of measurement tools suited to the task at hand and being sure that the individual entrusted with the task has sufficient experience, it is also important to consider how the information collected will be used. If the "measurements" are carried out e.g. with the *Tebo-dorn* shown in Figure 18.8, then the information collected can "only" be used to accept or reject an item. If on the other hand it is a matter of measurements carried out with a vernier gauge, then it will be possible to compare the results with the original dimensions of the object and possibly also with the results of intermediate condition monitoring measurements.

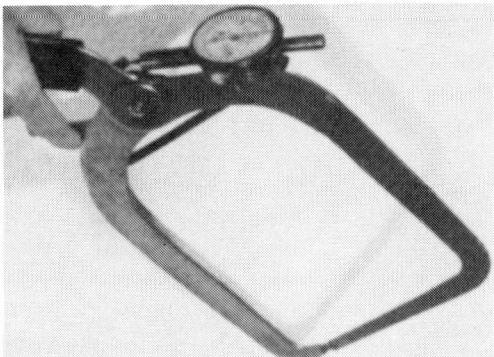

Figure 18.9. Mechanical measuring device with gauge.

In order to perform such comparisons, the measurement results from previous data collection must be preserved in a systematic manner so that results can be compared. In practice this means that a data registration and storage system must be established. Such a system can be established using a manual archive, or it can be computerized.

The purpose of comparing measurements carried out at different times is of course based upon the desire to be able to identify and to compute the rate of the breakdown process to which a given item is exposed and to thereby determine appropriate times for future measurements and to optimize preventive maintenance.

Training required/desired

There is no specific training course required within this special area of NDE technology. However, training in this area is included in all courses of instruction for mechanical engineering craftsmen in technical schools. A background as a craftsman and/or relevant experience will, therefore, be an important asset to an individual to whom the performance of these measurements is entrusted.

Anders Korsbæk

References

1) *API Guide for inspection of refinery equipment*, Chapter IV, Inspection Tools.

Chapter 19
NDE method combination

NDE principle

It is not possible to provide a technical description of a *particular* principle in the case of the combination of NDE methods, for by definition this chapter deals with the use of several selected NDE methods applied together in order to describe the condition of a material or a component.

However, in each individual case it is a question of defining the general guiding principles which are applicable in connection with the description of the condition of a particular system. Here it is a matter of determining those material or component parameters which will be characteristic for the deterioration expected in a given – known – environment and service for a given structure of a given material.

Thorough knowledge of the character of the material deterioration phenomena in question in relation to the environment and service is, therefore, extremely important for the optimum choice of the NDE methods which are to be included in the examination.

This in turn requires thorough knowledge of the measuring capabilities and limitations of the NDE methods available.

Development of the method

The development of combined NDE programmes began years ago when it became apparent that just checking the integrity of a welding joint by means of radiography during the production process was not adequate, because this method alone could not dependably reveal for instance lack of fusion, surface cracks, etc. in all cases.

Therefore, the examination was extended to include ultrasonic angle probe checks to reveal possible lack of fusion and dye penetrant examination or magnetic particle testing or, most recently, eddy current checks to locate possible surface cracks.

The evolution described here took place mainly in connection with the manufacturing control of welding joints. The methods as well as acceptance and rejection criteria in this area are today very well defined and generally covered in standards and norms, as the manufacturing process has a profound influence upon the method selected.

The situation is somewhat different for condition monitoring in general. Here knowledge of the material deterioration processes are, as mentioned earlier, decisive for which NDE methods are selected, and quite often rejection and acceptance criteria can not be defined beforehand as they depend on the actual service conditions including possible environmental influences etc.

Therefore, in the condition monitoring area choice of NDE methods will often require the development of procedures in each individual case in combination with the definition of rejection and acceptance criteria.

The system work technique is in this case a valuable tool and perhaps the only method which can provide a completely precise picture of the procedures.

An example is shown in Figure 19.2 which illustrates the use of the system work technique. Descriptions of this type can, as required, be directly computerized or used "manually".

Areas of application and practical applications

Combinations of NDE methods are employed in the areas mentioned in the following sections:

Component condition monitoring
In these cases it is a question of condition monitoring of an entire component – a motor, pump, pressure vessel, pipe, gearbox, etc. – undertaken perhaps even though no defects are suspected.

Such examinations, where several NDE methods may be used, should be employed on vital components as a preliminary check, providing a basis for subsequent NDE checks. In the oil and gas industry such preliminary checks are termed "base line inspections".

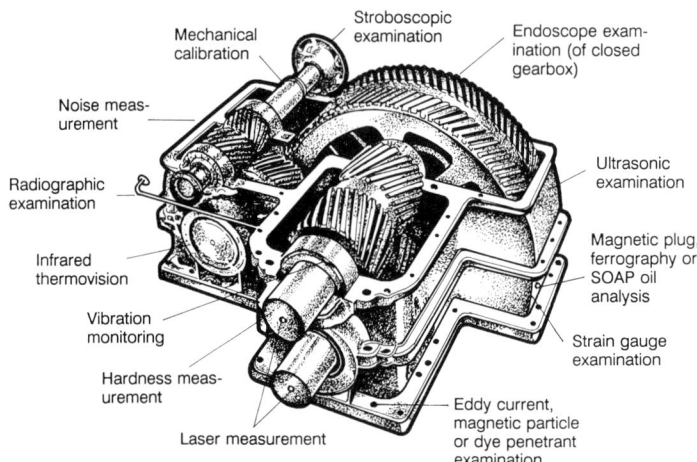

Figure 19.1. A gearbox with an indication of the NDE methods which might be used for component condition monitoring checks.

Figure 19.1 shows a gearbox with an indication of the NDE methods which might be used on each component part. The many NDE methods are, however, only shown here as an illustration of the various possible choices.

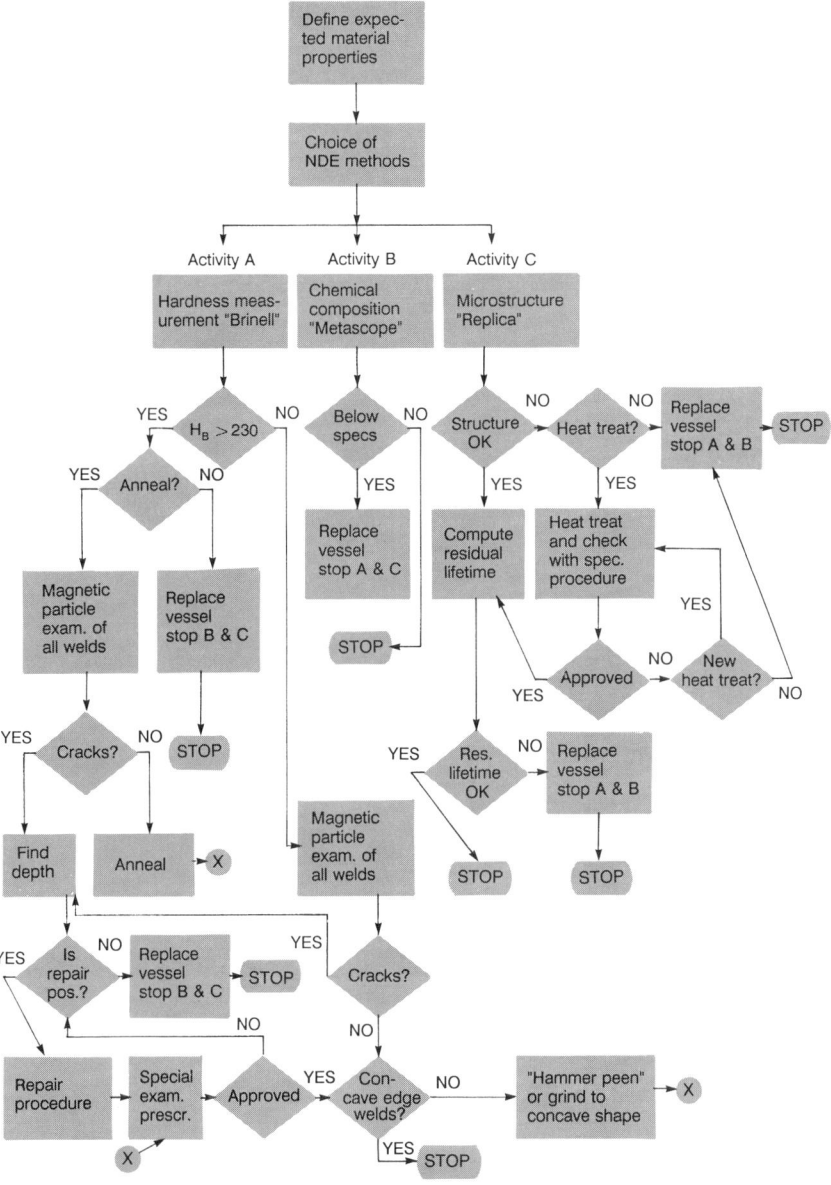

Figure 19.2. Condition monitoring programme for temper embrittlement in 1¼ Cr - ½ Mo material in heavy-wall pressure vessels (see also the text).

Material condition monitoring
Here is a case of the description of the condition of a particular material, e.g. in connection with inter-granular corrosion, stress corrosion, spheroidization, fatigue, creep, temper embrittlement, etc.

In such cases material and process related NDE parameters must quite often be combined in order to provide a complete description of the condition of a material.

Figure 19.2 shows an example of a programme for checking the temper embrittlement of a low-alloy ferritic steel.

The programme indicated might be supplemented with ultrasonic attenuation measurements, where the attenuation in the "used" material is checked in comparison to defect-free, new material. Another additional supplement might be acoustic emission measurements for the detection of possible internal micro-cracks, as the replica check indicated in Figure 19.2 will only reveal micro-cracks in the surface.

Integrated NDE inspection
NDE method combinations are also used for specialized testing equipment where the NDE equipment forms an integral part, e.g. in the "intelligent pig" used for checking pipelines which is shown in Figure 19.3.

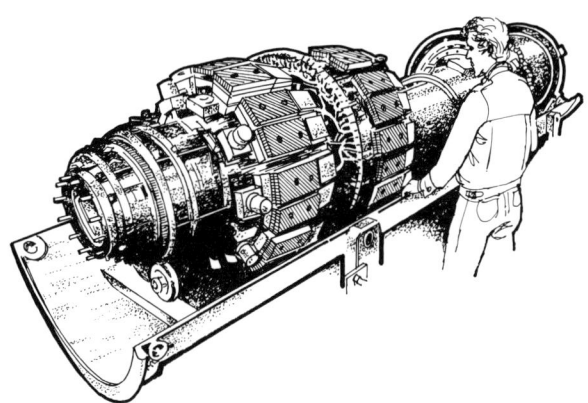

Figure 19.3. "Intelligent pig" for NDE examination of pipelines. This equipment has been developed by the British Gas Corporation.

The following NDE methods are integrated in this equipment which has been developed by the British Gas Corporation:

- A special type of magnetic leak field examination (Magnetic Flux Leakage) used to check loss of metal.

- Ultrasonic (Elastic wave) equipment used to detect stress corrosion, fatigue and laminations.

- Mechanical calibration for checking the geometry of pipelines.

All data gathered are then processed in a central computer unit.

Specialized inspection equipment of this type is very expensive to acquire and to use. It is, therefore, only used in very large oil and gas piping systems.

There are also simpler and less expensive "pigs" which can only be used to check for and to localize leaks and blockages in pipelines. Leak checking in this case can be performed by means of acoustic emission, for example. When an AE sensor is integrated in the "pig", the "pig" will automatically stop at a leakage. Localizing leaks or blockages can be done using an external geiger counter and a radioactive source built into the "pig".

In its simplest form the "pig" can just be an isotope ball which is used to localize blockages (see the chapter on *Isotope techniques*).

Predictive NDE programmes
Finally, NDE method combinations can be used in predictive NDE programmes. In such programmes NDE measurements may in some cases be performed continuously. This is true for example of some types of vibration monitoring, temperature and pressure measurements, etc.

In most situations, however, such NDE measurements are – for practical or economic reasons – only carried out periodically or non-periodically, depending upon the NDE programme in use.

In such cases it will often be desirable to carry out some form of check measurement, e.g. of the technical operational parameters which cause material degradation. Such measurements are thus indirect and relative with respect to the material degradation, and they should in any case *only* be used for correction of inspection and repair schedules (residual lifetimes) which have been computed on the basis of direct, absolute measurements.

In order to compute this correction it must be possible to mathematically calculate the relative effect of the operational parameters upon the degradation of materials.

Figure 19.4 shows an example of a combination of both direct and indirect NDE methods. In this case the subject of the examination is a piping system where the direct wall thickness measurements are performed by means of ultrasonic scanning and/or radiography, while the indirect measurement of the aggressiveness of the corrosive medium is performed by means of an electrical resistance probe (ER probe – see Chapter 8 on **ER probes**).

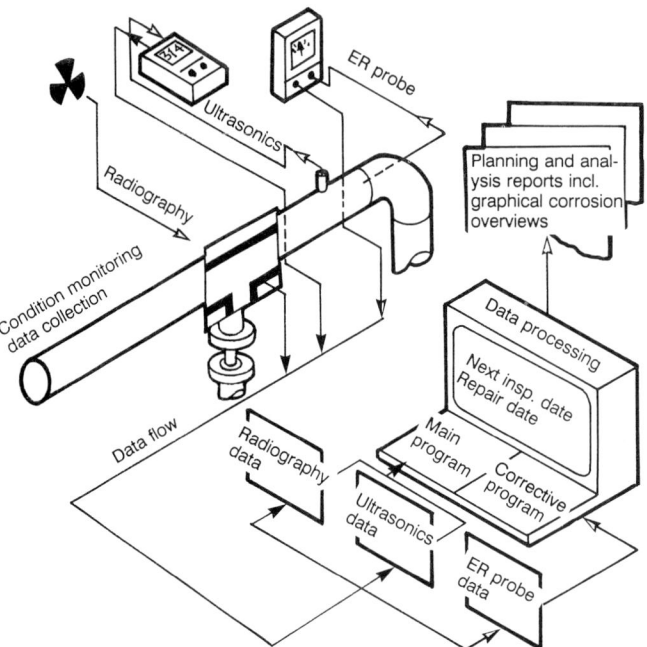

Figure 19.4. NDE method combinations, datalogging and treatment in a predictive corrosion/erosion monitoring programme developed by SVEJSECENTRALEN – The Danish Welding Institute.

In such programmes the data processing, which mainly aims to determine the time for the next inspection – with a certain safety margin – and computation of the time for the next repair, can be performed manually or using computer programs.

The cost/benefit factor in connection with the predictive programmes is often very favourable. Documented examples exist where this ratio is 1 : 24 or even better.

It is, however, recommended that a cost/benefit prognosis be carried out before the implementation of a predictive NDE programme and that possible findings are successively recorded and – if possible – capitalized in a log book or other easily accessible recording system in order to document the value of the programme.

Future developments
Today work is being carried out in many places with combinations of NDE methods, method programme integration, integration of technical and administrative systems, development of data transmission systems and the development of predictive, optimizing NDE programmes.

An example of such a system which has been developed is the Danish National Railways' system for automatic, track-based detection of defects in the wheels of railroad cars. The system is continually being extended by Caltronic A/S.

Training required/desired

Specific training requirements for NDE method combinations do not exist at the moment. However, the individual who is to establish such a programme must master the necessary disciplines in order that the result will be a reliable and suitable basis for making decisions.

Therefore, thorough knowledge of the many NDE methods and their potential applications and limitations is necessary. Furthermore, knowledge of which deterioration mechanisms may affect a given material must be present to form the background for making optimum choices of NDE methods.

Finally, experience in the use of the system work technique will be an important asset in connection with the development of NDE procedures.

A specific training programme for NDE technicians has, however, been developed, and the implementation of the programme is expected to take place in early 1990.

K. G. Bøving

References

1) *Here's experience of British Gas with on-line inspection.* **Technology**, Jan. 31 and Feb. 7, 1983.

2) *MOCS-W T 2.0 The predictive computer program for maintenance which assures optimum benefits from plan condition monitoring.* K.G. Bøving, The Danish Welding Institute, Publication no. 85.33.

3) *NDE methods and their use for condition monitoring purposes in maintenance programs.* K.G. Bøving, The Danish Welding Institute, Publication no. 84.18. Revised, expanded special publication from **Svejsning**, no. 4, 1984.

4) *Grundsätzliche Überlegungen der Instandhaltungsprinzipien von Anlagen und ihre Anwendung im Rahmen der unternehmerischen Möglichkeiten.* (Principles and Management Tools for Plant Reliability). K.G. Bøving, **Maschinenschaden** no. 1, 1987.

5) *New Procedures in Non-destructive Testing.* P. Höller, Springer-Verlag, Berlin, Heidelberg, New York, 1983.

6) *Non-destructive inspection and quality control.* **Metals Handbook** no. 11, American Society for Metals.

7) *Methoden zur Überwachung des Anlagenzustandes.* (Methods of monitoring plant conditions). K.G. Bøving, **Maschinenschaden**, no. 1, 1989.

Chapter 20
NDE methods under development

Known methods are improved

Many of the well known NDE methods, e.g. radiography, ultrasonics and eddy currents, are at present evolving very rapidly, particularly with respect to data recording, data handling and data presentation. These developments have as their goal, among other things, to make the methods faster, more quantitative and less dependent upon the operator.

New methods are developed

In addition the growing need for dependable quantitative inspection techniques has inspired many NDE technicians and researchers to try to develop or adapt methods which have not previously been used for NDE. Many of the new methods have potential capabilities for condition monitoring even though some of them seem to be too expensive and/or too complex.

In this chapter six new or relatively new methods and their potential capabilities for NDE will be discussed. The six methods are indicated in Figure 20.1.

NDE method terminology	Abbreviation
Stress pattern analysis by thermal emission	SPATE
Pulsed video thermography	PVT
Moiré contour mapping	MCM
Holographic interferometry	HI
Computerized tomography	CT
Positron annihilation	PA

Figure 20.1. New NDE method technologies.

SPATE

Stress pattern analysis by thermal emission

NDE principle

The method is based upon the thermoelastic effect which was discovered by Lord Kelvin in 1853. The thermoelastic effect is an expression of the fact that the temperature of a body changes when the body is subjected to an elastic deformation. The temperature change at any given point depends upon the total mechanical stress at that point. The greater the stress the greater the temperature change upon deformation. Therefore, if a body or a part is subjected to small cyclic stresses, it should in theory be possible to map the original mechanical stresses in the object. The possibility is now a reality thanks to modern infrared measuring equipment and data technology.

Development of the method

SPATE has been developed in England by the firm Ometron. The measurement setup is shown schematically in Figure 20.2. A transducer loads the subject with a periodic force causing small temperature differences, a few hundredths of a degree Celsius at the most. These small temperature differences are recorded by an infrared detector which is placed in a scanner camera. The camera is controlled by a microcomputer and scans the surface of the subject. The two channels make it possible to analyze the temperature signal from a small region and to localize the area. In order to – among other things – suppress noise signals a correlator is used to ensure that only temperature signals with the same frequency as the load signal are recorded. The relative phase between the temperature signal and the load signal is also recorded making it possible to distinguish between compression and tension. The computer converts the signals to a colour image of the distribution of mechanical stresses in the test subject. Different colours correspond to different strain values.

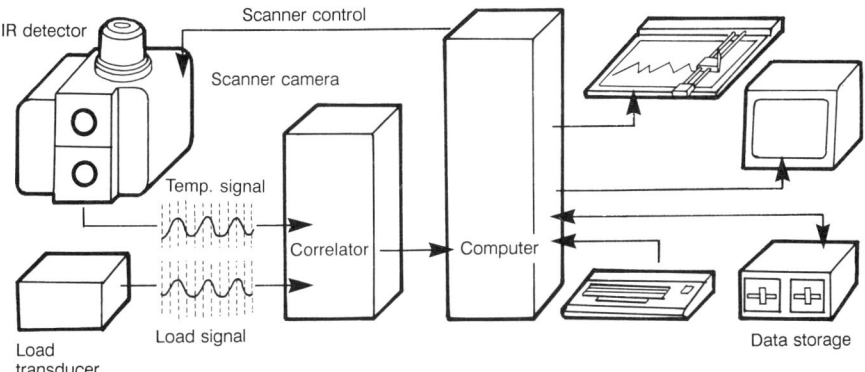

Figure 20.2. Main components in the SPATE system (Reference 1).

The SPATE method has many advantages compared to the traditional techniques for measuring stresses. The most apparent are that the method does not require physical contact with the object, that it is fast and that it is very sensitive. If the infrared detector is cooled by means of liquid nitrogen, the equipment can measure temperature differences of a thousandth of a degree at a distance of 10 metres. In steel this corresponds to strains of only one newton per square millimetre.

The greatest disadvantage of the method is the price. The price of the equipment in 1986 in England was £100,000.

Areas of application

The measuring equipment has been used in England, among other things, to map the concentration of strains in a radar mast, to examine pressurized tanks and parts of railroad cars. The method is well-suited to performing tension measurements on welded joints e.g. in the off-shore industry. The method can be used with almost all solid materials, for only very modest requirements are placed on the elasticity of the material. It is also expected that the method will be able to be applied to the mapping of stresses in concrete.

The stress mapping is very detailed because the equipment scans the subject partitioned into measurement regions of only one square millimetre. Because it is possible to scan surfaces with a width of several metres, the method will most certainly be used extensively in the future.

PVT

Pulsed video thermography

NDE principle

This technique is a thermal method based upon the following principle: A brief heat impulse acts upon the surface of an object under test causing transient temperature changes in the object as the heat impulse diffuses into it. The diffusion and thereby the temperature change is strongly dependent upon the presence of internal defects in the test object. The method can be used to localize defects in thin metal samples, in composite materials or at the interface between a coating and the base material. The method has, just as SPATE, only become possible thanks to modern infrared measuring equipment and computer technology.

Development of the method

The method, which has been developed at Harwell in England, is illustrated in Figure 20.3. A heat impulse with a duration of just a few milliseconds and covering a considerable surface of the object is generated by means of a large flash tube. The illuminated area or the corresponding area on the back of the test object is scanned by means of a thermovision camera. In this manner a "frame-by-frame" sequence of the temperature distribution pattern can be recorded. These patterns reveal details beneath the surface which affect the inward diffusion of heat. The sensitivity of the method depends upon the thermal properties of the object and on the depth of the defects, but the sensitivity of the method can be computed by means of a model for heat diffusion.

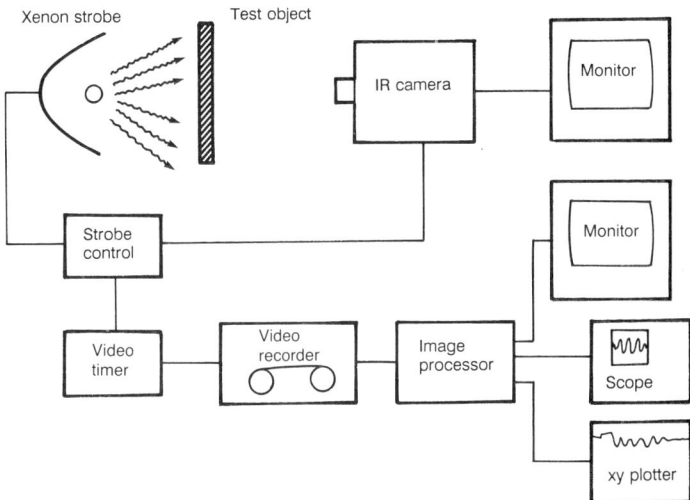

Figure 20.3. Main components of the PVT *system (Reference 3).*

Areas of application

The PVT method has turned out to be particularly effective in connection with the examination of fibre-reinforced components, honeycomb structures, metal-on-metal glued constructions, and coatings. PVT is particularly sensitive when one is searching for delaminations, density variations, thickness variations, bonding defects, and defects in composite materials.

MCM

Moiré contour mapping

NDE principle

The method is based upon the *moiré effect* which in essence is the following: Two relatively dense systems of curves or lines which overlap create an interference pattern called a moiré pattern. The word moiré is derived from the French word for a silk garment in which the threads are arranged in such a way that they form such interference patterns. The interference phenomenon is due to a simple shadow effect and is also sometimes called *geometrical* or *mechanical interference*.

MCM can be carried out with the so-called shadow-moiré method which is illustrated in Figure 20.4. This method only requires simple equipment, i.e. one or possibly two point sources of light, an optical grating and a camera. The grating is placed in front of the object, illuminated so that it casts a shadow on the surface of the object, which should preferably have a light, non-specular (diffuse) surface. Due to the overlap of the grating with the shadow of the grating on the surface of the object, the view from the camera is a pattern of moiré stripes which under certain experimental conditions are contour lines, i.e. curves which join points which are at a fixed distance from the plane of the grating. By means of photography a simple contour line image of the object is obtained.

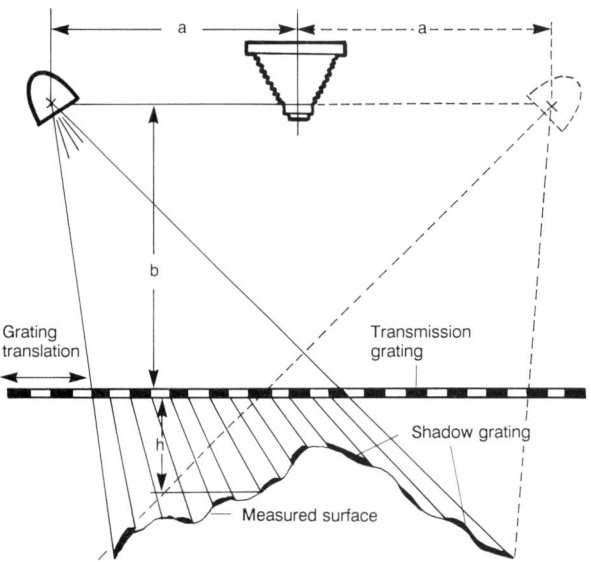

Figure 20.4. The shadow moiré technique for mapping of contour lines. The light source(s) and the camera aperture are all at the same distance b from the transmission grating.

Development of the method

One of the first NDE applications of the moiré technique was moiré strain analysis which was developed in the late 1940s, while the first publications concerning the shadow moiré technique appeared around 1970. At first pictures were taken with a stationary grating and one-sided illumination. Takasaki at the University of Shizuoka in Japan improved the technique shortly thereafter by introducing the translation of the grating under the picture-taking process (exposure times were several seconds) and "shadowless" illumination by using two light sources placed symmetrically on either side of the camera.

Figure 20.5 shows an example of shadow moiré pictures of an object with a relatively complicated surface topography. These three images illustrate the improvements which Takasaki introduced. When photographs are taken with a fixed grating (Figure 20.5a), an annoying side effect is generated, namely interference is visible between the imaged grating lines and the closely packed moiré stripes, i.e. a sort of second-order moiré.

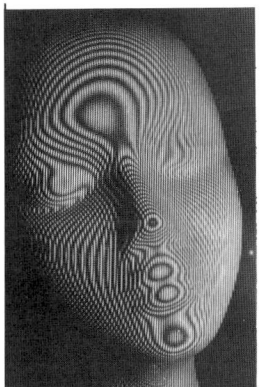 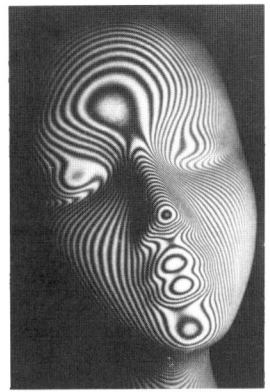 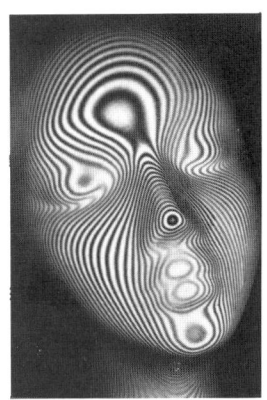

Figure 20.5. Shadow moiré exposures of the head of a plastic manikin: (a) stationary grating, one-sided illumination; (b) grating translated with constant speed, one-sided illumination; (c) grating translated with constant speed, double-sided illumination. Height differential between neighbouring curves $D \approx 1.65$ mm.

This effect completely disappears when the exposure is made with the grating in motion (Figure 20.5b), for the grating lines are smeared out during the exposure while the primary moiré pattern is not changed. The one-sided illumination can cause prominent features on the surface of an object to cast shadows on certain areas of the surface to be mapped (Figures 20.5a,b). This can be avoided by using "shadowless" illumination, i.e. by illuminating the object with two light sources placed symmetrically with respect to the camera (see Figure 20.4). If the two light sources and the camera aperture are equidistant from the grating, then the moiré pattern will be the same no matter whether the object is illuminated from the left or the right (Figure 20.5c).

The exposure conditions for the pictures in Figure 20.5 were such that the moiré stripes are contour lines with an approximately constant height difference between neighbouring lines of $\Delta h \approx (b/a)\,d$, where b is the common distance between the grating and the light sources and camera aperture, a is the distance between the light source and the camera aperture, and d is the distance between the lines of the grating (see Figure 20.4).

For contour mapping of larger objects or objects which can not remain fixed during an exposure, a projection moiré technique has been developed, see Figure 20.6. The method is used for example by the Meat Research Institute, Bristol, England, for making contour images of cattle to study their growth and the development of the animals under various conditions of feeding and environment. In principle the method consists of superimposing the image of a grating projected on a flat white screen and the image of the same grating projected onto the object (e.g. a cow painted with a chalk mixture in suspension).

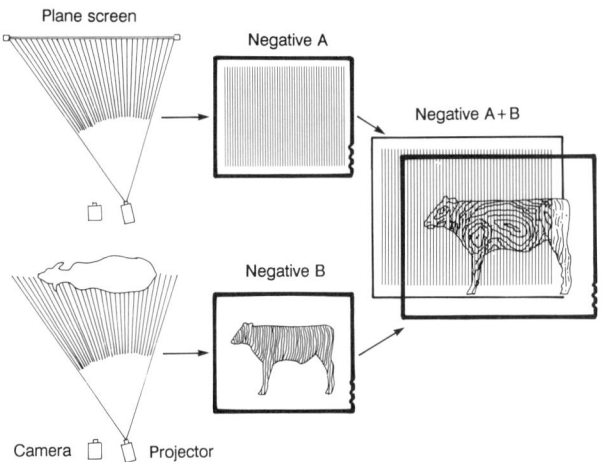

Figure 20.6. The projection moiré technique for producing contour line mappings of larger and/or moving test objects (Reference 6).

The deformation of the projected grating is a function of the distance of the surface points from the reference plane (the screen), and the moiré stripes represent the contour lines. When flash tubes are used, the projection moiré method can be used for objects in motion. The latest development in MCM is the introduction of a video camera and digital image processing techniques.

Areas of application

If the topography of an irregular surface is to be characterized, it is convenient to use contour line mapping, and in many cases this can be achieved in a simple manner with MCM. With respect to NDE such a mapping can be useful in the following situations: (1) production of form-critical parts, (2) measurement of displacements normal to the surface, (3) checking for changes in form over time. Both points (2) and (3) are of interest in connection with maintenance checks. MCM methods can handle a wide range of object sizes. The only requirement which must be met is that the surface of the object must be fairly smooth, light and diffuse. For small and medium sized test objects good precision can be achieved using the simple moiré shadow technique. The development of video and image processing techniques will presumably create new, interesting applications for MCM.

HI

Holographic interferometry

NDE principle

Photography is a method for recording the intensity (the square of the amplitude) of light waves. Holography is, in contrast, a method used for recording an entire optical wave front both with respect to amplitude as well as phase with a view to later reconstruction. The completed exposure is called a *hologram* (derived from the Greek *holos* meaning "whole"). In contrast to conventional photography which records a three dimensional object as a two-dimensional picture, the hologram preserves the true three-dimensional character of the object. The holographic recording is achieved in principle as shown in Figure 20.7a.

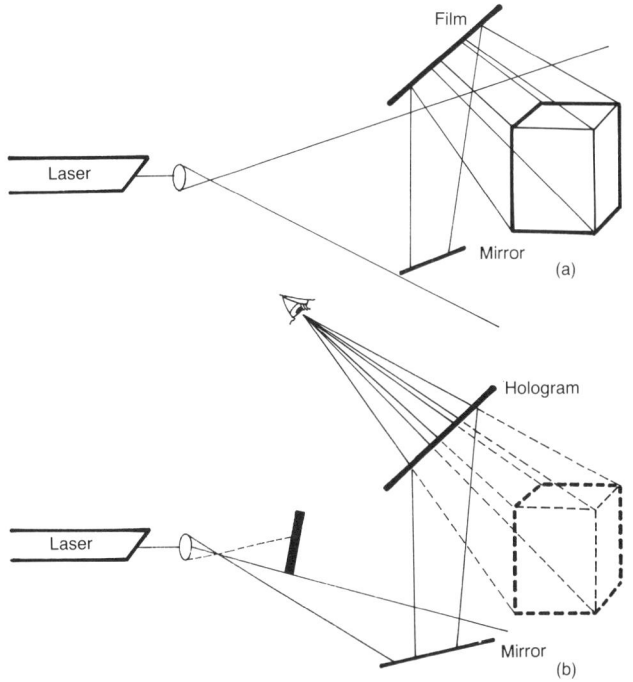

Figure 20.7. Holography. (a) Exposure of the hologram, (b) reconstruction of the image by diffraction of the reference radiation by the hologram.

Light from a laser is split by means of appropriate optics into two waves, one of which is used to illuminate the object, the other is used as a reference wave. An interference pattern is formed between the reference wavefront and the object wavefront on the holographic film. The hologram, i.e. the developed and fixed film, appears as a rule as a dense and complicated system of interference lines which have no resemblance at all to the object which it represents.

Figure 20.7b illustrates how the hologram is illuminated with a reference wave and how the object wave is reconstructed by means of diffraction. An observer who looks at the hologram will, therefore, see a virtual three-dimensional image of the object behind the hologram.

Because interference patterns are additive, several exposures can be made on the same film with subsequent reconstruction of several images which interfere with one another. This is the principle behind holographic interferometry (HI) which can be used for NDE. There are three HI methods which have turned out to be particularly useful for recording microscopic displacements on a test surface, namely: (a) the *double exposure method*, (b) the *simultaneous method*, and (c) the *time averaging method*.

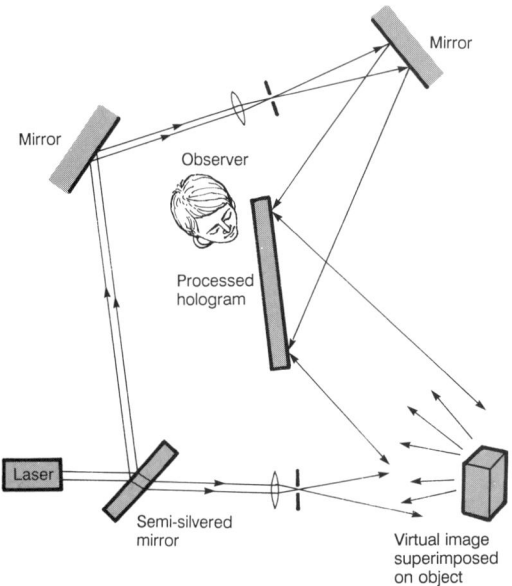

Figure 20.8. Holographic simultaneity method (real time HI).

- When the *double exposure method* is used (referred to as *double exposure HI* or *frozen fringe HI*), holograms of an object in two slightly different states are recorded on the same film. When reconstruction is carried out an image is created with interference stripes superimposed on those areas of the object which have been displaced between the two exposures. The interference picture can be regarded as a moiré pattern created by the two superimposed holograms.

- With the *simultaneous method* (*real time HI*) a hologram is recorded by a single exposure of an object in a desired reference state. After developing and fixing the photographic film it is returned to its original position so that the real object overlaps its holographic image. If the object surface is now displaced in relation to its reference position, interference lines (real time HI) will be observed, see Figure 20.8. A continuous registration of the time dependence of the displacements can be achieved by filming the hologram image.

- The *time averaging method* (*time average HI*) is particularly suitable for the examination of periodic oscillations and vibration types with small amplitudes. In this case the hologram is recorded on a film during a time interval which is large compared to the period of the oscillation under examination. The resulting hologram can be regarded as a superposition of an entire set of holograms. The interference lines in the reconstructed image reveals anti-nodes of the oscillation.

All three HI methods provide interference curves which represent points with the same displacement. If the illumination and observation are nearly parallel with the displacement direction, then the displacement between two neighbouring interference curves is about half the wavelength of the laser light. In other cases the sensitivity is less.

Development of the method

The holographic technique was discovered by Dennis Gabor in 1947, a discovery for which he received the Nobel Prize in 1971. However, it achieved practical significance only after the helium-neon laser was discovered in 1962. A prerequisite for doing holography is, as indicated, that the object and reference waves interfere in spite of relatively large differences in path length. This places such restrictions on the coherence of the light that only laser light can fulfil them.

Holographic film must possess a high resolution, and it is, therefore, fine grained and rather insensitive. When using an ordinary, low-powered laser, long exposure times are required. Serious stability problems can therefore arise when using HI in industrial surroundings. Today these problems can be solved by using pulsed, high-energy lasers. As an alternative to photographic film, a special holography camera with thermoplastic film has been developed. This film is processed dry on site and is thus particularly suitable in connection with real time HI where precise placement of the hologram can otherwise be a serious problem. Thermoplastic film also has the advantage that it can be flattened out and reused.

It can be advantageous to perform the analysis of HI holograms using a video camera attached to a computer with appropriate image processing facilities.

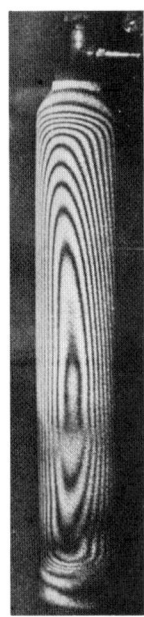

Figure 20.9 (a) Reconstructed image of the pressure tank from a double hologram exposure at tank pressures of 150 atm. and 145 atm. respectively. (b) The experimental setup during the exposures of the double hologram. Laser (A), pressure tank painted white (B), mirror for reference wave (C) and film (D), for exposure of the hologram. (Reference 7).

Areas of application

The relatively simple double exposure method has found many industrial applications. The deformation of an object between two states, e.g. unstressed and stressed, can reveal stress concentrations and defects in the object. As examples deformation measurements of machine tools in operation and of radio-telescope antennas acted upon by wind loads can be cited. Deformation of the test object between two loaded states can also reveal defects. For example, anomalies in the interference stripe pattern in the reconstructed image from a double hologram of an automobile tyre taken with two different air pressures in the tyre can reveal binding defects between the various layers of rubber and canvas. Figure 20.9 shows an example of how pressure tanks are examined. The picture of the pressure tank is reconstructed from a double hologram taken with 150 atmospheres and 145 atmospheres of pressure respectively. Anomalies in the interference stripe pattern reveal possible defects or weak areas in the pressure tank.

The time averaging method (time average HI) is used for the analysis of oscillations. Figure 20.10 is a reconstruction of a time averaging hologram of a turbine blade which has been acoustically excited to about 20 kHz. The dark interference stripes correspond to lines of nodes.

Figure 20.10. Reconstruction from a time averaging hologram of a turbine blade which has been acoustically excited to a resonant frequency of about 20 kHz. (Reference 8).

The introduction of video and image processing technology will, just as in the case of the optical moiré technique, enhance the capabilities of HI and provide it with a wider range of applications.

CT

Computerized tomography

NDE principle

Computerized tomography using ionizing radiation can be characterized as cross section radiography. The principle, roughly speaking, is illustrated in Figure 20.11. X-rays or gamma radiation from a point source is collimated to a flat, fan-shaped beam which penetrates a slice in the cross section of the object under examination. The intensity of the transmitted radiation is recorded by a linear arrangement of many small detectors or a single, linear, position-sensitive detector. The intensity at each point of the detector arrangement is a measure of the total absorption (or attenuation) of the radiation in the corresponding beam direction. In this manner a projection with an absorption profile of the cross section of interest is obtained. During the scanning procedure, whereby either the object alone or the source-detector arrangement are rotated step-wise about an object axis perpendicular to the cross section, a series of such projections through the same cross section is obtained.

On the basis of these absorption profiles and computer technology the absorption distribution in the cross section of interest can be mapped and displayed in the form of a grey scale or a colour scale on a monitor. These absorption values are generally, but not exactly, proportional to the mass density. The image is called a *tomogram* (derived from the Greek *tomos* = to cut, cf. *atomos* = indivisible).

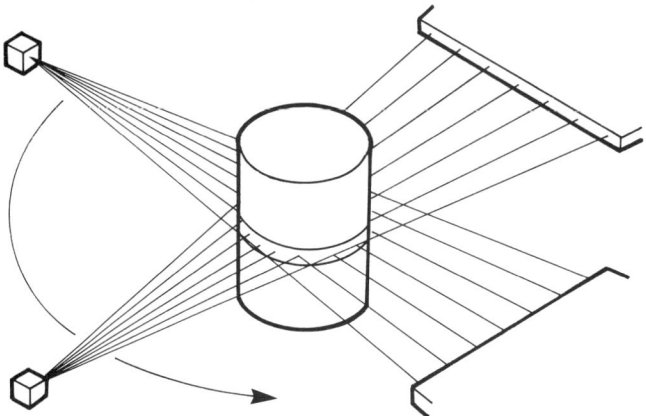

Figure 20.11. Tomography. The scanning procedure, whereby a radiation source and detector array are rotated stepwise, e.g. in increments of a degree at a time, around the object, is illustated. An alternative method is to allow the source and the detectors to remain fixed and to perform the stepwise rotation with the object.

The quality of the tomogram depends mostly on how precisely the cross section is scanned and on the uncertainties in the individual measurements. During a particular exposure there is a lower limit on the uncertainty determined by the statistical nature of the radiation (fluctuations in the stream of photons). In a well-dimensioned CT system the uncertainty approaches these statistical limits. In medical tomography one can typically achieve a spatial resolution of 1 mm and a relative density resolution of 0.02% for single cross section scans exposed for less than one minute.

Development of the method

CT with X-rays was first achieved by Hounsfield in 1971 (Nobel Prize, 1979) at the Central Research Laboratories E.M.I., Middlesex, England. The first commercial CT scanner for medical diagnosis required an exposure time of four minutes to perform a single cross section scan. It was used as a brain scanner, for the brain is one of the few human organs which can "stand still" during such a long exposure time. Today the fastest CT systems can perform a single cross section scan in less than a second. The medical CT systems represented an enormous advance over X-ray diagnosis. This success also inspired the development of industrial applications.

The first CT system designed specifically for industrial NDE applications was already developed in 1974 at the Lawrence Livermore Laboratory in the US. Since that time other institutions and industrial firms in the US, England, West Germany, France, the Soviet Union and Japan have proceeded to develop industrial CT scanners.

In connection with industrial applications of CT it is not generally necessary to achieve the extremely precise density resolution of 0.02%, while on the other hand a geometrical resolution better than one millimetre will often be desirable, e.g. for revealing micro-cracks. It is not necessary, as is the case with medical CT, to be so concerned about limiting the radiation dose or exposure time, making it possible to use simpler scanning and detector systems. The industrial CT scanners which are on the market today cost about £200,000-500,000, which is actually a factor five or ten less expensive than medical CT scanners. In the following we will briefly discuss some CT systems for NDE and some of their applications.

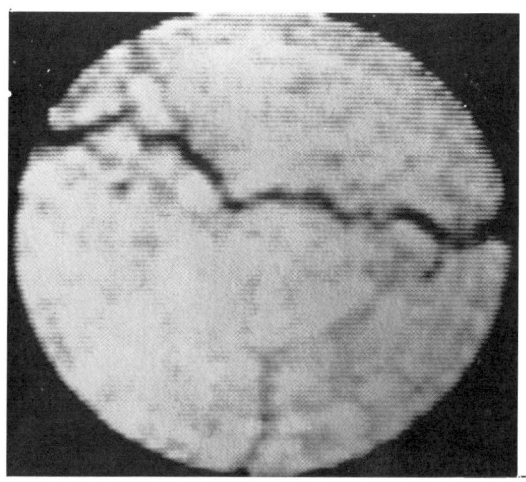

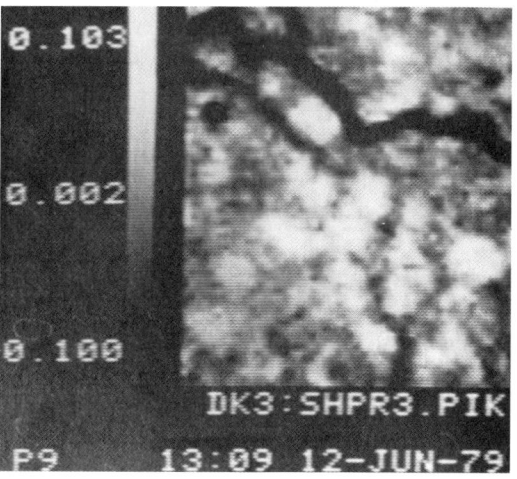

Figure 20.12. Tomogram of a concrete cylinder 15 cm in diameter. Recorded with a SMS scanner. A 1.6 curie CS-137 source was used. Single slice was exposed for 40 minutes. (a) Overview reconstruction of the entire cross section (1.6 mm between horizontal lines) and (b) a finer reconstruction of a portion of the same cross section (0.8 mm between horizontal lines). The white regions are stone, the grey areas are cement, and the dark regions are cracks or pores (Reference 12).

Areas of application

Scientific Measurements Systems (SMS) in Austin, Texas, in the USA has developed a CT system based on gamma sources, e.g. Cs-137 or Ir-192. A total of 31 detectors are arranged in a fan-shaped geometry, and for each projection the detector arrangements rotate in small steps over a small angular interval to fill out the spaces between the detectors. The various projection directions are attained by step-wise rotation of the object under test.

Figure 20.13. The CT scanner at BAM in Berlin. The test object on the object table is a 250 mm diameter glass cylinder for storing radioactive waste. The glass cylinder is inside a 7 mm thick steel container. During exposure of the tomogram the object is rotated step-wise on its axis (Reference 14).

The SMS apparatus has been used for the examination of logs to be used for telegraph poles and electrical wires. Annual rings, knots, rot, insect damage, and moisture content could be clearly observed. The apparatus has also been used successfully to map cross sections in concrete pillars. The distribution of stone aggregates and cracks is clearly visible in the tomogram of Figure 20.12. In spite of the fact that the pixel (picture element) size was only a square millimetre, cracks with a width of only a tenth of a millimetre could be observed. Such cracks reduce the absorption within a pixel by 10% which can easily be distinguished from the statistical image noise, particularly when the crack runs through several pixels.

Bundesanstalt für Materialprüfung (BAM), Berlin, has constructed a powerful and very flexible CT machine (Figure 20.13), which was brought into service in 1982. The BAM apparatus can be used either with gamma sources (isotopes) or X-ray tubes.

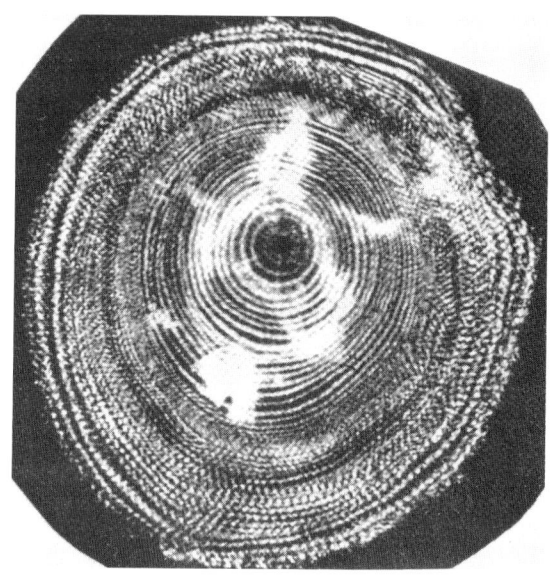

Figure 20.14. (a) A tomogram of a cross section through a living pine tree taken with a CT scanner at the University of Tokyo. Recorded using 47.5 kW, 20 mA, X-radiation. A total of 90 projections (2 degree rotation steps) and 1200 measurements per projection, i.e. a total of 90 × 1200 = 108,000 intensity measurements. (b) Photograph of the same cross section (Reference 13).

X-rays or gamma radiation is recorded just as in the SMS apparatus by means of an arrangement consisting of 31 scintillation detectors. It can handle a test object with a mass of up to 1000 kilograms, and the maximum scanning circle diameter is 100 centimetres. The BAM was the first non-military CT service centre outside the USA which offers routine inspections to all customers on a commercial basis.

The BAM equipment has been used to take tomograms of test objects representing a wide range of absorption coefficients, e.g. embalmed heads, ceramic insulators, massive glass rods, cement columns, and thick-walled steel pipes.

The growing concern for deforestation has created a need for a transportable CT scanner for the examination of living trees. Such a system has been built in West Germany (University of Marburg) and in Japan. The German apparatus uses a radioactive isotope (Am-241) as the radiation source, while the Japanese apparatus is equipped with an X-ray tube. The exposure of a tomogram is achieved by allowing the radiation source and the detector arrangement to move step-wise around the tree on a small platform affixed to the tree. To make the system as light and inexpensive as possible the detector arrangement consists of only three scintillation detectors, which means that the exposure time must be correspondingly longer: up to 10 hours exposure is necessary to achieve a tomogram of the highest quality. Figure 20.14 shows an example of a tomogram recorded using the Japanese equipment.

One of the advantages which CT possesses compared with conventional radiography is that the probability of detecting narrow cracks is much greater. This is due to the fact that the test object is irradiated from all angles.

CT has many potential applications for NDE in condition monitoring. Whether or not these capabilities will be utilized is to a large degree a question of economics, and this depends upon the further development and upon price reductions of industrial CT systems.

PA

Positron annihilation

NDE principle

Positrons (particles with the same mass as electrons but with a positive charge) have turned out to be very useful for NDE by using them as very sensitive probes for revealing defects in solid materials. Positrons are emitted during the radioactive decay of certain radioactive isotopes, e.g. Na-22 (emits positrons with energies up to 0.54 MeV) and Ge-68 (positrons with energies up to 1.89 MeV).

Positrons which are sent from a radioactive isotope into a test object have a penetration depth of up to about 1 to 1.5 mm in metal and up to about 2 mm in fibre-reinforced plastic materials. The positrons quickly lose their kinetic energy and become thermal (i.e. their energy is reduced to the kinetic energy due to temperature, about 0.03 eV) in about 20 picoseconds (20×10^{-12}s). After this there are two options for the further fate of the positron, both of which are relevant for NDE. After thermalization a positron will, both in metallic and non-metallic materials, diffuse with an average lifetime of about 200 picoseconds, until it collides with an electron. The particles will be annihilated and be transformed completely into two gamma quanta (gamma photons) each with an energy close to 0.51 MeV (see Figure 20.15). In non-metallic materials an electron and a positron can form positronium which is a short-lived hydrogen-like union of the two particles. If, however, the electron and the positron have parallel spins, i.e. "rotate" in the same direction about parallel axes, the positronium "atom" has a much longer lifetime than a normally annihilating positron, namely a lifetime corresponding to a half life of about two nanoseconds (2×10^{-9}s). This will normally be the case for about three-quarters of the positronium "atoms".

The lifetime of a positron is somewhat increased if it diffuses into a defect in a crystal lattice. This effect can reveal beginning fatigue in metals before any crack can be observed. Positron annihilation is the most sensitive method for detection of the atomic displacements which are connected with plastic deformation damage in regions where fatigue cracks are developing.

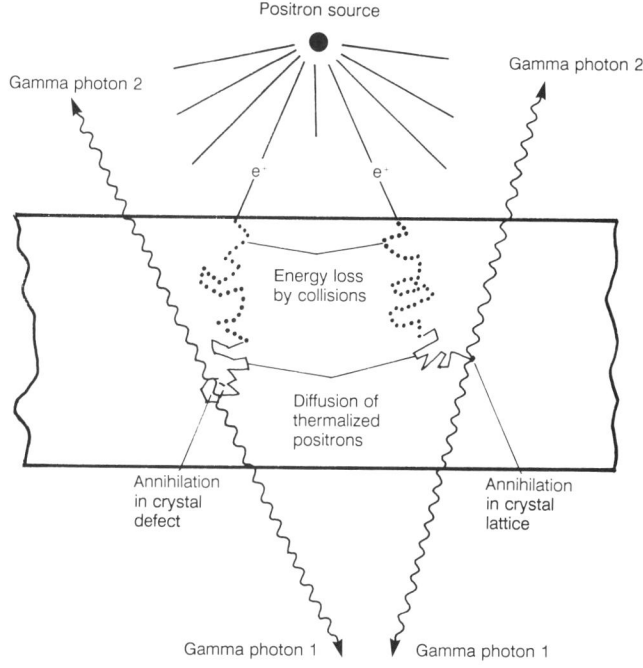

Figure 20.15. The life histories of positrons in solid, crystalline materials.

As an example of the use of positronium, the measurement of the moisture content in polymers and composite materials can be mentioned, because water molecules reduce the average lifetimes of the positronium. The loosely bound electrons in polar molecules such as water will "pluck" the positron from a positronium atom with greater probability than the tightly bound electrons in the base material.

Development of the method

The existence of positrons was theoretically predicted first and then experimentally verified in the early 1930s.

The positron annihilation processes were studied with methods known from experimental nuclear physics, i.e. lifetime measurements, measurements of electron impulses and measurements of the energy distribution of the emitted annihilation radiation by means of energy-sensitive gamma detectors. The means by which the various positron parameters are measured is shown schematically in Figure 20.16. During NDE applications positrons are injected into the surface of the object from an ordinary radioisotope holder or from a radioactive foil. Only weak sources are needed with activities of up to one millicurie, so the method requires only a minimum of protection from radiation.

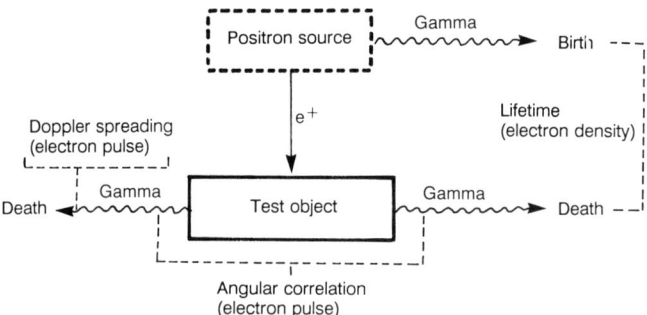

Figure 20.16. Overview of the measurement of positron parameters. Registration of the emitted gamma radiation is the fundamental principle involved in the PA technique (Reference 16).

Areas of application

PA has been used in materials science for the study of crystal defects in metals since the late 1950s. This area of research has been dramatically expanded and intensified after the discovery of the so-called *voids*, which are vacancy concentrations occurring due to intense exposure to neutron radiation. The presence and development of voids, which are small, microscopic cavities, has a great influence on the mechanical properties of metals. PA turned out to be the most sensitive method for the detection of voids, and it has, therefore, been natural to attempt to use PA for other NDE purposes.

PA has been used to record the development of fatigue in titanium alloys, for the study of the front edge of a growing crack and for tracing the resulting phase changes in the surrounding material. PA will be able to be used to study the damage with occurs as the result of explosion-forming of metals. The measurement of the lifetime of positronium appears to be a particularly promising technique for the detection of moisture in the surface of fibre-reinforced plastic materials. The method is apparently independent of how the fibres are laid and at which temperature the polymer is hardened. And the method is unaffected by the relative humidity of the material, its temperature and the character of the surface.

A. Lindegaard-Andersen

References

SPATE
1) *Thermoelastic stress analysis.* Bill Cummings, **Engineering**, Feb. 1985.
   [This is a four page article which explains the physical principle in an easily understandable way and describes the construction of the system as well as applications and development options.]

2) *Dynamic effect during vibrothermographic NDE of composites.* S.S. Russell and E.G. Henneke, **NDT International**, vol. 17, no. 1, Feb. 1984, pp. 19-25.
   [A rather theoretical article about a method closely related to SPATE.]

PET
3) *New tests find industry's small faults.* **New Scientist**, 10, Sep. 1986, p. 36.
   [Brief discussion of PET and other newer methods such as PA.]

4) *The development and the present status of moiré topography.* Hiroshi Takasake, **Optica Acta**, vol. 26, no. 8, pp. 1008-1019, 1979.
   [A good, readable article about the development of MCM up to 1979.]

5) *Theoretical aspects and practical applications of moiré topography.* S.S. Xenofos and C.H. Jones, **Phys. Med. Biol.**, vol. 24, no. 2, pp. 250-261, 1979.
   [Thorough mathematical analysis of the method. The applications discussed are medical in character including the registration of physical deformities due to various illnesses.]

6) *Recording the shape of animals by a moiré method.* C.A. Miles and B.S. Speight, **Journal of Physics E, Scientific Instruments**, vol. 8, no. 7, pp. 772-776, Nov. 1975.
   [Good article about the experimental conditions and quantitative aspects of the projection moiré technique.]

HI
7) *Holografin hittar dolda materialfel (Holography locates hidden material defects)*, Hans Bjelkhagen, **Teknisk Tidsskrift**, no. 18, pp. 19-21, Nov. 1975.
   [Very readable overview article about industrial applications of HI.]

8) *Holographic non-destructive testing.* Editor Robert K. Erf., Academic Press, New York 1974.
[A complete treatise on the various holographic NDT methods and applications up to 1974.]

9) *Holographic detection of defects under the surface of solid objects.* **American Journal of Physics**, vol. 51, no. 11, pp. 984-987, 1983.
[Describes a simple technique which by heating a test object locally can defect defects below the surface.]

10) **DOPS-NYT**. Published by the Danish Optical Society, vol. 1, no. 2, Apr. 1986.
[This issue has holography as the theme and contains several contributions about HI.]

11) *Computer reconstructed X-ray imaging.* G.N. Hounsfield, **Phil. Trans. Royal Soc. London**, vol. A292, pp. 223-232, 1979.
[Well-illustrated and pedagogically prepared article which reviews the principle, applications and development of the first three generations of CT scanners.]

12) *Examination of concrete by computerized tomography.* I.L. Morgan, H. Ellinger, Klinksiek and J.N. Thompson, **American Concrete Institute Journal**, pp. 23-27, Jan.-Feb. 1980.
[Brief technical description of an industrial CT system and its applications.]

13) *Computed tomography of use on live trees.* M. Onoe, J. Wen Tsao, H. Yamada and M. Yoshimatsu, **Materials Evaluation**, vol. 41, pp. 748-749, May 1983.
[Brief description of a special Japanese CT system.]

14) *Recent developments in the industrial application of computerized tomography with ionizing radiation.* P. Reimers, W.B. Gilboy and J. Goebbels, **NDT International**, vol. 17, no. 4, pp. 197-207, Aug. 1984.
[Overview article on the development of industrial CT until 1984. The article contains 58 references to the literature.]

PA

15) *Positron annihilation.* C.F. Coleman and A.E. Hughes, publ. in **Research Techniques in Non-destructive Testing**, vol. III, Ed. R.S. Sharpe, Academic Press, 1977.

16) *Crystal defects studied by positrons*, Kurt Petersen, Doctoral dissertation, Danish Technical University (DTH), Polyteknisk Forlag, 1979.

Chapter 21

Noise measurements

NDE principle

Noise or *sound* is due to elastic pressure vibrations in the air. A microphone is used to perform an objective measurement of these pressure vibrations. The collected/measured pressure vibrations are handled by the noise measurement equipment (often called a *noise meter*). The noise meter can be equipped with various filters making it possible to emulate the sensitivity of the human ear. In addition other filters can be used, e.g. octave band filters, 1/3-octave filters or narrow-band filters. Noise meters can also be equipped with integrating circuits so that noise load/noise doses over shorter or longer periods of time can be measured. The noise measuring device and its components are illustrated schematically in Figures 21.1 and 21.2.

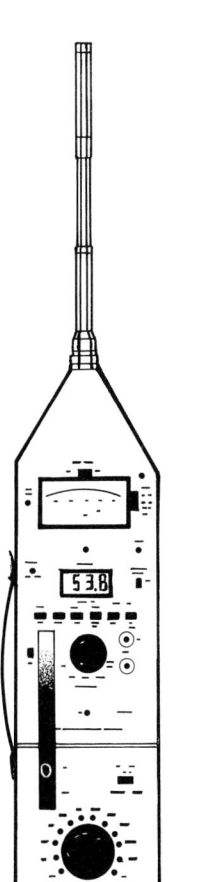
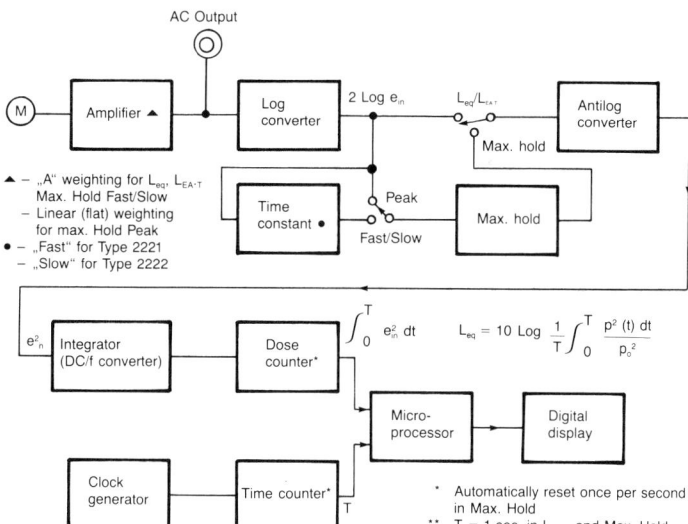

Figure 21.1. Noise meter.   Figure 21.2. Schematic diagram of the noise meter.

Sensitivity
The noise meter is constructed in such a manner that its sensitivity to pressure and frequency changes corresponds to what the human ear can detect. Sound pressure

levels from about 20 dB to 130 dB over 20 μPa and frequency changes from about 20 Hz to 20,000 Hz can be detected. The lower pressure level limit corresponds to the whisper of leaves in a light wind, and the upper level corresponds to the sound produced by a jet aircraft on a take-off run. See also Figure 21.3.

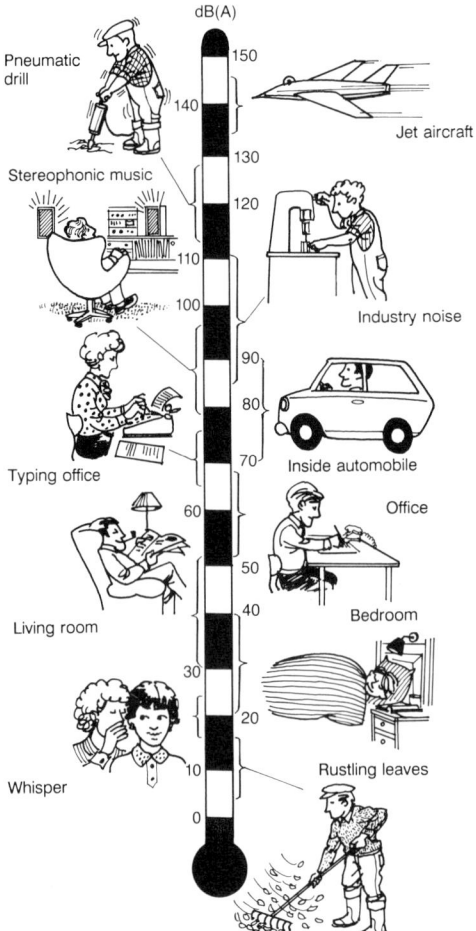

Figure 21.3. Examples of noise levels in the units dB(A) over 20 μPa.

Calibration
The noise meter is calibrated against a reference sound source with a known pressure level and frequency.

Advantages

- The noise meter is portable.

- The noise meter is flexible with the option of choosing different microphones and selecting various internal and/or external filters.

- The noise meter provides immediate results.

Limitations

- The measurements are sensitive to irrelevant noises and background noise.

- The microphones have a very limited operating temperature range.

Development of the method

The noise meter has been developed over the years, and recent developments in electronics and computer technology have led to further improvements and refinements. The noise meter can usually be provided with an array of options for performing various acoustic measurements.

Areas of application

The noise meter is particularly useful when demonstrating whether or not specific official requirements regarding noise levels have been fulfilled. Guidance in this connection is provided from national and local authorities. Rather stringent requirements are placed upon the measurements and particularly to the boundary conditions under which they are carried out, and strict requirements are also incumbent upon the operator to perform careful execution of the measurements and to document them.

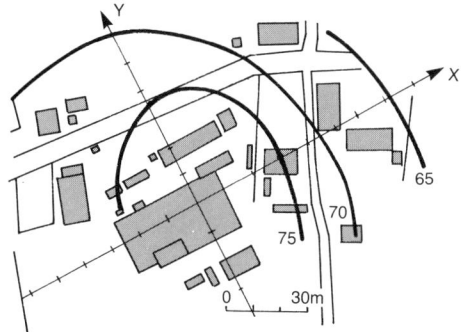

Figure 21.4. Sound source mapping.

The noise meter can also be used for mapping the noise levels within a particular area or working environment, *noise mapping*, see e.g. Figures 21.4 and 21.5. Another application is checking whether or not a shipment of machines and equipment adheres to given requirements.

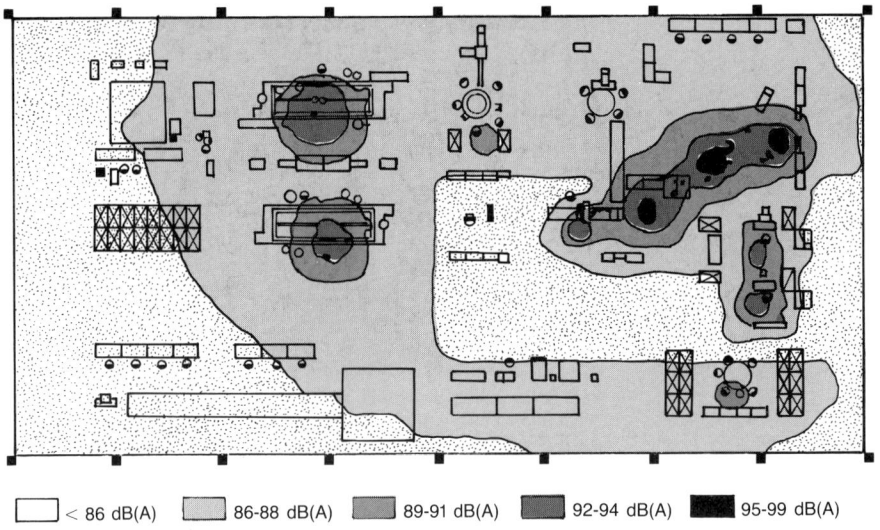

☐ < 86 dB(A)   ▨ 86-88 dB(A)   ▦ 89-91 dB(A)   ▨ 92-94 dB(A)   ■ 95-99 dB(A)

Figure 21.5. Noise source mapping of an indoor environment.

One of the most widespread areas of application is preventive maintenance. Noise spectra are measured at given locations at appropriate time intervals and then compared. In this manner it is possible to discover and correct defects which may have turned up, before they can lead to serious breakdowns.

In order to perform reproducible measurements, an array of documents specifying standardized measurement procedures are available to the measurement technician. See the references at the end of this chapter.

Practical applications

As mentioned above, the noise meter can be used for many different tasks. For example see Figure 21.6, monitoring of a rotary blower where a check measurement revealed a defective bearing. After repair the noise level returned to the normal level.

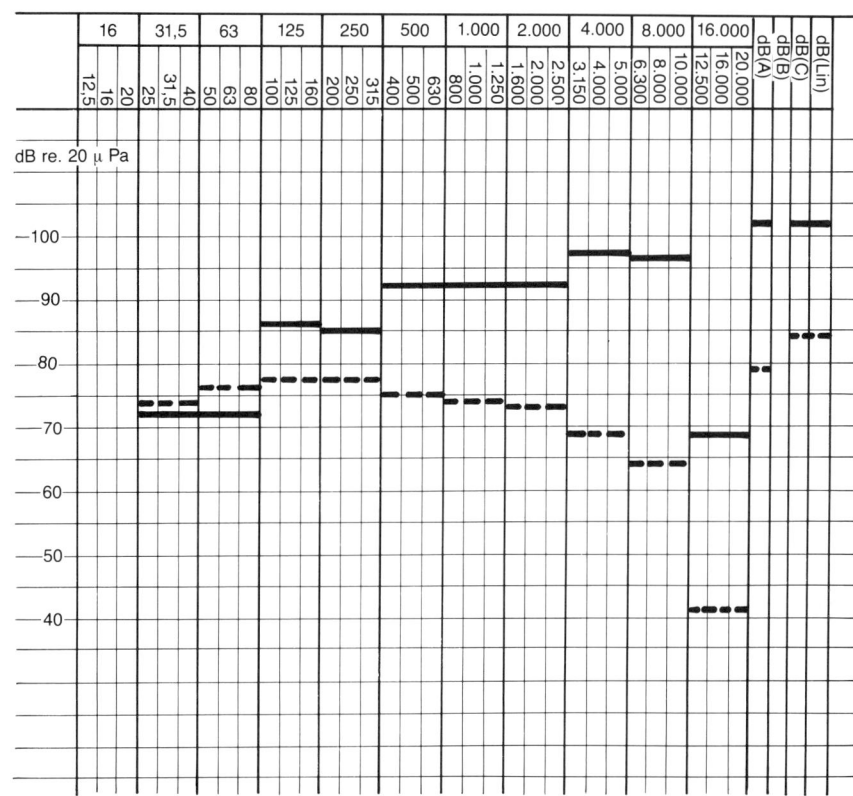

Figure 21.6. Check measurements, octave band analysis and broad band measurements: L_A, L_C and L_U.
-- with defective bearing,
--- with bearing repaired.

Figure 21.7 shows the result of the introduction of a noise abatement screen on a compressor in a pump- and compressor-room.

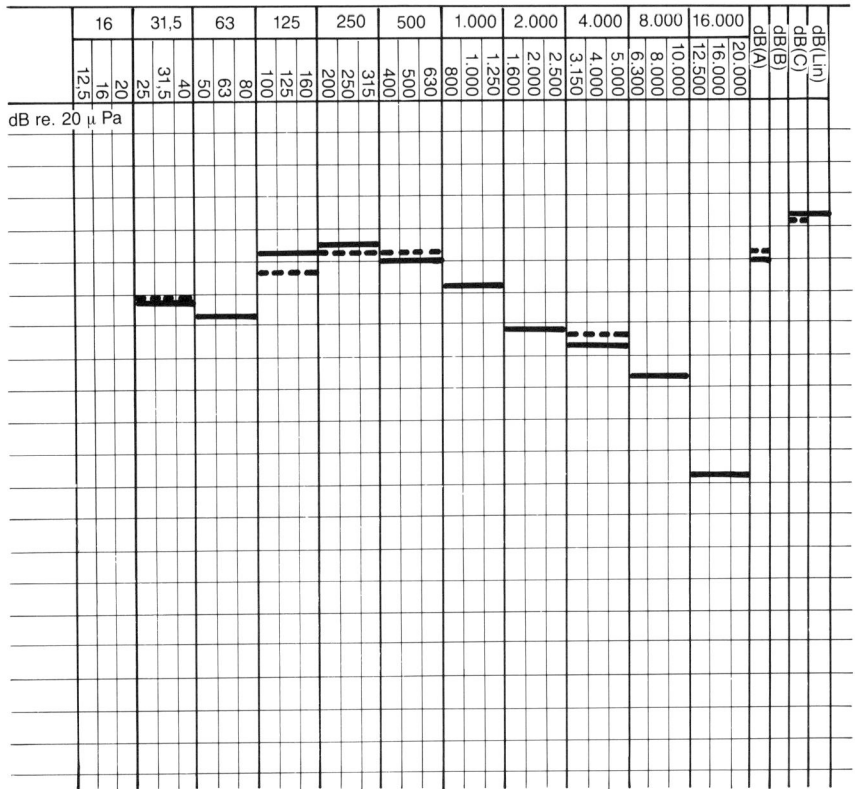

Figure 21.7. Result of noise isolation in a compressor-room: octave band analysis and broad band measurements: L_A, L_C *and* L_U.
-- noise levels in diffuse sound field without noise abatement of compressors,
--- noise levels in diffuse sound field with noise abatement of compressors.

Figure 21.8 shows a delivery check measurement performed on the ventilation system installed in the cab of a fork-lift truck.

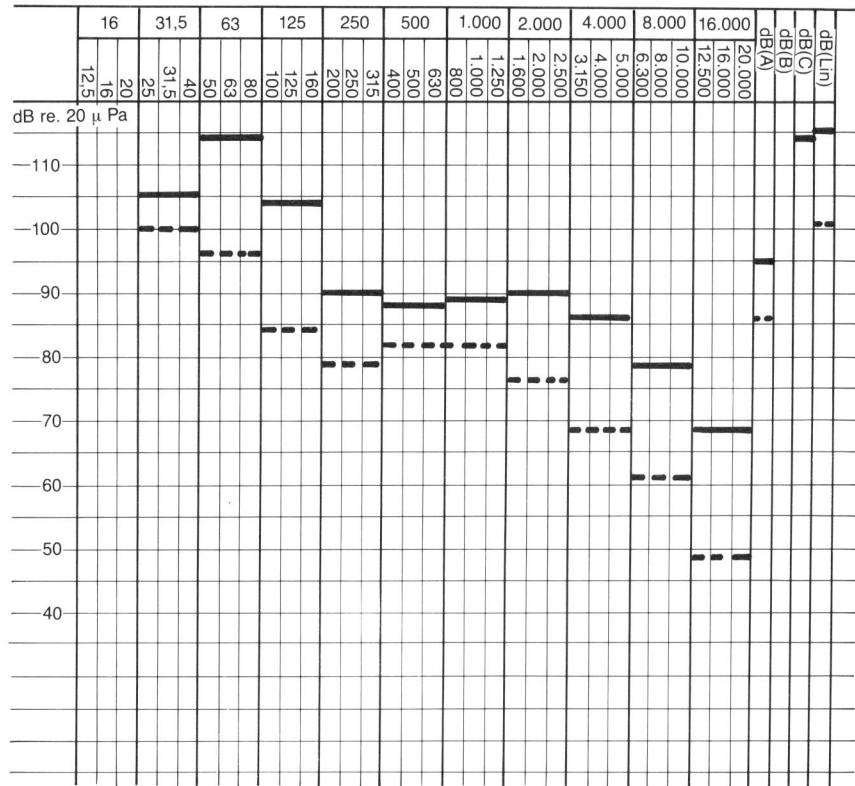

Figure 21.8. Noise levels measured in the cab of a fork-lift truck: octave band analysis and broad-band measurements: L_A, L_C and L_U.
-- *without ventilation system in operation,*
--- *with ventilation system operating.*

Training required/desired

In order to attain sufficient benefit from the measurements it is important to provide the operator with thorough instruction, perhaps supplemented with a special training course.

If the measurements are to be used as documentation for the authorities, then the measurements should be carried out by an authorized laboratory, unless some other arrangement has been made.

Lene Mikkelsen
B.W. Kristensen

References

1) *Report no. 32 from Acoustic Laboratory, ... Environmental noise from industrial plants. General prediction method.* The Danish Academy of Sciences, 1982.

2) ISO 3741-42-43-44-45-46, *Determination of sound power levels of noise sources.*

3) *ISO Standard Handbook 4*, Acoustics, vibration and shock, 1985.

4) *Acoustic noise measurements*, 5th edition, June 1988, Brüel and Kjær A/S, Denmark.

5) *Noise control*, 2nd edition, 1986, Brüel and Kjær A/S, Denmark.

Chapter 22

Optical pattern recognition

NDE principle

Optical pattern recognition is a technique which is based upon the use of a video camera and a computer with the ability to store images.

The computer carries out operations upon and recognizes images which have been acquired by means of the video camera.

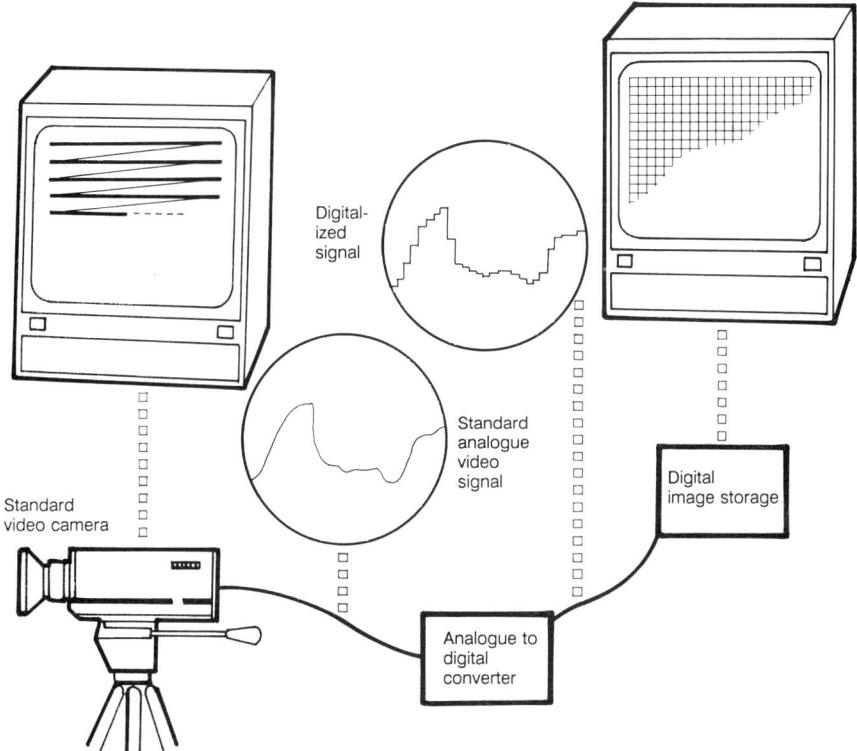

Figure 22.1. The analogue video signal from the camera is digitized at full video speed, i.e. 25 frames per second. The digital information is placed in the image storage area of the computer and can be displayed on a monitor if desired.

The analogue video signal from the camera is digitized and is placed in the image storage area of the computer (see Figure 22.1).

A CCD (charge coupled device) camera is often employed (see Figure 22.1a) because it is more stable and dimensionally constant than a tube-based camera.

Digitization is carried out at full video speed. Information in the image storage area of the computer can be renewed 25 times each second.

Figure 22.1a. A CCD camera is often used. CCD stands for "Charge Coupled Device". It means that the image is captured by a chip which consists of a large matrix with light-sensitive elements. The light charge in each element is "clocked" one element at a time.

The image storage area is organized like a large matrix with a typical size being 512 x 512 points (*pixels* = picture elements). For each pixel contains information about the light level at that location. This information is stored as a digital value between the values 0 and 255. The value 0 corresponds to totally dark, and 255 corresponds to completely white.

One can imagine the information in the image storage area to be the information acquired from an enormous array of photocells which measure the light level in e.g. 512 x 512 points. This corresponds to a total of 262,144 photocells.

The computer can address each memory cell and read out the light level stored in every single pixel represented in the storage area. This means that it can find appropriate black/white transitions in the image, measure distances, and identify and check contours.

Development of the method

Historically digital image processing and optical pattern recognition have been developed by research institutes and similar institutions, but in recent years complete systems with camera and computer have been taken into serious use in industry.

The computer technology and the camera technology is quite advanced and operates perfectly well with the present-day resolution and light sensitivity. The technical data in these areas are constantly being improved.

There can be a big difference in the prices of cameras depending upon the resolution, light sensitivity and sharpness of focus which they can deliver. The price of the computer components in such systems has been decreasing in recent years. Complete turn-key systems for optical pattern recognition can be acquired for prices from about £7,000 and up.

Areas of application

One factor which is very decisive for whether a measurement can be performed with sufficient certainty and precision is the illumination of the object. The illumination must emphasize those details in the picture which are important for the measurements desired.

The limitations of the systems often lie in the fact that it is not possible to emphasize precisely those details in an image which are important. If the computer can not find the necessary information in the image, then it cannot carry out the measurement.

In many measuring setups background illumination is used to create silhouette images. An example with a silhouette image is shown in Figure 22.2.

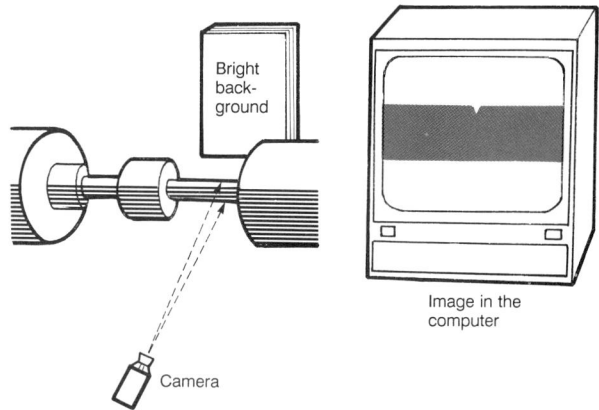

Figure 22.2. An illuminating plate forms the background and creates a silhouette image of the axle. In this manner a dependable measurement of the axle contours can be achieved.

In silhouette images the contours of the object appear quite sharply with the illuminating light source behind the object and the front of the object quite dark.

Optical pattern recognition

On the other hand details inside the object will normally be lost or appear with very low contrast.

The system in Figure 22.2 is monitoring an axle. The illuminating plate in the background causes the axle itself to appear dark with a good dark/light transition, which sharply defines the edge contour of the axle. The image in the computer is shown with greater detail in Figure 22.2a.

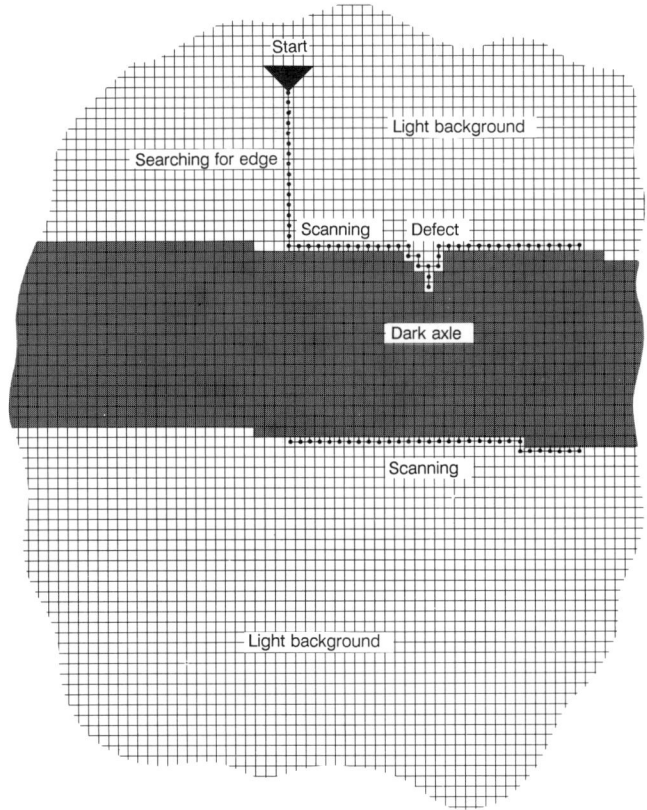

Figure 22.2a. The figure shows a section of the image storage area containing the silhouette image of an axle. The resolution appearing in this illustration is about one tenth of the resolution which is normally used. The computer program starts the analysis at a pixel in the illuminated region and searches downward until finding the axle then continues scanning along the axle.

It is apparent that the computer program starts out in the light region for example and searches downward until the transition to the axle is found. Then scanning is continued along the axle until possible defects are located.

By also scanning on the opposite side of the axle the diameter in pixels can be measured, and this can be converted to an absolute measure in millimetres.

In Figure 22.3 an example is shown with forward illumination. Such illumination is employed when defects are to be found on the front surface or contours on the front surface of the object are to be measured.

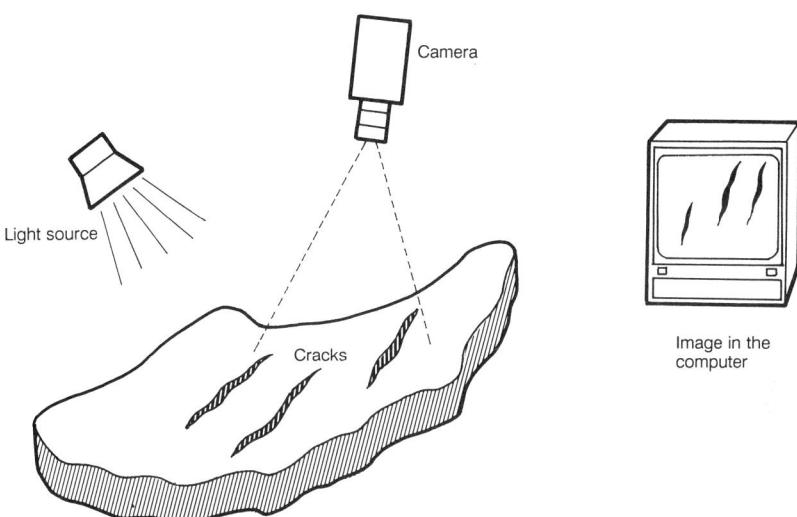

Figure 22.3. With oblique, forward illumination on a somewhat specular surface, cracks and other irregularities will appear as dark, shadowy areas in the image. These defects will be detected by the computer program as regions with lower light intensity values in the corresponding pixels in the image storage area.

In many cases it can be advantageous to use shadow effects. Figure 22.4 shows a type of three-dimensional measurement. The third dimension (the height of the object) is observed by allowing a stripe of light (e.g. from a laser) to strike the object from an angle different from the one from which the camera views the object. In this manner the position of the beam of light in the image storage area of the computer can represent the height variation of the object along the stripe of light.

The object to be measured and checked will often be moving. In order to avoid an unclear image, it is necessary to freeze the object in motion. This can be achieved by using strobe lights for illumination, intensely illuminating the object for example 10 μs of the 40 ms over which the camera integrates the exposure. In this manner a very sharp image can be obtained even though the object is moving at a high speed or rotating rapidly. The strobe light illumination must be synchronized with the image capture process.

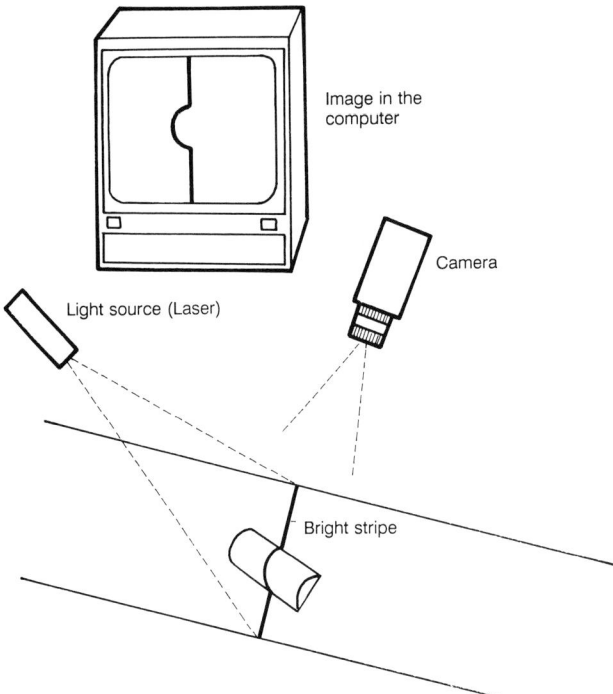

Figure 22.4. The figure illustrates a technique for three-dimensional measurement. The height variation of a beam of light moving across the subject is seen in the camera, because the camera sees the beam from a different angle than the angle from which the light is projected.

Another technique for freezing moving objects is the use of very fast cameras which can "gate" the picture into the system in a very short time interval. This is usually a more expensive solution, but it can be necessary in certain situations where it is not possible to make the stroboscopic illumination dominate, e.g. when performing measurements in daylight.

The frequency with which measurements are carried out in an optical pattern recognition system will typically be 10-15 measurements per second. This corresponds to 1/25 s for the exposure and 1/25 s for the computer to handle the image information, which is often more than adequate for the computer. This is possible because the computer does not examine all of the pixels in the image but concentrates on those in the most interesting and critical regions.

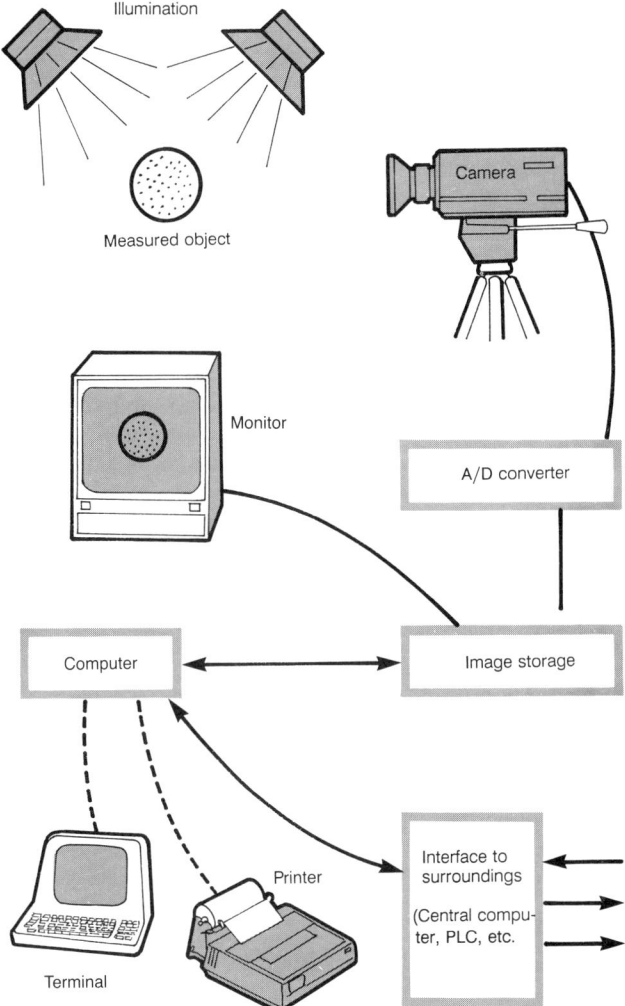

Figure 22.5. The figure shows a complete system for optical pattern recognition consisting of:

1) *illumination which emphasizes important details,*
2) *one or more cameras,*
3) *an analogue to digital converter,*
4) *an image storage area,*
5) *a computer which analyses the information in the image storage area,*
6) *a monitor for displaying the image,*
7) *a printer for printout of results and statistics,*
8) *a terminal through which the user can communicate with the system, and*
9) *an interface with communications facilities to other control systems, regulation systems, central computers, etc.*

Optical pattern recognition

Practical applications

A practical example from the piping industry is the measurement and control of pitting in the X-ray pictures of pipes.

Figures 22.6 and 22.6a show a pipe without pitting and with pitting respectively.

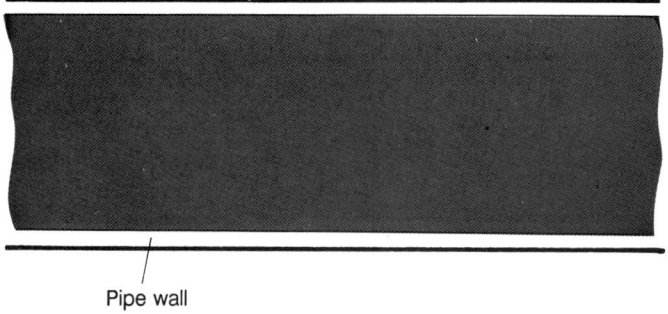
Pipe wall

Figure 22.6. This is an X-ray photograph of a pipe with no pitting. The light level is roughly constant across the entire pipe surface.

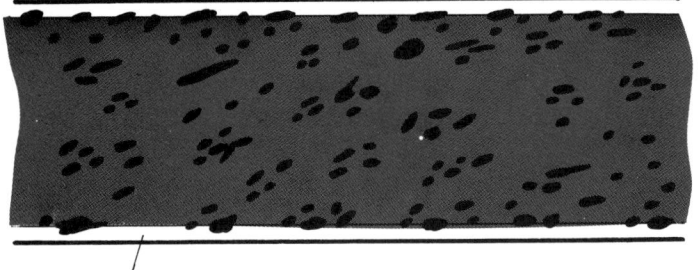
Pipe wall

Figure 22.6a. This X-ray photograph shows a pipe with serious pitting. The light level varies substantially across the surface, and the contour of the pipe is uneven.

The system detects pitting partly due to the variations in light intensity across the surface of the pipe (the pits appear as dark blotches), and partly due to variations in the diameter of the pipe, since the pipe contour becomes quite uneven when some of the pits cause the diameter to be increased.

Training required/desired

It is not essential that the user of an optical pattern recognition system possess a great deal of knowledge about computer technology and programming. The system can be purchased as a turn-key unit, programmed to a particular task. On the other hand it is usually very important for the user to understand the importance of illumination so that the image quality in the measurement setup is always optimized. No matter how intelligently the pattern recognition system may have been programmed, it will never be able to measure and recognize features which do not appear in the image.

Jesper Nilausen

References

Primarily due to the fact that optical pattern recognition has first been utilized within the past few years there is not yet an extensive literature describing the installation and use of systems for optical pattern recognition. But there are a number of books, cited here, describing the components and operation of such systems:

1) *Vision in man and machine.* D. Levine, McGraw-Hill Book Company, New York, 1985.

2) *Digital image processing.* Rafael C. Gonzales & Paul Wintz, Addison-Wesley Publishing Company, New York, 1987.

3) *Digital image processing (a practical primer).* Gregory A. Baxes, Cascade Press, New York, 1988.

4) **Robot technology:** *Interaction with the environment*, volume 2, Philippe Coiffet, Kogan Page, London, 1983.

Chapter 23

P-scan

NDE principle

P-scan is an ultrasonic system with automatic documentation which is used to perform automatic or manual ultrasonic examination tasks. The purpose of such examinations can be to locate defects in an object or to measure the thickness of a material. The use of P-scan requires knowledge of the ultrasonic methodology which is discussed in Chapter 34 on *Ultrasonics*.

The P-scan system differs from other ultrasonic systems in that it stores all data while the examinations are carried out and displays it in a so-called P-scan image on a monitor display.

The P-scan image is a projection image of the object under examination, e.g. an image of a weld or part of a pipe. In the P-scan image the defects which have been found are automatically shown at their correct locations, or the thicknesses of the material for the entire area which has been scanned are automatically displayed.

The results of the P-scan examination can be saved on disk and presented as hard copy from a printer.

Development of the method

The P-scan system has been developed as a result of the desire for and the requirement for direct documentation of ultrasonic examination results. *Direct documentation* means reporting which is independent of the subjective evaluation of signals by the operator. In the field of medical technology equipment has been developed to perform ultrasonic examinations of the human body, where the result has the form of a cross section, a so-called B-scan image.

This development was extended to the examination of metallic objects, and in the mid-1970s the first P-scan equipment for the examination of welds was put into service.

The P-scan system has been developed and manufactured in Denmark. In several other countries ultrasonic systems have been developed with varying degrees of automation and documentation. This has occurred for example in the US, England, France, Japan and Spain.

In 1985 a new generation of the P-scan system appeared: battery-operated and portable. All technical details in this chapter refer to this new generation of P-scan systems.

The P-scan equipment

The P-scan system consists of a number of components. The most important of these is the P-scan apparatus itself (see Figure 23.1), which is the central unit in the system. The P-scan apparatus is portable and battery-powered, containing the ultrasonic pulser/receiver, the data processor, the data screen, two disk drives and a printer.

Figure 23.1. The P-scan apparatus is battery-powered, weighs 18 kg and contains the ultrasonic pulser-receiver, the data processor, display screen, two disk drives and a printer.

The ultrasonic pulser-receiver

The ultrasonic pulser-receiver produces the electrical pulses which in turn generate ultrasonic pulses in the transducer or the probes. It also receives electrical pulses from the probes. These pulses contain all the information received by the probes as echoes or reflected ultrasonic waves. This information may be considered to be the primary results of the examination. In the ultrasonic receiver the pulses are amplified and are then transferred to the data processor for further treatment.

Data processor

The data processor has many functions. It controls the ultrasonic module, it receives and processes the ultrasonic echo signals and combines them with the corresponding probe position data. It produces the P-scan images which appear on the screen during examination. Finally, it controls the disk drives and the printer as well as the automatic scanner if this is used for the examination.

Disks
The data disks in the P-scan system are used for storage of examination results and the corresponding procedure data. The disks are used in another computer with colour graphics and colour printer for further treatment, evaluation and presentation of the examination results.

Data screen
The data screen displays the P-scan images while the examination is being carried out. On the screen the operator may also follow the ultrasonic signal in A-scan presentation as on a conventional ultrasonic instrument. The content of the P-scan image is discussed in a later section of this chapter.

Printer
The main P-scan unit contains a small printer. It is used for printout of procedure data or examination results in case this is required on site during or immediately after the examination. The final reporting is produced on a colour printer which is connected to an external computer.

Scanners
For continuous generation of probe position data, scanners or probe manipulators are used as an important component in the P-scan system. There are two types: *automatic scanners* and *manual scanners*.

An *automatic scanner* is a piece of equipment which causes the probes to be moved in a pre-defined pattern. It is controlled either by the computer or by means of a separate control unit. An automatic scanner mounted on a pipe can be seen in Figure 23.2.

With a manual scanner the probe is moved by hand as in the case of a conventional ultrasonic examination. The probe is mounted in the scanner, which continuously sends position data to the computer of the P-scan system where it is combined with the ultrasonic data to form P-scan images. Figure 23.3 shows a picture of such a manual scanner.

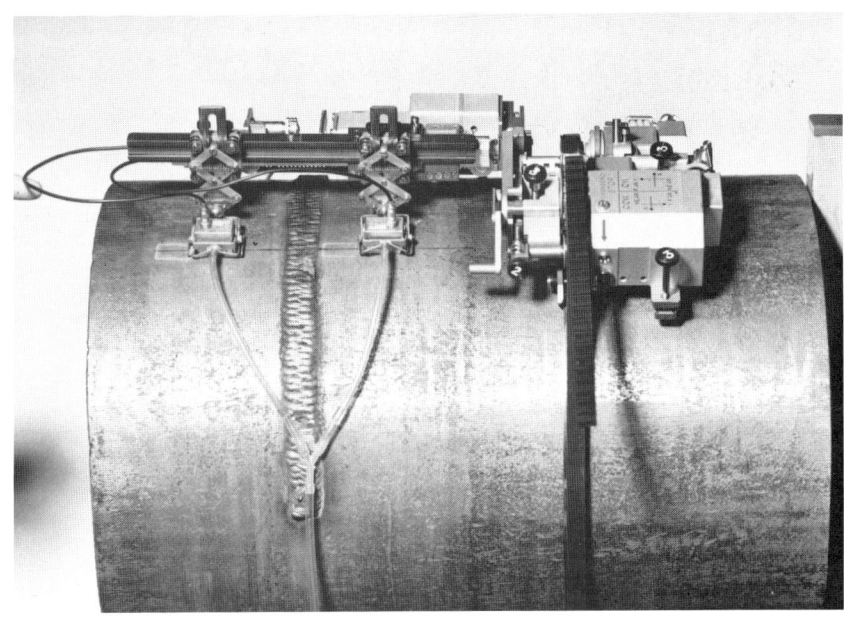

Figure 23.2. An automatic scanner mounted on a pipe for examination of a weld.

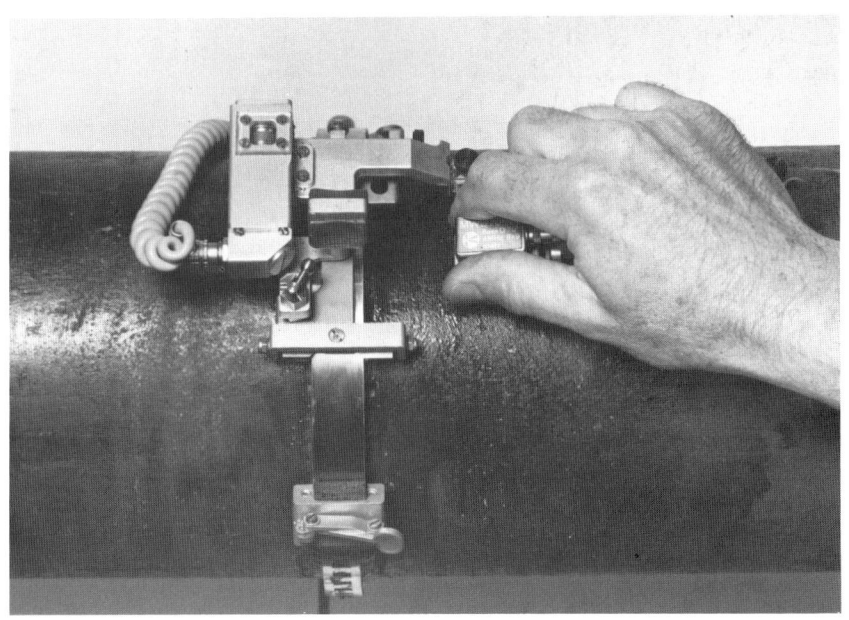

Figure 23.3. A manual scanner mounted on a pipe for examination of a weld.

The P-scan image
As mentioned earlier the P-scan system constructs an image on the monitor screen during an examination. The *P* in the name P-scan stands for *projection*. The P-scan system shows the results of the examination in the form of projection images.

Figures 23.4 and 23.4a illustrate the sort of projection images which one might obtain on the screen in connection with the examination of a butt weld.

Figures 23.5 and 23.5a show projection images obtained when measuring the thickness of a pipe. This image is also called a T-scan image, where the *T* stands for *thickness*.

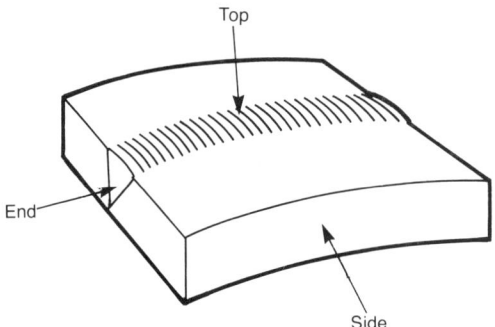

Figure 23.4. The locations of weld defects are shown by means of three projection images of the volume examined. The three images are called TOP, SIDE and END views. In the P-scan different colours are used to indicate from which direction the defect is indicated.

The external computer
As mentioned the P-scan results are stored on disks. These disks can be transferred to an **IBM PC AT** or compatible computer where the results can be displayed on the monitor, evaluated and processed as desired. The program supports colour graphics, which can either be used to separate indications from the different sound directions (weld joint program) or to display various material thicknesses in various colours.

The complete documentation consists of P-scan images in colour prepared as printouts from the external computer.

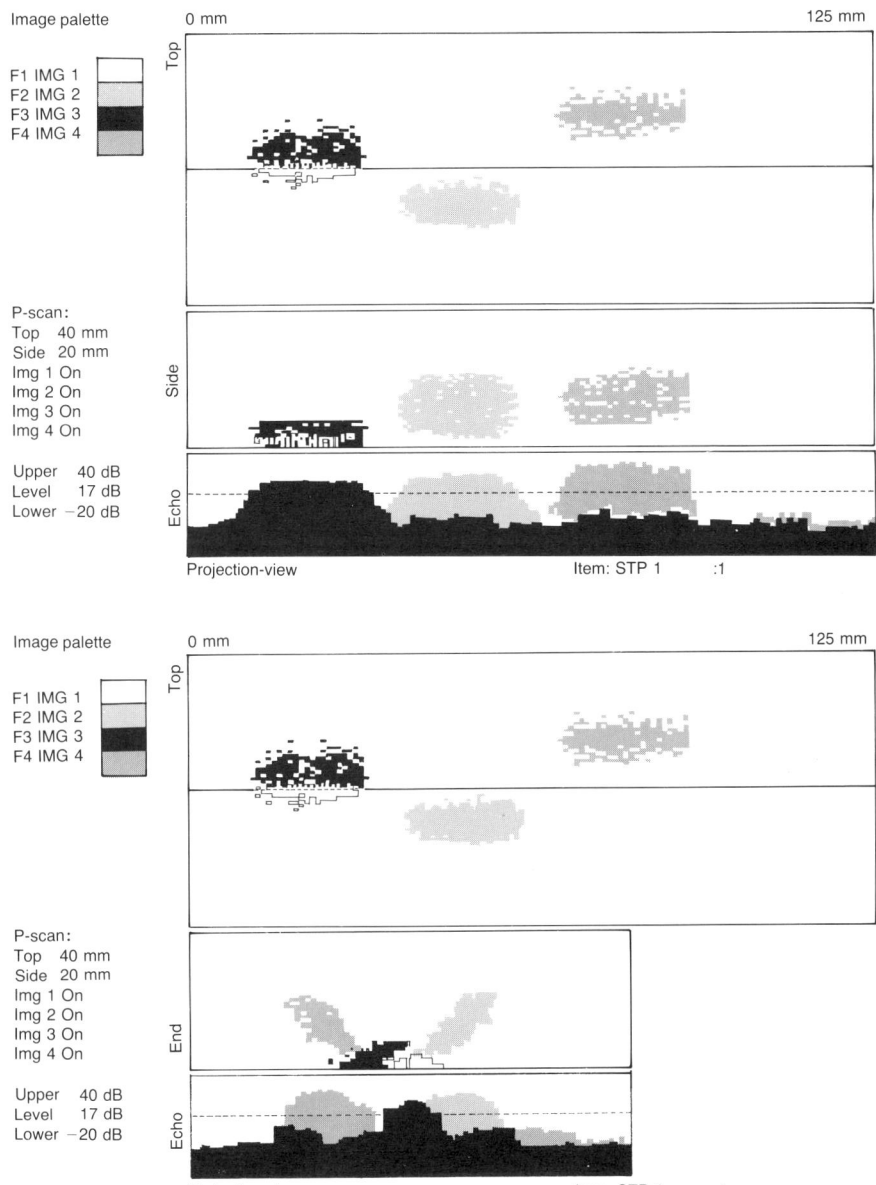

Figure 23.4a. P-scan images of a weld as shown in Figure 23.4.

Areas of application

P-scan technology is used both for in-service inspection and for production control.

In-service inspection of welds and materials using P-scan can be quite relevant in cases where there is a need to map an object and/or to provide detailed documentation.

P-scan is also appropriate in cases where it is possible to automatically scan an object and thereby carry out an examination quickly.

Examination of welds
As an example we can mention examination of welds in structures where the risk of fatigue cracking exists. P-scan can also be used to check welds with a risk of creeping cracks and stress corrosion cracking.

The properties of the P-scan system in this connection were first utilized to examine critical welds in nuclear power plants where it has been necessary to map existing cracks or other defects and to follow their development during the lifetime of the components.

The characteristics of the P-scan system with respect to thorough documentation of defects in welds is now also used for piping systems and pressure tanks in petrochemical and other chemical plants, in conventional power plants and on welds in steel structures such as platforms, bridges, buildings, storage tanks, submarines and ships.

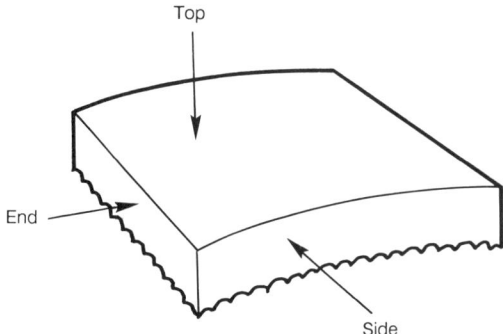

Figure 23.5. The mapping of material thicknesses is shown as three projection images or cross sections: Eight colours are used to illustrate various material thicknesses.

Greater benefit is obtained from a P-scan examination if a P-scan report acquired during production inspection is available. The results of such a report can form the basis for the evaluation of later examinations undertaken during the lifetime of the component.

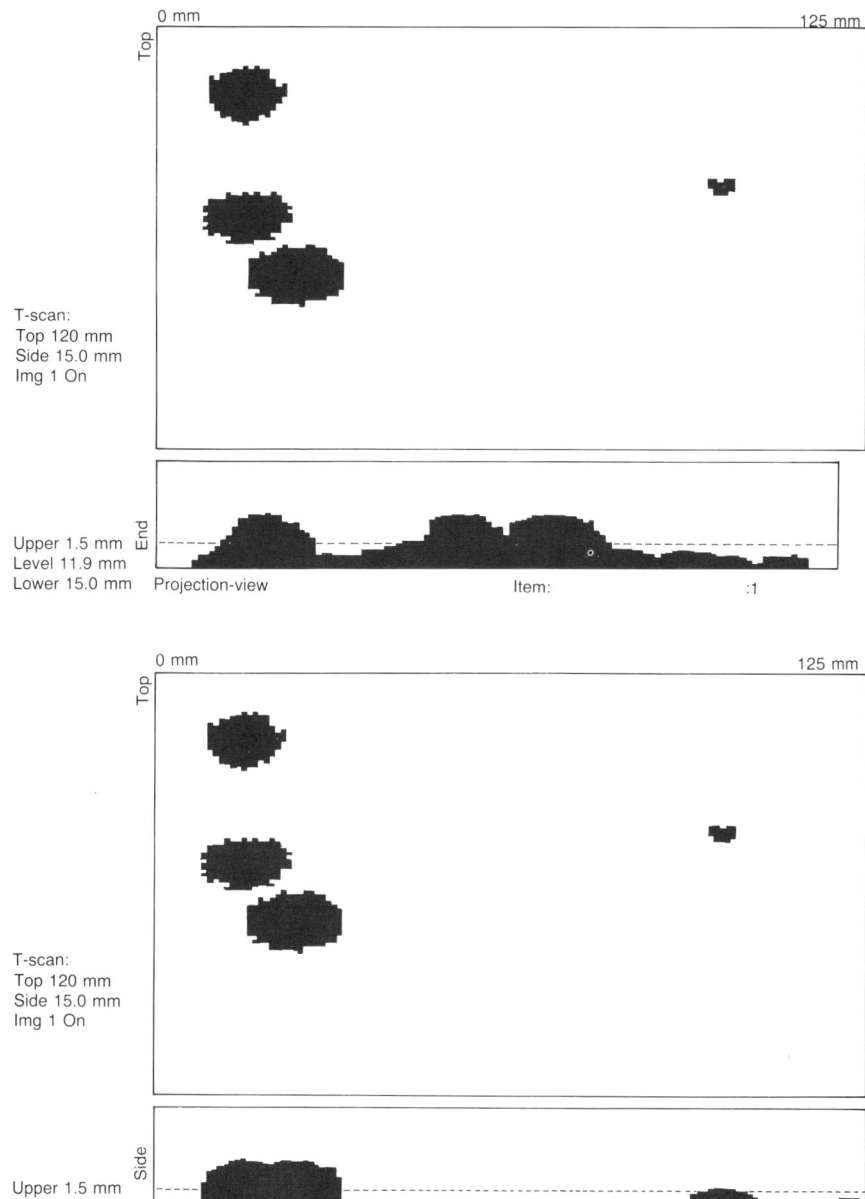

Figure 23.5a. T-scan images of the object shown in Figure 23.5.

Thickness measurements

P-scan is used to measure the wall thicknesses of piping systems, pressure vessels and storage tanks as a form of in-service inspection. The P-scan system is particularly well suited to locate and measure corrosion and erosion damage. The

P-scan

thinnest part of an object is found by scanning the entire area which is under examination and then locating the thinnest material thickness on the P-scan image.

By performing thickness measurements using conventional ultrasonic equipment or digital instruments the thickness is only determined at selected points. There is, therefore, a significant risk of not finding the minimum wall thickness, particularly in case pitting or erosion damage is present.

Practical considerations
As in the case of conventional ultrasonic examinations certain practical requirements must be satisfied by the object to be examined or measured.

The surface of the subject must be reasonably smooth and clean. Local irregularities in the form of weld spatter, holes in the surface or splotches of paint will preclude measurement of certain areas. Smooth, adhering paint is no hindrance to ultrasonic measurements. However, energy losses in the paint layer must be accounted for when measuring echo amplitudes. The paint thickness must also be taken into account when performing thickness measurements.

Deviations from the theoretical shape of the surface (plane, cylindrical or spherical) can give problems when performing mechanical scanning, for it is not always possible for a mechanical scanner to maintain good probe contact, e.g. on a wavy surface, even though it is otherwise suitably smooth and clean.

Mechanical scanning imposes even greater demands on accessibility than manual scanning. In addition to the region on the surface where the probe must move, the mechanical or automatic scanner needs room to move about. Therefore, the position of pipe nozzles, supports, scaffolding and other possible hindrances must be taken into account when planning an examination.

Thickness measurements of steel plates, e.g. in containers, can be disturbed or impeded if the plate has many areas with slag inclusions or actual lamination.

Water, continuously applied to the object, is the coupling medium between the probe and the object during mechanical scanning. After use the water will run down over the object. In such cases consideration must be given to whether or not it will be necessary to remove it.

P-scan examinations with a manual scanner can be carried out on battery operation, while mechanical scanning will require a power supply on site for operation of the scanner and the water pump.

Practical applications

Figure 23.6 illustrates the automatic scanning of a piping system in a nuclear power plant. The purpose of this examination is to map the material thickness at and around elbows in the system.

Figure 23.2 shows a P-scan examination of a piping system in the laboratory. Weld joints are examined with two angle probes mounted in an automatic scanner. Figure 23.4a shows the result of the examination of part of the weld. The P-scan image shows a root defect and two internal defects in the weld.

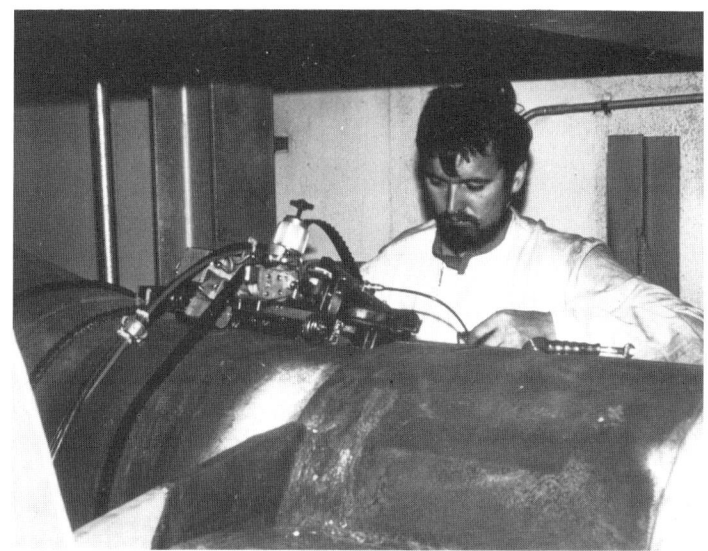

Figure 23.6. An automatic scanner in operation measuring the thickness of a piping system in a nuclear power plant.

Training required/desired

A P-scan operator must possess intimate knowledge of the use of ultrasonics as an inspection method, e.g. corresponding to Nordtest level 2 or CSWIP 3.6. A Nordtest or corresponding certificate will often be required in connection with the examination of welds. Beyond the training in ultrasonics, special training in the use of the P-scan system is required.

During automatic scanning two operators will normally be needed to operate the system. The minimum requirement on operator number two is that he or she has been trained in the use of P-scan equipment, and that he or she has adequate knowledge of ultrasonics.

Norman Thomsen

References

1) *Recent advances in the documentation of ultrasonic weld inspection.* Sven Erik Iversen, The Danish Welding Institute, 1982.

2) *P-scan, an improved technique for documentation of manual and automatic ultrasonic inspection of welds.* Svend A. Lund, Sven Erik Iversen. The Danish Welding Institute, 1985.

3) Willy D. Kristensen, et al.: *SUPERsaft ultrasonic image reconstruction,* **12th WCNDT World Conference on Non-destructive Testing,** Amsterdam, Holland, April 23-28, 1989.

4) X. Edelmann: *Ultrasonic testing of austenitic stainless steel components.* **Proceedings of the 4th European NDT Conference,** London, England, Sept. 1987.

5) K.V. Rasmussen and C. Christensen: *Automated ultrasonic mapping of corrosion and hydrogen cracking.* **12th WCNDT World Conference on Non-destructive Testing,** Amsterdam, Holland, April 23-28, 1989.

6) The Danish Welding Institute: P-scan system description and brochures.

Chapter 24
Pinhole detection

NDE principle

Pinholes (pores) in insulating coatings on electrically conductive substrates can be detected simply by connecting the substrate to one pole of an electric power source and searching the coating surface with a counter electrode connected to the opposite pole. When a pinhole is encountered, a current will pass through the detection device which can then emit an audible or visible signal indicating the presence of a defect.

A thin coating can be searched using a sponge dipped in water which has been treated to reduce surface tension using a weak soap solution as a wetting agent. Direct current sources at voltages from 9 V to 90 V are used. See Figure 24.1.

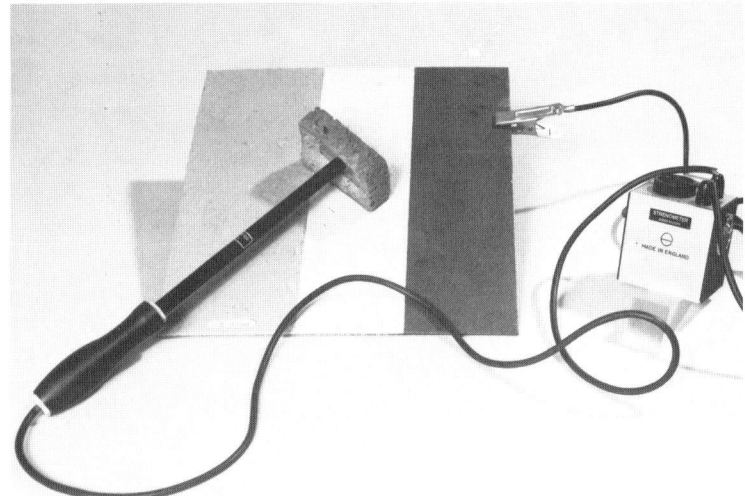

Figure 24.1. Low voltage pinhole detector.

For thicker coatings the usual procedure is to use dry conditions and voltages which are so high that they ionize the air within the pinholes so that it becomes conducting. Defect currents cause sparks which can be utilized in the dark to help localize the pores. By using even higher voltages the coating itself can be ionized so that a hole is burned in it, thus revealing weak places in an apparently homogeneous film. This latter purpose has caused such apparatus to be referred to as a holiday detector (see Figure 24.2). Test voltages of up to 30-40 kV are used.

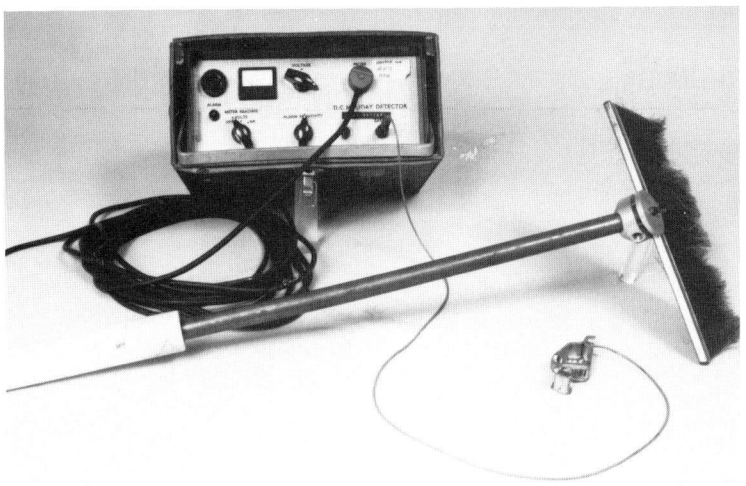

Figure 24.2. High voltage DC pinhole and holiday detector.

A common characteristic for both test methods is that the protective coating must have a certain insulation resistance in order for the method to be successful. Metal coatings, most metal pigmented coatings and black products which have a content of "carbon black" above a certain limit cannot be examined in this manner. A constant error signal will continuously be recorded in such cases so that the presence of actual defects is concealed.

The generation of sparks during high voltage testing produces a certain burning at the place of the defect.

A prerequisite for the application of both methods is that the substrate is dry and generally also clean. Dirt will often cause an interfering electrical conductivity on painted surfaces at relative humidities above c. 70% so that a constant error signal occurs.

Development of the method

Many years ago it was realized that the ability of paints and plastic films to protect against corrosion in an aggressive environment is dependent both upon their resistance to water vapour diffusion as well as to the penetration of water through pores or weak areas. Direct water penetration was of course particularly damaging because it initiated a deterioration process (Figure 24.3). But experience from high voltage technology suggested that one could regard the protective coatings as insulators and test their integrity by exposing them to a high voltage difference.

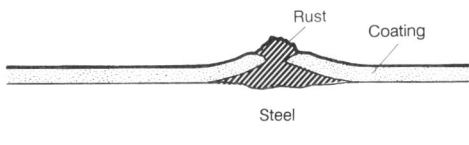

Figure 24.3. Underrusting of a pinhole penetrating the coating.

The first transportable equipment for pinhole detection was developed with induction apparatus as a model. Later automotive ignition systems were emulated with a small electric motor activating contact points in the low-voltage primary circuit of a transformer causing high voltage pulses in the secondary circuit.

Gradually pure high voltage AC equipment was developed both as battery-powered and AC-powered versions, but all of these devices had the common property that the test voltage was not well-defined. This was due, among other things, to the electrical capacitance in the measuring circuit. The desired degree of reproducibility could only be achieved using DC equipment. These have now been developed to the point where a low voltage battery supply is converted to an alternating current by a solid-state generator, and the alternating current is transformed to a high voltage, rectified and filtered.

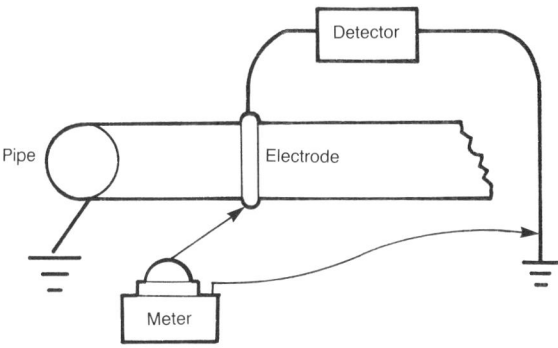

Figure 24.4. Pinhole detection using pulses. The measurement of the effective test voltage is performed using a special high voltage voltmeter.

Direct current devices can – in contrast to pulse and AC equipment – be adjusted continuously, and the output voltage can be measured directly using an ordinary voltmeter. The measurement of the effective test voltage, particularly when using pulsed equipment, is rather complicated (see Figure 24.4), and voltage adjustment must usually be made in steps.

Pinhole detection

Areas of application

In many cases where a coating is used to protect against corrosion it is important that the coating be free of pores or that it only has a limited number of penetrating pores. In particularly demanding situations weak (thin) areas must also be avoided.

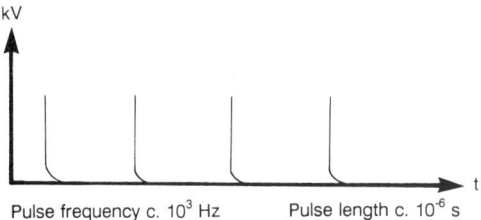

Figure 24.5. Pinhole detection with pulses. Voltage vs. time.

Checking for pores can be carried out non-destructively using low-voltage direct current via a liquid film or by using high-voltage direct current, pulsed (see Figure 24.5) or alternating current.

Low voltage testing
The method is used mostly to demonstrate the presence of a certain level of quality of painting systems used in aggressive to very aggressive atmospheres.

The method is suited for use with insulating coatings with thicknesses up to about 300 μm using a 9 V voltage source and up to about 500 μm with a 90 V source.

As mentioned, only penetrating pores are registered, and the selectivity is not particularly good. The sponge must be lifted and moved after each pinhole is located, because an error signal will occur over a considerable distance in a continuous moisture film. Lifting the sponge entails the risk that pores will not be detected.

The sensitivity of the method is not particularly well-defined, but pinholes which are invisible to the naked eye can be detected.

When using this method it is common practice to tolerate a certain number of pores on a painted surface, e.g. 3-10 pores per metre on plate and profile edges and 3-10 pores per square metre surface area.

High voltage testing
The method is used when the coating thickness of an insulating layer exceeds 500 μm and when particularly high demands are made on the integrity and the homogeneity of the surface. The latter application is the most important, and a completely defect free surface at a given test voltage is (usually) required.

Coatings which are tested are typically internal corrosion protection layers in pipes, valves and tanks against media, water included, which are aggressive towards steel (see Figure 24.6). They may be manufactured from unsaturated polyester, two component epoxy, hard rubber, natural rubber, etc.

Figure 24.6. Searching for pinholes in the bottom coating of an oil tank.

In the piping industry it is standard practice that all steel pipes which are to be buried in the ground or placed under water are tested in connection with the surface treatment which they receive in production facilities and immediately before being installed as well. The latter check is particularly important in connection with protective coatings applied on-site to welded joints.

Test voltage
The test voltage which is used may be given in a standard, e.g. 25 kV for PE coatings of steel pipes intended for use underground and manufactured according to the DIN 30670 standard, but it is often specified as a certain percentage the breakdown voltage for a perfectly prepared coating, e.g. 30 or 50%. For thick coatings other DIN standards refer to a rule of thumb which specifies the test voltage as 5 kV + 5 kV/mm coating thickness.

Voltage type
Ideally, direct current is used, but the method is often undependable when working outdoors in inclement weather because the creep currents on dirty surfaces cause a constant alarm signal to occur.

It has turned out that very short voltage pulses succumb to this source of error at much higher relative humidities, so that equipment using voltage pulses is in fact predominant in the piping industry in spite of the difficulties with calibration.

Probe types
The ideal probe is spherical, but due to the need to increase productivity it is not used. The next best solution is the blade probe consisting of conductive rubber with rounded edges, but this type can be difficult to handle because it does not effectively follow discontinuities in the coated surface. Therefore, brush probes are quite widely used. They are composed of thin, elastic metal threads, in spite of the fact that it is difficult to define the test conditions. Corona discharge phenomena at the tips of the wire threads can lead to untimely ionization of the air and the coating, so the action of the probe corresponds to that of a higher and unknown test voltage.

Rolling spiral springs are used as probes for long pipes (see Figure 24.7).

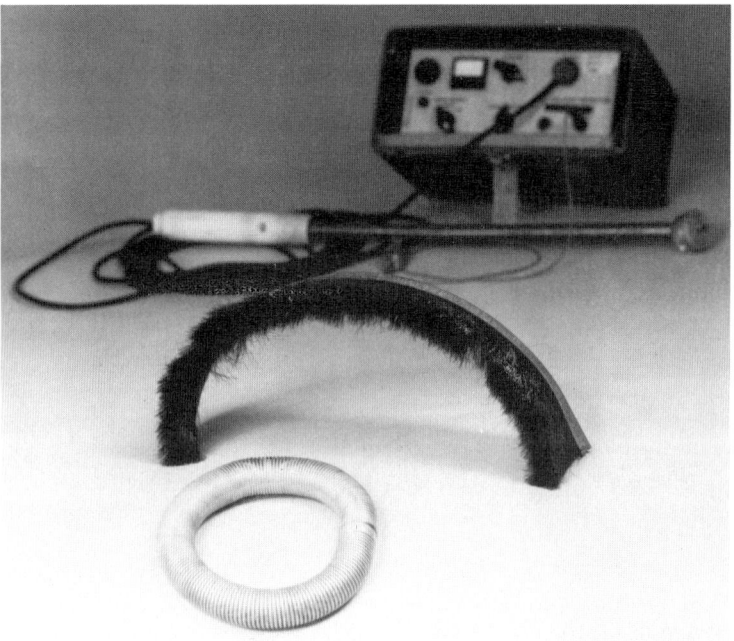

Figure 24.7. Spiral spring probe and brush probe for pinhole testing on pipes.

Deterioration of material
The mildly destructive character of the test – it makes existing defects worse due to burn damage – makes it essential to repair all defects which are detected.

All coating types which contain chlorine and fluorine are thermally decomposed when sparks are formed, producing sulphuric and hydrochloric acid. They can initiate adverse corrosion processes down on the metal surface. It is, therefore, not advisable to undertake high voltage tests on coatings which contain chlorine and fluorine compounds. Examples of this type of compound include: PVC, Chloroprene (Neoprene) and Teflon coatings.

Practical applications

In addition to the applications already cited, checking bitumen insulating layers on bridges with steel deck plates can be mentioned. On a large bridge in Denmark with more than 30,000 square metres of driving surface, a 3 mm insulating layer was initially checked visually for pores. It was very time-consuming and quite ineffective. Testing with a high voltage pulsed apparatus dramatically improved both capacity and quality.

Training required/desired

No particular requirements for training or certification exist for this area.

It is, however, recommended that the operator develop and maintain his or her practical experience. Furthermore, an understanding of the characteristics of various surface layer treatments is important in order to achieve a dependable interpretation of the results of an examination.

Jørgen Møller

References

1) *Recommendations for the corrosion protection of steel structures*, 1st edition, February 1982, DS-Recommendation DS/R 454, Translation Edition , June 1984.

Chapter 25

Pressure testing

NDE principle

Pressure testing is carried out by applying a usually predetermined pressure which must often be held for a predetermined period until the pressure is again reduced.

For safety reasons liquids (e.g. water or oil) are preferred as pressure media during testing. Because liquids are almost incompressible, damage to the surroundings will generally be significantly less in case of rupture than the damage which can be caused by the explosion of the test object using compressed gas.

In connection with liquid pressure testing procedures it is thus very important that air is bled from the system while liquid is forced into the system to be tested.

The principles used are generally quite simple and independent of specialized equipment. The main principles involved in liquid pressure testing are shown in Figure 25.1.

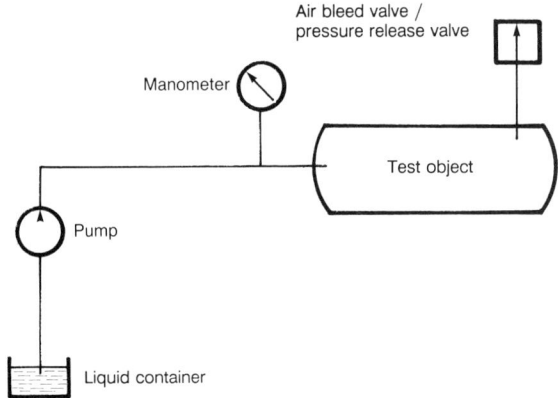

Figure 25.1. Test setup for the performance of liquid pressure testing.

On the basis of a pressure test it is possible to determine whether or not a test component can withstand the desired pressure, which as a safety precaution normally is set somewhat higher than the maximum working pressure for the component.

In connection with pressure testing, leakage checks and, where possible, visual inspection of the test object are usually carried out.

Development of the method

Motivated by the desire to combine liquid pressure testing with a simple but effective form of leak detection, the setup shown in Figure 25.1 can be extended by placing a valve as indicated in Figure 25.2. If the valve is closed when the desired pressure has been applied to the component under test, then any pressure drop due to leakage or volume changes of the now closed system (e.g. plastic deformation) can be detected in the form of a pressure change.

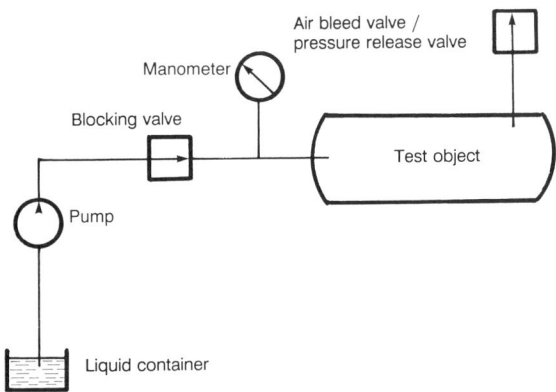

Figure 25.2. Extended test setup.

The leak testing procedure can of course be carried out in other ways, and some of these methods are discussed in Chapter 14, **Leak testing**. During liquid pressure testing a method such as coating the component with chalk can be effective, because the applied layer of chalk changes colour when wet.

If the pressure test is undertaken with a gas as the pressure medium, then leak testing can be carried out by lowering the component into a liquid during testing. The component can also be wetted with a soap solution or similar mixture which forms bubbles due to escaping air (leak detection fluid is widely available commercially), or leaks can be detected using a probe which is sensitive to a particular gas.

The volume expansion which occurs in connection with a pressure test can be determined by checking the dimensions of a component before pressure has been applied and while under pressure, but even more precise data can be obtained by means of the test setup shown in Figure 25.3 (water jacket volumetric expansion test).

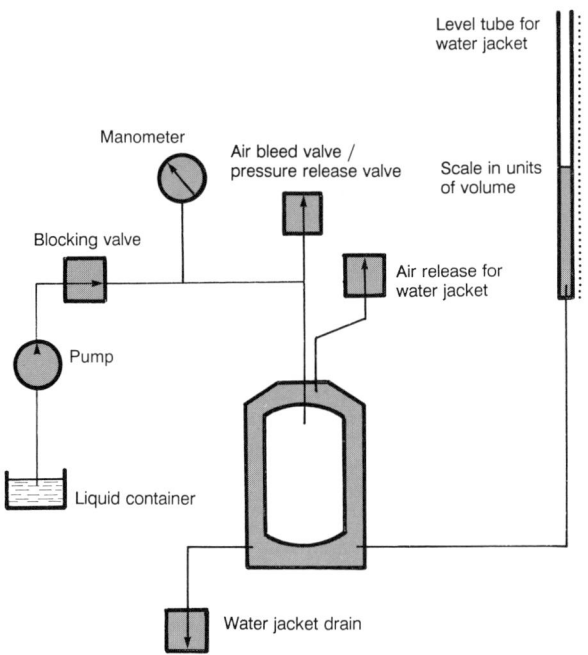

Figure 25.3. Water jacket volumetric expansion test.

Here a test subject has been lowered into a liquid-filled container ventilated by means of a valve located at the top of the container. The container is connected with an open expansion tube with a scale marked in units of volume. By reading the liquid level in the expansion tube before, during and after the test, the elastic expansion of the component and any residual expansion can be calculated.

A special form of pressure testing is *vacuum testing* or testing with external pressure, and can be relevant in the examination of hoses and pipes. This sort of testing is designed to check the resistance of the component to collapse due to pressure.

In connection with many pressure tests it is common to perform a visual inspection, where the extent of any corrosion, occurrence of cracks or other defects can be checked and the prospects for continued use of the component can be evaluated.

Such evaluations can lead to the use of other non-destructive examination techniques before making any decision regarding repair or replacement. For example material thickness measurement using ultrasonics can provide useful information in such cases, and this data can be compared with the required minimum thickness.

National authorities usually specify requirements regarding pressure testing of pressurized systems both before they are put into service and at given intervals thereafter or after repairs are performed on an operational system, and the test pressures are usually set equal to the most extreme operating pressure anticipated multiplied by a factor which can vary depending upon which safety risk might be associated with a breakdown in the system.

Areas of application

Pressure testing is a useful and handy NDE method for the examination of most pressure bearing components.

Individual components such as oil level glasses, temperature sensor fittings, etc. can be pressure tested independently of the systems in which they will be a part, but if special requirements are in force regarding the integrity of complete systems, then such auxiliary components should be mounted in place during the examination of the entire system.

As mentioned earlier the requirements concerning pressurized containers and piping systems are usually set by the authorities, and the requirements which they set should be regarded as minimum requirements.

Due to considerations of economic operation and safety, it can be advisable to carry out pressure testing and inspections (both internal and external) at shorter intervals than required by the authorities, and in particular it should be noted that careful inspection carried out by a knowledgeable person can make a significant contribution to maintaining a high level of safety and to the judicious planning of repairs or replacement of pressurized components.

Practical applications

Common practice is as mentioned often determined by the authorities, and the regulations are often classified according to the construction or application involved, such as stationary containers, transportable containers and piping installations, etc.

For reasons of economy transportable containers, of which there are many in circulation, are pressure tested in fixed setups with up to 12 containers at a time.

Training required/desired

With respect to carrying out the pressure test itself no special requirements are incumbent on the operator, but in order to attain a responsible and satisfactory evaluation of the results of the examination and to make decisions concerning the possible need for supplementary tests, considerable experience in this area is needed. It is also commonly required that a competent authority or particularly qualified individuals appointed by the authorities are authorized to perform these tasks.

Kjeld Grønfeldt

References

1) *Technische Regeln Drukgase* (Technical regulations for pressurized gases) (TRG).

2) British Standard BS 5045: *Specification for transportable gas containers.*

3) British Standard BS 5430: *Specification for periodic inspection, testing and maintenance of transportable gas containers (excluding dissolved acetylene containers).*

Chapter 26

Radiography

NDE principle

X-rays or gamma radiation is used to perform a *radiographic examination.* X-rays and gamma radiation have the property that they can penetrate solid materials. Both are forms of electromagnetic radiation, like visible light.

X-radiation

X-radiation is produced by accelerating electrons up to very high energies (50-400 keV) and then causing them to decelerate. When they collide with a target material the electrons are slowed down, and their kinetic energy is liberated, among other things, as X-radiation ("Bremsstralung" = radiation due to braking). Some of the electrons interact with electrons in the target and knock electrons out of their "orbits" around atomic nuclei, i.e. from a particular energy level. When an electron is missing, electrons at higher energy levels will perform quantum jumps down to the energy level of the missing electron. Due to such quantum jumps X-radiation (characteristic radiation) is emitted.

Gamma radiation

Some of the atomic nuclei in a radioactive material are unstable and tend to decay to a more stable condition. During this radioactive decay process the atomic nuclei emit energy in the form of radiation and/or elementary particles. The electromagnetic radiation which is emitted is called *gamma radiation.*

In order to examine an object it is irradiated with X-rays or gamma radiation. The radiation will be absorbed in the object to varying degrees depending upon the thickness of the object, the composition of the material and the wavelength of the radiation.

That portion of the radiation which penetrates the object can be registered by recording it on a film. Just as visible light, X-rays and gamma radiation can expose film. The more radiation penetrating the object and striking the film, the darker the film appears when developed.

By examining differences in exposure of the film, it is thus possible to determine differences in the thickness and composition of the object. This means that one can see whether or not corrosion is present in pipes, because varying material thicknesses will attenuate the radiation differently. When the exposed object is concrete it is possible to observe the steel reinforcing rods, because various materials attenuate the radiation differently and thus expose the film accordingly.

Development of the method

In 1895 Wilhelm Conrad Röntgen discovered that the electrical discharge in a cathode ray tube could cause a fluorescent screen to glow. Experiments showed that a previously unknown form of radiation was generated, which Röntgen called *X-rays*.

An electrical discharge occurs in the gas-filled cathode ray tube. During discharge positive ions are created which strike the cathode. The collision causes the liberation of electrons which generate X-radiation when they are stopped by the glass wall of the tube.

Due to the size of the electron beam striking the glass wall of the tube, the cathode ray tube produces an image which is not perfectly sharp. The cathode ray tube was rapidly improved. A metal anode (tungsten) was introduced. The metal anode can withstand greater stresses and is also more efficient at producing X-rays. The cathode was constructed as a concave lens so that the electrons could be focused toward the anode. In this manner better resolution was achieved.

In 1913 the cathode was replaced with a filament as the electron source, and air was removed from the tube to eliminate discharges. This principle is still in use today.

In an X-ray tube energies of up to 50-400 keV can be achieved, corresponding to X-rays which can penetrate up to 50 mm thicknesses of steel. X-radiation is also created in accelerators. In a linear accelerator (up to 20 MeV) electrons are accelerated by a running electromagnetic wave. In a betatron (up to 30 MeV) electrons are injected into a circular path where they are accelerated by a magnetic field which increases in strength.

The future will surely yield substantial improvements in image quality. The development of advanced computer technology will make it possible to record three-dimensional images, cross sections and to perform many operations on images.

Areas of application

The future will provide us with increased usage of composite materials, plastics, light metals and new means of joining materials (adhesive bonding, electron beam welding, laser welding, plasma coating).

Microfocus

Within the past few years X-ray tubes have been developed with a resolution which is about 100 times better than ordinary X-ray tubes. This is achieved by focusing the electrons from the filament on a very small area (the focus). The electrons are focused by means of a magnetic field. The system is called *microfocus* (see Figure 26.9).

Radiography with microfocus opens new possibilities for locating defects in very small details: down to 10 μm, e.g. structures in metals and fibres in composite materials.

Microfocus is also well suited in situations where the need for real time imaging exists. One can obtain real time imaging of objects at rest as well as objects in motion. They must not, however, move faster than 50 cm/minute.

Microfocus combined with real time imaging is an extension of the systems used in airports to X-ray baggage.

Radiography can be used with all materials and is independent of the magnetic and electrical properties of the material. There is, therefore, no doubt that radiography will also be a widely used method of condition monitoring in the future.

Radiography can be used in any situation when one wishes to view the interior of a subject. It can be used to check an object for internal faults and construction defects (e.g. faulty welding, soldering, riveting or adhesive bonding). It can also be used to examine what is inside an object, if something has become stuck inside a hollow object or is not in its proper position. One can also perform measurements of size, e.g. measurements of the material thickness of pipes and cylinders in hydraulic systems. Using special stroboscopic X-ray tubes it is also possible to examine objects which are moving rapidly, e.g. motors.

The limitations imposed upon the use of radiography are set in part by the type of material through which the radiation must penetrate and in part by the material thickness. The thicker the material the radiation must pass through, the more scattering (i.e. the more noise) is generated in the object. At some point with a sufficiently thick object, depending on the material, the signal-to-noise ratio becomes too small for a usable image to be obtained.

In the case of steel, defects corresponding to about 2% of the penetration thickness can be discerned.

In practice there is yet another limitation. One must have access to both sides of an object, for the film must be placed on one side and radiation must be transmitted from the other.

This handbook is only concerned with condition monitoring. Therefore, two methods will be discussed, where radiography can be used in condition monitoring, namely *radiography on-stream inspection* and *stereo radiography*.

Guidance concerning the use of radiography in the production monitoring of welded joints is given in ISO 1106, for example.

Radiographic on-stream inspection (ROSI)

The use of radiography, ultrasonics, etc. while a processing system is in operation is classified as *on-stream inspection*. The classification is also used even though the system is shut down, e.g. for the examination of pipes in a power plant boiler.

Piping systems are an integral part of all processing plants, oil refineries, petrochemical plants, power plants, etc. Pipes are subject to corrosion, wear, sedimentation and clogging with solid particles. Pipes can also be clogged due to icing and breaks in internal components.

Part of ROSI examinations is the measurement of the material thickness of piping. This examination is carried out by performing *tangential imaging*.

Tangential imaging

Figure 26.1 shows the setup. The film is placed inside a light-tight cassette mounted on the pipe. In order to achieve a uniform exposure, the film should be placed at right angles to the central ray (the film must not be bent around the pipe). The source is placed on the other side at an appropriate distance.

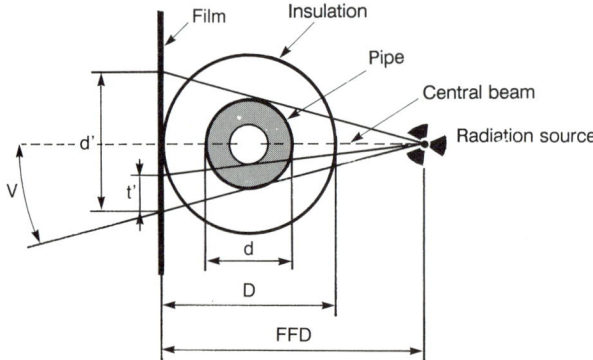

Figure 26.1. The setup for tangential imaging.

The apparent wall thickness t' is measured on the developed film. Since the film is placed some distance from the tangential plane of the pipe, the wall thickness is enlarged on the film. The measured thickness t' must, therefore, be reduced by a factor R. The real thickness t is obtained from the following expression:

$$t = R t'$$

The factor R can be obtained from the following expression:

$$R = \sqrt{1 + \left(\frac{D-d}{2 \cdot FFD}\right)^2 - \frac{D}{FFD}} \quad \text{(with insulation)}$$

where *FFD* = Film Focus Distance.

If the pipe has no insulation and the film is placed right next to the pipe, that is

$D = d$, then R can be computed from the following expression:

$$R = \sqrt{1 - \frac{d}{FFD}} \quad \text{(without insulation)}$$

Normally the apparent pipe diameter d' can be measured directly on the film, and R can be replaced with the simple expression

$$R = (d/d') \cos v$$

For FFD values greater than three times the actual pipe diameter d, the value of $\cos v$ is very nearly equal to unity, and one obtains:

$$R = d/d'$$

If the source is placed too close to the pipe, the image on the film will not be sharp. Depending upon the physical size of the source, the source should be three to six times the pipe diameter from the film.

Measurement accuracy

For a normal exposure with reasonable sharpness one can figure on a measurement error which is less than ±0.3 mm. The measurement accuracy depends, however, on the material thickness penetrated by the radiation and whether or not the pipes are filled. In addition coatings can "reduce" possible corrosion.

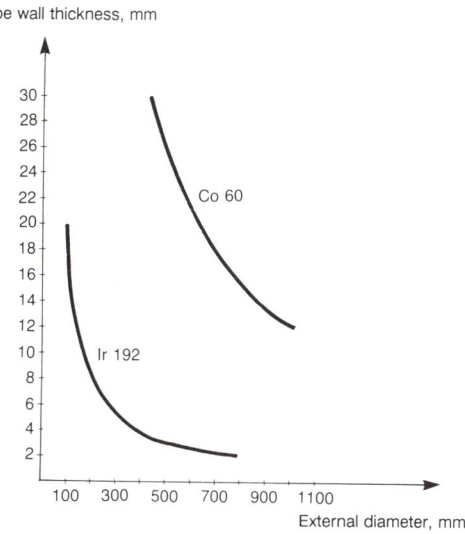

Figure 26.2. Limitations on material thickness vs. exterior pipe diameter for two radiation sources.

Radiography

Limitations

The isotopes Ir-192 and Co-60 emit radiation which can penetrate up to 80 mm and 150 mm of steel, respectively. The diameter of the pipe also places limits on the material thicknesses which can be penetrated to achieve tangential imaging. See Figure 26.2.

Sources of error

It is not possible for the eye to distinguish clearly between both light and dark areas of the film at once. High contrast film or film which has been overexposed will contain light and dark sections. It can be difficult to observe the material thickness on such films. X-ray film usually has emulsion on both sides, and one can, therefore, sometimes improve the readability of the film by scraping off the emulsion on one side.

Stereo radiography

When using radiography a two-dimensional image (length and breadth) is usually obtained. By means of stereo-radiographic exposure a three-dimensional image can be achieved. That is one can perform depth measurements, e.g. to determine the placement of steel reinforcement within concrete.

The technique involves making two exposures on the very same film. The source must be displaced a distance a between the two exposures, but the film must not be moved.

A lead marker is placed on the irradiated side of the object. By measuring the displacement c of this mark on the film, it is possible to avoid measuring the material thickness of the object. This can often be difficult or impossible, e.g. because of floors or walls.

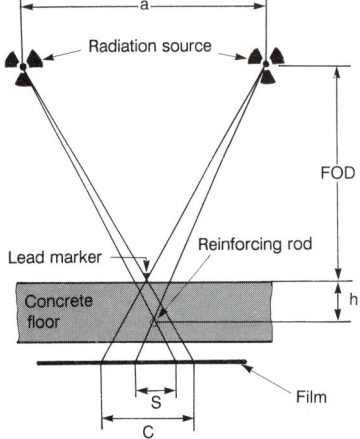

Figure 26.3. Geometry of the setup for stereo radiography.

The displacement *a* of the source and the distance *FOD* (Focus Object Distance) from the source and perpendicularly down to the object is measured on the spot; *s* and *c* are measured on the film, where *s* is the displacement of the steel reinforcement for which the depth *h* is to be determined. The quantity *h* is determined by means of the following expression:

$$h = FOD\,(c - s)/(a + s)$$

When measuring *c* and *s* one must be sure that *a*, *c* and *s* all lie in the same cross sectional plane, i.e. *s* and *c* must be measured in the same direction as one has displaced the radiation source (see Figure 26.4). The quantities *s* and *c* need only be measured perpendicularly to the image of the steel reinforcing rods in the case where the radiation source is displaced perpendicularly to the reinforcing rods but still in the same plane.

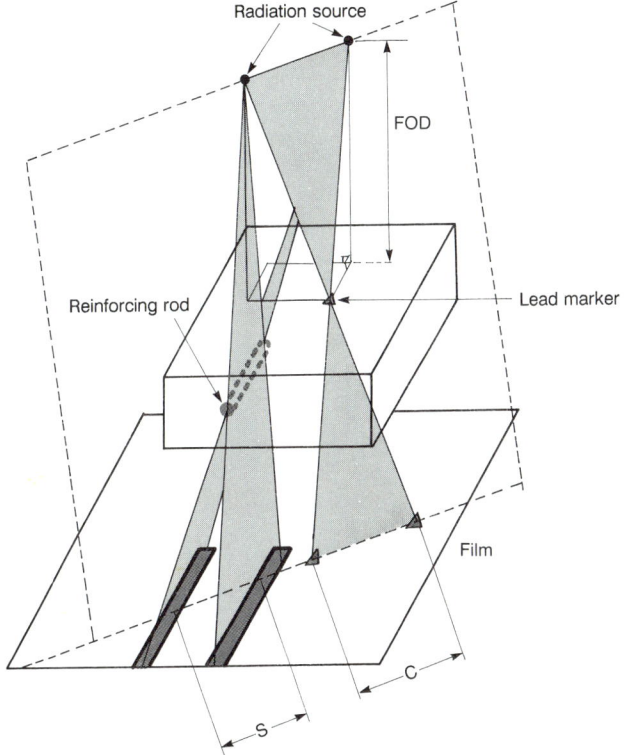

Figure 26.4. Three-dimensional perspective view of the measurement geometry.

In certain situations it can be advantageous to expose two films, one for each position of the source. In this case two lead markers must be placed on the film side of the object. When two films are used, the films must be in close contact with the object so that the lead markers are not imaged in displaced positions on the films. The lead markers serve as reference points.

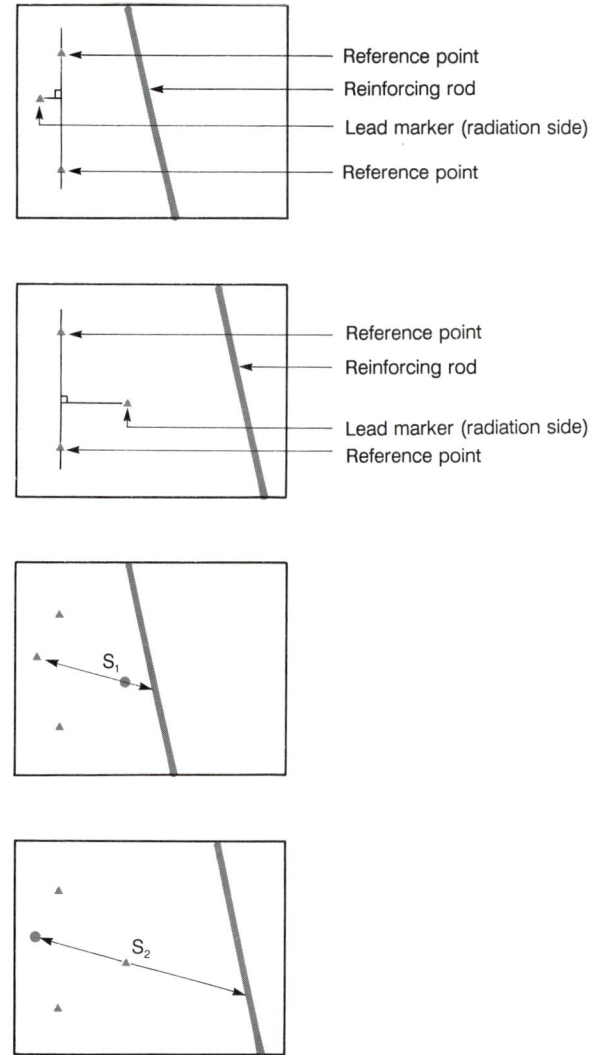

Figure 26.5. Determining the location of reinforcing rods in concrete using two X-ray films exposed with the radiation source in two different positions.

The placement of the images of the lead marker which is mapped from the radiation side of the object in relation to the fixed points is measured on both films, see Figure 26.5. The placement is transferred from one film to the other; c can now be measured directly on both films; s is obtained from the expression

$$s = s_2 - s_1$$

where s_1 and s_2 can be measured either on film 1 or film 2; now h can be computed as if one had used one film.

Measurement accuracy

The measurement accuracy of the result is quite dependent upon how careful one is about the geometry of the setup and the measurements. For example, it is a prerequisite that the source is displaced parallel to the plane of the film. The source must be displaced sufficiently so that s and c are large enough to make the effects of measurement inaccuracy small. The quantities s and c should be more than 50 mm. In practice the measurement accuracy is about ±5%.

In order to achieve the correct setup and thus the best result, the personnel performing the condition monitoring should be clearly informed concerning what is to be examined and why, and which factors can cause errors.

Practical applications

Stereo radiography is used in particular to locate reinforcing rods in concrete constructions (concrete floors, beams, balconies, consoles, etc.). The method is used both with new as well as older constructions where there is a need to evaluate condition and safety. The method is reasonably fast and inexpensive. It is sufficiently accurate to make the measurements can be usable as a basis for computing the strengths of constructions. An important advantage with a stereographic exposure is that it is unnecessary to break up concrete to determine how the reinforcing rods are positioned. Previously, physically breaking up the object was the only method available. This has led to problems, because it can be difficult to perform a good, durable repair.

The radiation from Iridium-192 can penetrate up to 300 mm of concrete. If it is necessary to penetrate greater thicknesses, e.g. concrete bridges, one can use Cobalt-60, which can penetrate up to about 1 metre of concrete.

Sources of error

In constructions where reinforcing rods are close together or the rods are present in several layers, it is possible to misinterpret the film. It is, therefore, important that the individual who interprets the film has access to construction diagrams. The two film method can be used to help avoid misinterpretations.

Radiation safety

Both ROSI and stereographic exposures are normally performed using radioactive isotopes. These sources emit very penetrating radiation. X-radiation and gamma radiation are health hazards. It is, therefore, necessary to close off the area when making exposures to avoid exposing the public to radiation. The distance from the barrier to the exposure site will typically be about 20 metres for Iridium-192 and 50 metres for Cobalt-60.

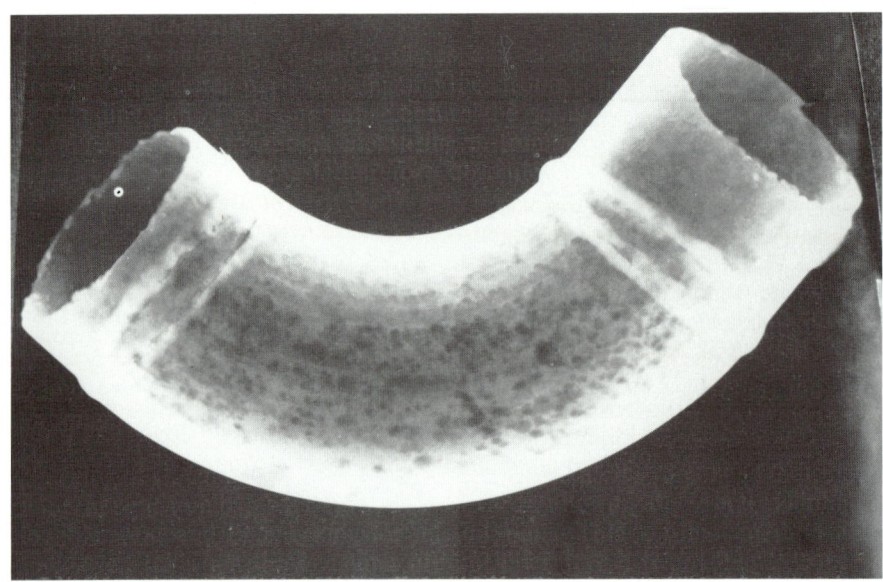

Figure 26.6. Example of pitting in a 6" pipe bend.

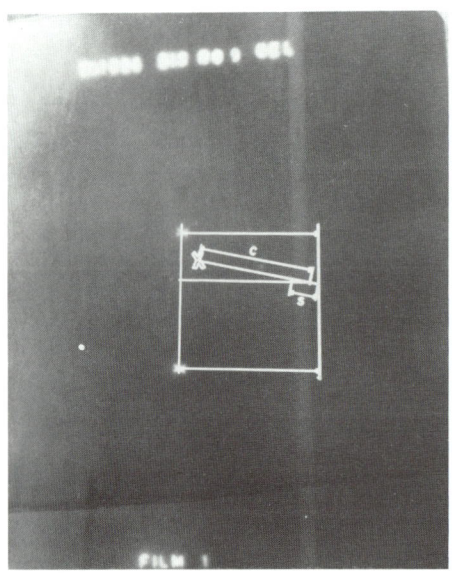

Figure 26.7a. Stereo exposure with two films on 20 cm concrete. Film 2.

Companies which perform radiography must have their equipment approved by an appropriate government agency. Similarly, personnel who operate X-ray or gamma radiation equipment must carry exposure films and must be qualified to work with radioactive sources. Government agencies often offer courses of instruction in radiological hygiene.

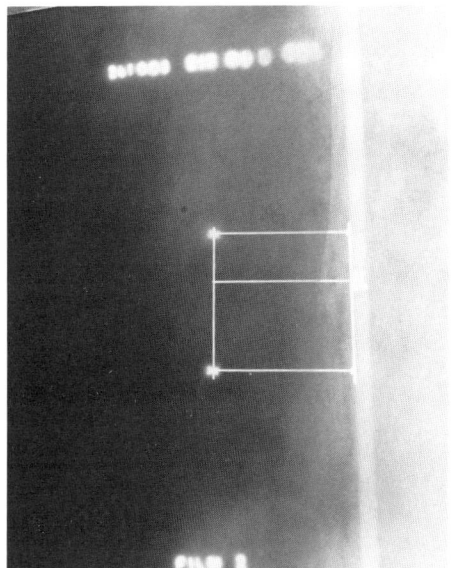

Figure 26.7b. Stereo exposure of 20 cm concrete. Film 1.

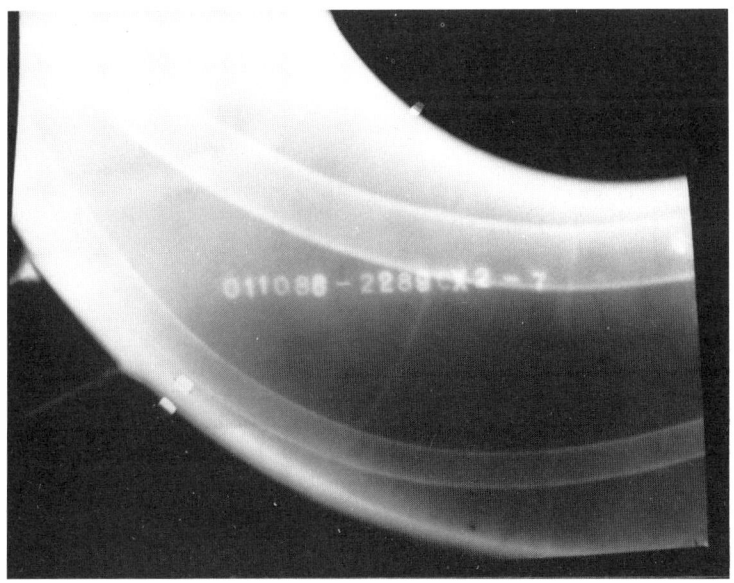

Figure 26.8. Example of the use of radiography to measure the material thickness of an insulated 150 mm (6") production refinery pipe used for steam transport.

Radiography 273

Training required/desired

Radiography is used in the field of production monitoring. In such situations it is normally required that operators must be qualified according to the rules laid down by NORDTEST (valid throughout the Scandinavian countries). This means that the operator must submit to a test to document that he or she has adequate theoretical and practical knowledge and is competent to evaluate a film for weld defects. Furthermore, he or she must have worked regularly with radiography for at least one year. In order to take this test the operator must have attended a variety of courses for a total of about four weeks.

It is advisable that individuals who are to perform condition monitoring tasks undergo a similar training programme.

Gert Lassen

Figure 26.9. Advanced X-ray examination using micro-focus. The first industrial equipment in Scandinavia is in operation at The Danish Welding Institute. It opens completely new possibilities, e.g. for locating microscopic defects.

References

1) *Industrial radiography*. Agfa-Gevaert Corporation.

2) *Radiography in modern industry*. Kodak Corporation.

3) *Grobstrukturprüfung mittels Röntgen - und Gammastrahlung.* Egon Becker, Deutsche Verlag für Grundstoffindustrie - Leipzig, 1983.

4) *Industrial radiology, theory and practice*, Applied Science Publishers, London and New Jersey, 1982.

5) **Metals Handbook**, volume 11, *Non-destructive inspection and quality control*, American Society for Metals, Metals Park, Ohio 44073, USA, 1976.

Chapter 27

Replica technique

NDE principle

Non-destructive material examination by means of the *replica technique* is a *metallographic* method which is based upon the examination of the microstructure of the surface of the metal. Transportable equipment is used for the examination.

The metallographic equipment which is used for prior preparation of the surface consists of portable grinding and polishing equipment.

Before commencing with the grinding process, proper care should be exercised in the case of critical material thicknesses to reduce damage to the component under examination. The removal of material by grinding during a non-destructive examination should be on the order of 0.1 to 0.4 mm depending upon the "past history" of the component under test, as one must be careful to remove any oxide scale or decarbonized surface layer.

During the subsequent fine grinding process, grinding paper no. 50, 120, 220 and no. 400 can be used in that order. After grinding, the region to be examined is carefully cleaned with cotton and alcohol then dried completely using a hot air dryer. During the grinding processes it is important to avoid unnecessary heating or cold deformation of the surface. Thus only light pressure should be used on the tools during the processes. After grinding the surface with the various grades of grinding paper, grinding grain remnants should be removed from the surface before each new, finer-grade grinding paper is used.

After the final grinding operation has been carried out, the surface must be polished. This can be done in the following ways:

- Electrolytic polishing can be performed using an electrolyte and transportable electropolishing equipment.

- Mechanical polishing can be performed using polishing discs and diamond paste or spray (particle diameters from 7 μm down to 1 μm). When doing the polishing manually the same equipment is used as in the prior fine grinding process.

After polishing, the surface is cleaned again carefully with cotton and alcohol and dried thoroughly with a hot air dryer. In particular after electropolishing it is important to perform the cleaning process quickly and carefully to avoid permitting the normally aggressive electrolyte to cause corrosive attacks on the newly polished metal surface. The choice of electrolyte depends upon which type of material one is working with. Ready-mixed electrolytes for various types of materials can normally be purchased from companies which deal with metallographic equipment.

The subsequent etching process should be performed using the etching reagent specified for the material under examination. The etching reagent, etching time and etching temperature depend on which material one is working with and which information one is seeking concerning the material. After etching, the surface should be cleaned carefully with cotton and alcohol and carefully dried using the hot air dryer so that the surface is now free of dust, moisture, grease and is undamaged.

After this, an impression of the metallographically produced material structure is made. The *replica technique* is used for this purpose.

A replica consists in principle of a plastic foil which must have a minimum thickness of 0.06 mm. An appropriate solvent is applied to the foil, then the softened foil is pressed against the location to be examined. When it has hardened, the foil is removed (during this operation one must be careful that the foil is not damaged by bending, fingerprints, etc.). See Figure 27.1.

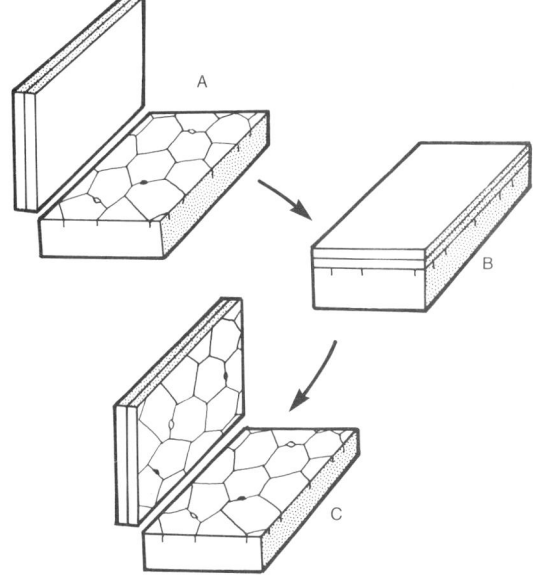

Figure 27.1. "Transfer" of surface structure by means of a replica.
A. *The prepared surface, ground, polished and etched.*
B. *The softened replica is pressed firmly onto the surface.*
C. *After hardening the replica can be removed, and the "mirror image" of the microstructure can be examined under a microscope.*

The structure impression on the replica can then be observed in a microscope either by transmitted light or by reflected light, for the foil manufacturer may have provided a reflective coating on the reverse side. Examination and the evaluation of the structure are conducted here in the same manner as with ordinary metallographic examinations.

If it is desired that the evaluation be undertaken on the spot, portable metal microscopes can be used.

If larger magnifications of the structure are desired with a clear differentiation of the various structural components (e.g. carbides and cavities), then replicas can be studied in a scanning electron microscope (SEM) by first evaporating a reflective, conductive material such as aluminium, gold, silver or carbon onto the surface of the replica.

The evaporation of the reflecting and conductive materials onto the surface can also provide significant improvements in the structure image to be examined when examinations are performed with a light optical microscope.

Development of the method

The metallographic preparation of the sample which is undertaken prior to the production of a replica has undergone a relatively slow evolution since the late 1940s from being a purely manual process to a more mechanized and automated process. In Denmark this development has been initiated primarily by Dr. E. Knuth-Winterfeldt who developed an electrolytic polishing apparatus suitable for short-term polishing of metallographic specimen subjects. Further development of this equipment led to the development in 1955 of transportable equipment for performing non-destructive electrolytic polishing. The equipment has been commercialized and marketed by the company Struers K/S. A parallel evolution took place of equipment for wet grinding based upon water-tolerant silicon carbide grinding paper both for use in the laboratory and as transportable equipment for performing metallographic examinations.

A replica was developed by Struers K/S in the early 1960s to supplement the transportable metallographic equipment developed by this company. This replica consisted of an aluminium foil to which a special type of acetate foil was affixed.

The use of the replica technique has been relatively limited in scope until the late 1970s to conventional on-site metallographic examinations.

In 1976 the technique began to be employed both in Denmark and Germany in connection with the monitoring of components which are stressed by high temperatures.

Areas of application

The replica method can in principle be used for the examination of all metallic materials, but the technique is used most frequently for the examination of steel, and here in particular for checking the microstructure of materials and for crack type determination and crack detection.

But since the replica method in principle is simply the transfer of the topography of the surface of a metal to a plastic foil, the method can also be used for more untraditional examinations. These will not be discussed in detail here but only mentioned for completeness.

- *Hardness measurement.* To document hardness measurements carried out in the field, the impressions can be transferred to a replica.

- *Wear examinations.* The evaluation of wear scratches and marks can be carried out using a replica taken at the actual locations. The fineness of a grinding process can be checked in a similar way.

- *Profile measurements.* Large topographic level differences can be measured and evaluated after casting using special two-component plastics which provide an exact impression with much detail.

Reproducibility when using the replica method is on the order of 0.1 μm.

Microstructure examinations

The fact that the replica method can be used to examine microstructures has to do with the circumstance that the replica is evaluated with a light optical microscope or a scanning electron microscope in which the structure of the material is apparent.

It is often appropriate to use the method to check whether a particular microstructure has been "damaged" during the operation of a component or whether the given mode of operation is inappropriate to the microstructure.

Typical examples of this are the breakdown of a material structure spheroidization after a long period of operation at high temperature, actual overheating due to inappropriate operation or perhaps due to the action of fire. In these cases the microstructure of the material is changed, and on the basis of previous experience an approximate operating temperature can be indicated.

In connection with the evaluation of structures the method can be used with stainless steel, for the ordinary 18/8 types are sensitive to temperatures over about 400°C. At these temperatures a gradual breakdown of the structure occurs, so that the possibility of crack formation is present due to intercrystalline corrosion.

Crack type determination

Because each type of crack has its own special characteristics, a type determination is usually possible by using the replica method. Because the method is restricted to very small areas, methods such as dye penetrant examination, magnetic particle and eddy current are often used to roughly determine the distribution and extent of cracks, whereupon replica examination can be used for type determination.

The following table shows some examples of what can be identified using the replica method.

Original defects	Welding defects	Operational defects
rolling taps[1]	lack of fusion	ductile fracture
casting defects	solidification cracks	brittle fracture
incorrect heat treatment	HAZ cracks[2]	fatigue
cracks	hydrogen cracks	thermal fatigue
	reheat cracks	corrosion fatigue
	pores/slag etc.	creep
		hydrogen damage
		stress corrosion
		intercrystalline corrosion
		liquid metal embrittlement

1) defects caused by excessive rolling of a material
2) HAZ = Heat Affected Zone

Type determination of cracks is very essential, for it is only in this manner that the cause of the crack can be identified. With the crack mechanism as a starting point, corrective actions such as reducing the level of stress and/or eliminating an actively corrosive environment can be evaluated. Corrective action should only be taken after the cause of cracking has been determined.

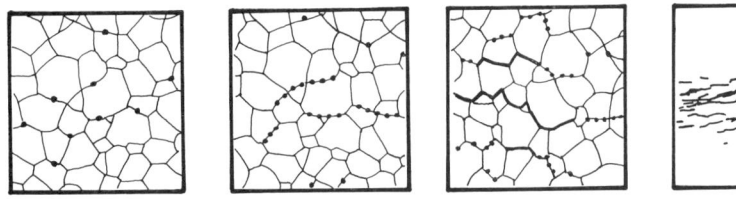

Figure 27.2. The development of creep cracks from scattered individual cavities in the grain boundaries to oriented cavities to actual macroscopic cracks.

Crack detection (creep)

The replica method is particularly well suited to revealing defects due to creep cracking. Creep cracks are formed in components and system parts due to long-term stresses at high temperatures (> c. 450°C). Thus it is particularly advantageous to check boiler components, steam pipes, turbine components and oven parts for creep damage by means of the replica method.

Because creep is by far the commonest cause of operational breakdowns or thermally stressed components in power stations, the application of the replica method is dominant in this energy sector, and it is used as an important tool in connection with the evaluation of residual lifetimes.

With the replica method one can recognize creep cracks at a much earlier stage than is possible with other NDE methods. This means, because creep cracks normally evolve relatively slowly, that there is plenty of time to plan repairs and/or replacements and thus avoid unplanned interruptions of service.

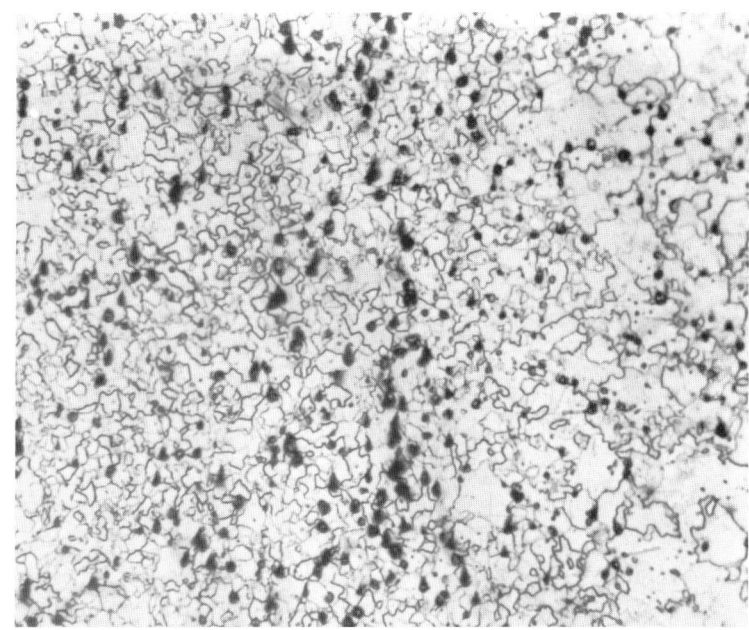

Figure 27.3. The early stages of creep cracking in the guise of cavities lying in the grain boundaries of the material seen with a visible light microscope (500 x).

Creep cracks are first recognizable using the replica method in the form of small holes (cavities) lying along the grain boundaries of the material. Under constant stress these gradually link together to form microcracks and then on to become visible macroscopic cracks. Figure 27.2 shows the progression of creep crack development in low-alloy carbon steel, while Figures 27.3 and 27.4 show photographs of creep-damaged materials seen in a visible light and in a scanning electron microscope, respectively.

Creep cracks start most often at the outer material surface and as a rule in connection with welded joints, material transitions or bends. The stresses can either be due to internal overpressure or due to bending moments in pipes, e.g. because of unsuitable pipe supports. Creep cracks often occur quite locally depending on local conditions of stress. The residual lifetime can, therefore, be quite different at various locations in a system.

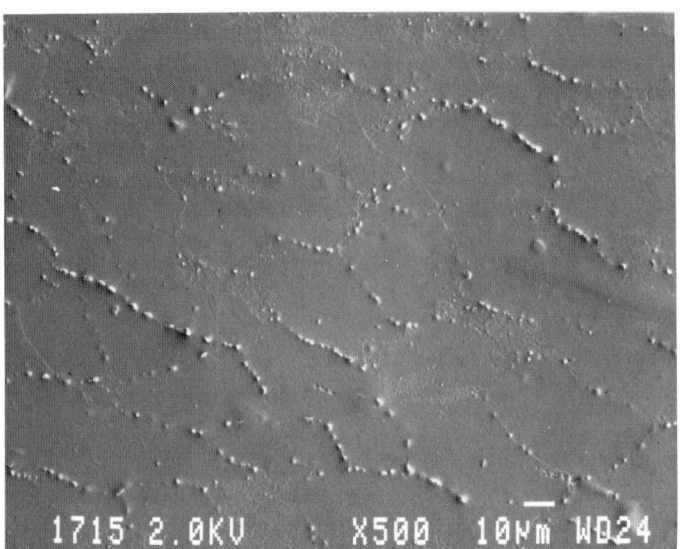

Figure 27.4. Early stages of creep cracks in the form of cavities along the grain boundaries of a material as viewed with a scanning electron microscope (500 x).

Limitations of the method

The replica method can only be used on surfaces and furthermore only on surfaces which are readily accessible. Thus it is not possible to perform replica examinations on the interior of pipes. The method cannot currently be used under water.

The replica method can be used in the temperature region from 5-40°C.

When using the replica method it is usually only possible to examine a limited region. Thus the electropolished region which is prepared only has a diameter of about 8 mm, see Figure 27.5.

Finally it should be mentioned that it is not possible to transfer colours using the replica technique.

Practical applications

As mentioned previously, the replica method is used predominantly for the examination of thermally stressed components, e.g. in power stations and in the chemical and petrochemical industries. In the following, two practical applications of the use of the replica method are presented.

A boiler manufacturer detected a leak during the pressure testing of a newly mounted boiler. The leak was traced to a pipe on the top of the boiler in a very inaccessible location near a spot where the pipe entered a header. Replacement of the pipe would have been a prohibitively complex process. Because it was desired that the cause of the leak be determined to evaluate the chance of finding similar defects in other pipes, a replica examination was initiated. The examination showed that a welder had been so unlucky as to burn through the pipe. His effort to cover the burn-through by welding and grinding resulted in penetrating slag encapsulation and porosities in the weld metal thus causing the leak. It could be assumed that the remaining pipes would not have similar defects. The pipe in question was repaired while checking with magnetoflux and then rewelded (and checked).

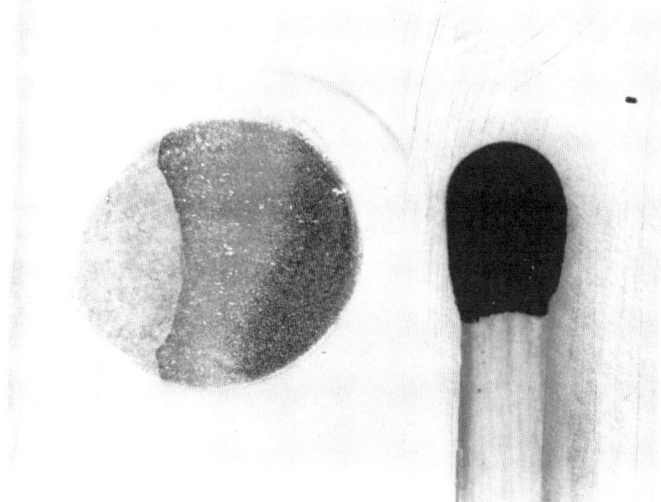

Figure 27.5. Electropolished spot up to a welding seam transferred to a replica. The spot covers part of the welding seam (left), the heat affected zone (HAZ) and some of the base material (5 x).

An oven with 40 vertical stainless steel pipes was heated directly by means of a centrally positioned oil burner in the bottom. In connection with an overhaul round cracks were observed in two of the pipes. The cracks were examined in the laboratory where it became apparent that intercrystalline stress corrosion was present. The microstructure of the material was also sensitized due to inappropriate heating. A replica examination of the remaining 38 pipes was initiated, revealing nine additional pipes with an inappropriate microstructure. These nine pipes were also replaced, and the oven was put back into service. Complete replacement of all the pipes was thus unnecessary.

Training required/desired

As with other metallographic examinations specific national or international requirements do not currently exist. The operator should have a technical laboratory background and special training at a metallographic laboratory in order to perform the actual metallographic work of grinding, polishing, etching and production of the replica. The work procedure should follow the guidelines indicated in standards such as DIN 54.150.

The evaluation of the replica should be performed by a specially trained, competent engineer with experience within the field of metallurgy. When evaluating creep stressed power plant components, the guidelines in TRD 508 and VGB-R509L should be followed.

Per B. Lundwigsen
Knud Erik Poulsen

References

1) DIN Standard 54.150. *Abdrückverfahren für die Oberflächenprüfung (Replica-Technik)*, (Reproduction procedures for surface testing (Replica Technique)), August 1977.

   This is a standard which contains a complete description of the procedure to be used when producing a replica.

2) ISO Standard 3057-1974 (E). *Non-destructive testing – Metallographic replica techniques of surface examination.*

   A standard which provides a brief, step-by-step description of the replica procedure.

3) TDR 508. *Oberflächengefügeuntersuchung zeitstandbeanspruchter Bauteile.*, (Surface texture examination of critical building components).

   This set of rules specifies how creep damage can be classified and proposes a schedule of periodic examination.

4) VGB-R 509 L: VGB-Richtlinie *Wiederkehrende Prüfungen an Rohrleitungsanlagen in fossilbefeuerten Wärmekraftwerken.* (Periodic testing of piping systems in fossil-fueled combined heating and power plants.) 1. Edition 1984.

   Guidelines for testing and evaluating test results in connection with residual lifetime evaluation of creep damaged components.

5) *Application of structural surface examination for the testing of power plant components subjected to creep stress.* W. Arnswald, R. Blum, B. Neubauer, K.E. Poulsen. **VGB Kraftwerkstechnik,** 59, no. 7, July 1979, pp. 537-548.

Practical experience with the use of the replica technique in connection with the examination of power plant components.

6) *Einsatz des Rasterelektronenmikroskops zur Früherkennung von Zeitstandschädigungen bei der zerstörungsfreien Gefügeuntersuchung.* (Use of the scanning electron microscope for early detection of defects by means of non-destructive structure examination). B. Neubauer, Lecture presented at the 8. meeting of the working group on scanning microscopy 11-12/10 1977. Berlin.

Description of the use of the scanning electron microscope for replica examination.

7) *Die Kombination von Abdrückverfahren (Replica-Technik) und Isostress-Methode und ihr Einsatz bei der Untersuchung von Materialien in dänischen Kraftwerken.* (The combination of replica technique and the isostress method and their use for the examination of materials in Danish power plants.) P.B. Ludwigsen. Der **Maschinenschaden 58** (1985) no. 6, pp 230-236.

A theoretical treatise illuminating such aspects as the detection limit for recognizing creep damage and creep crack growth using the replica method.

Chapter 28

SOAP, spectrometric oil analysis program

NDE principle

The *spectrometric oil analysis program* is a maintenance tool which is used to check the condition of oil lubricated mechanical systems (e.g. motors, gearboxes, hydraulic systems). The systems can be kept under surveillance without dismantling them. Abnormally worn components can be localized and replaced before a catastrophic failure occurs. The quantity and type of wear metals in samples of the lubricating oil is determined. The quantity can indicate something about the magnitude of the wear, and the type of wear metal can reveal which component is wearing out.

Wear metals

Wear metals are caused by the relative motion between metallic parts. The motion is accompanied by friction and wear on the surfaces which are in contact with one another. The metal particles (wear metals) are rubbed off due to friction and enter the lubricating oil. By measuring the quantity of wear metals in the lubricating oil, the degree of wear can be evaluated as being normal or abnormal. The wear metals have the same chemical composition as the components from which they come, and the type of wear metal can provide information about which parts are being worn.

Increased quantities of iron are quite common, since many parts are composed of iron, while an increase in the content of less common metals, such as silver, can often indicate precisely which component is being worn abnormally.

Wear profile

For systems which operate normally, wear metals are produced at a constant rate. This rate is the same for all normally operating systems of the same type.

A theoretical curve showing the concentration of wear metals as a function of time for a closed system without oil consumption is shown in Figure 28.1.

New or newly assembled systems have a tendency to produce metallic wear metals at a greater rate during the break-in period.

Every circumstance which increases the friction between the moving parts will increase the degree of wear and thereby increase the concentration of wear metals in the lubricating oil.

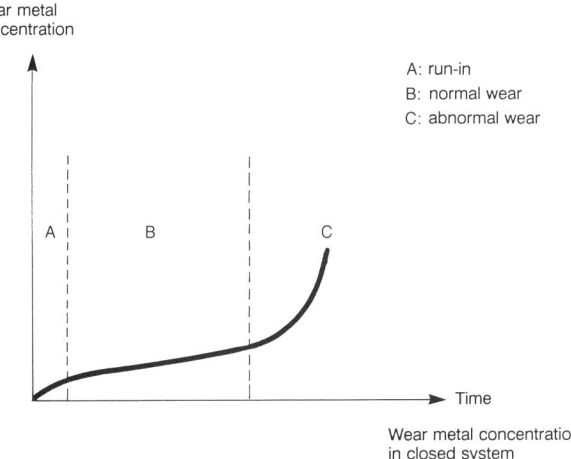

Figure 28.1. The concentration of wear metals in a closed system. A: break-in period, B: normal wear, C: abnormal wear.

If the abnormal condition is not discovered and rectified, it will often continue and accelerate with catastrophic failure as the result.

Wear metal concentration

In motors oil is consumed and new oil is added to replace it. If the new oil is added at the same rate as the oil is consumed, the concentration of metallic wear in the oil will vary as shown in Figure 28.2.

Oil is not generally added continuously, but at intervals, and the concentration of wear metals in the lubricating oil in a motor which is operating normally will vary as shown in Figure 28.3. If abnormal wear occurs, then the curve will increase above the uppermost dotted line.

Measurement of wear metal content

The quantity of wear metal can be measured in very small quantities using spectrometric analysis as described in the following sections.

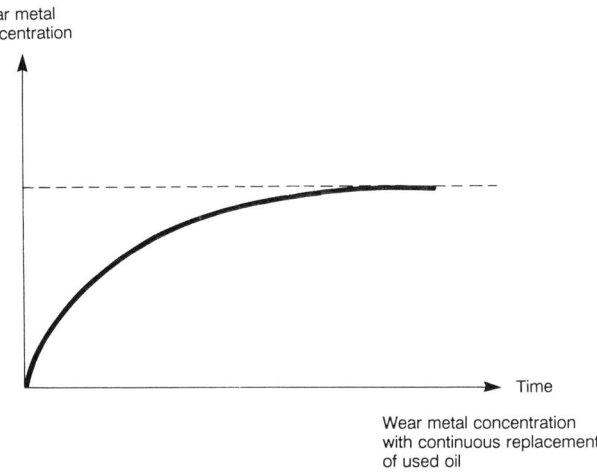

Figure 28.2. The wear metal concentration for continuous replacement of lubricant.

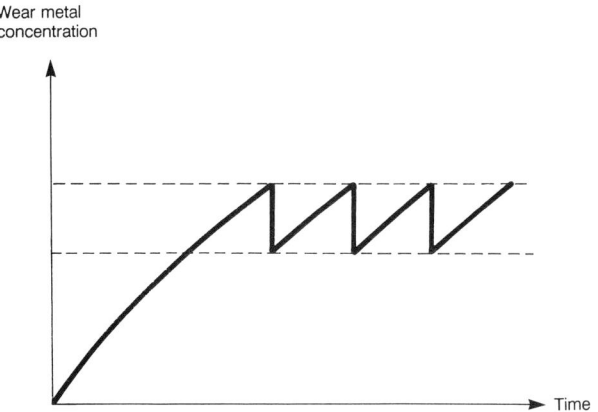

Figure 28.3. The effect of periodic replacement of consumed lubricating oil.

Emission spectroscopy

An emission spectrometer (Figure 28.4) is an optical instrument where the sample is "burned" in a spark between two electrodes. Energy is absorbed by the metals in the sample, and they emit light with wavelengths which are characteristic for each element in the sample. The intensity of the light is proportional to the concentration of the metal in the sample.

A newer variant of the classical emission spectrometer is the plasma emission spectrometer, e.g. ICP (*inductively coupled plasma emission spectrometer*), where the electrodes are replaced by a plasma in the form of ionized argon.

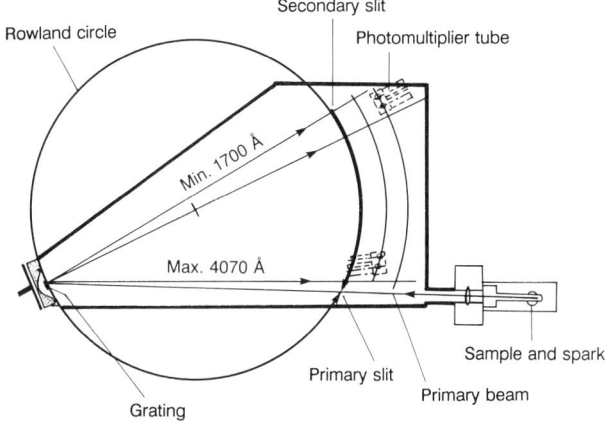

Figure 28.4. Emission spectrometry.

Atomic absorption spectrometry

In an atomic absorption spectrometer (Figure 28.5) the sample is burned in a gas flame, where the metal compounds are transformed into atoms which can absorb light at wavelengths which are characteristic for each metal. If one wishes, e.g. to determine the quantity of copper, then light with a wavelength characteristic for copper is sent through the flame, where the copper atoms absorb part of the light. The quantity of absorbed light is proportional with the quantity of copper in the sample.

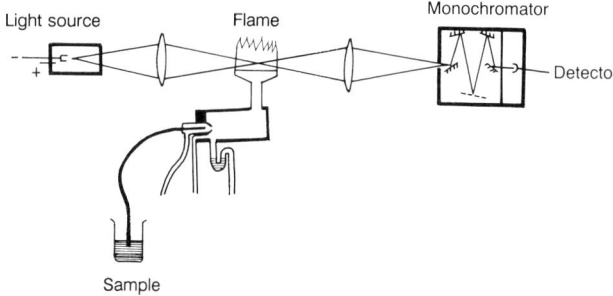

Figure 28.5. Atomic absorption spectrometry.

SOAP, spectrometric oil analysis program

Sampling and analysis

Samples should normally be taken while the oil is warm and before topping up with new oil. The sample is taken in a clean plastic bottle (50-100 ml is fine). It is best if possible to take the sample from above (the dip method) and somewhat above the bottom of the oil sump. See Figure 28.6. It is very important that the sample not be contaminated with foreign matter.

The sample is homogenized in the laboratory, and a portion of it is removed for analysis. Depending upon the viscosity of the sample and the method of analysis which is selected, it may be necessary to dilute the sample. When using atomic absorption spectroscopy the sample is diluted 5-20 times with an organic solvent.

In the laboratory all analysis results for a particular system (e.g. a motor) are recorded, and possibly graphs are drawn to indicate the evolution of the wear process. It is most practical if the laboratory responsible for evaluating the results also "sounds the alarm" if the wear metal concentration is abnormal and perhaps requests new samples or a more frequent sampling schedule for systems which are suspected to be defective.

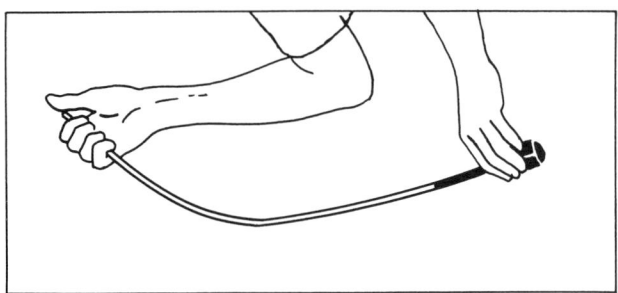

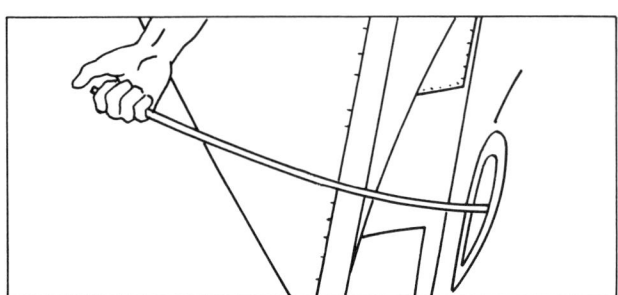

Figure 28.6. Sampling using the dip method.

Part of the idea behind SOAP is that the sample is analysed and the result is reported a short time after the sample is removed. Rapid analysis is also important because the samples do not have unlimited shelf-life. Metal particles can be deposited on the wall of the container, and small particles can clump together and become so large that they cannot be measured.

In fact an absolute measure of the metal content in the lubricating oil is not performed. Only particles under a certain size can be measured. In the case of atomic absorption the size of the particles measured is on the order of 0-10 μm. With emission spectrometry somewhat larger particles can be measured. It is, therefore, also important to use precisely the same method when comparing results.

SOAP program

When motors and other systems are to be checked using the SOAP method, a program must be developed. In this connection one must decide:

- Which elements are to be measured?

- What will be the sampling frequency?

- What criteria will be used for defining a condition as abnormal, i.e. at what wear metal concentrations must action be taken?

Here detailed knowledge of the construction of the test object must be available. What components comprise the system, and what is the chemical composition of the components? The construction material will often be cast iron or steel. In that case iron will be present in the lubricating oil. If stainless steel is present, then chromium and nickel will also be present. In sliding bearings tin and lead are present, and silver may have been used in hard solder joints. Furthermore, alloys of copper, aluminium, magnesium, etc. may often be found.

Sampling frequency

The shorter the time interval between sampling, the more effective SOAP can be. Thus for aircraft it would be ideal if a SOAP analysis were conducted after every landing. This is not possible, so the frequency of analysis must be determined based upon experience with the object of measurement and other practical considerations.

Criteria for abnormal wear

Experience is the determining factor for which wear metal concentrations justify closer examination of the object.

Manufacturers which recommend SOAP provide tables indicating normal and abnormal concentrations of wear metals and abnormally high increases of the concentration in a given period. An example of such a table is shown in Figure 28.7. The use of such tables requires careful adherence to oil top-up schedules and sampling plans.

Associated with the table is a key which indicates where information can be obtained about where in the system abnormal wear can be found. For the motor corresponding to the table in Figure 28.7 it is indicated, among other things, that an increased presence of the wear metals iron and copper combined with a lesser increase in silicon presumably indicates significant wear in a particular component in a flushing pump. If there in addition to iron, copper and silicon also is an increase in the magnesium content, then there can be wear in the reduction gear of an oil pump.

ATOMIC ABSORPTION, content in ppm

	Fe	Ag	Al	Cr	Cu	Mg
Abn. inc./10 h	5	2	2	2	2	4
Normal range	0-16	0-3	0-4	0-3	0-5	0-13
Marginal range	17-19	4	5	4	6	14-16
High range	20-24	5	6	5	7	17-19
Abnormal range	25+	6+	7+	6+	8+	20+

Average concentration of other elements:
Ni = 1, Pb = 4, Si = 5, Sn = 9, Ti = 1, Mo = 1

Figure 28.7. Aircraft manufacturer's guidelines concerning wear metal conditions in a certain motor.

Wear rate

In connection with the method discussed above based on tables supplied by the manufacturer one evaluates the condition of the system directly based upon the measured concentrations of wear metals in the lubricating oil. If the oil consumption is high, as in certain motors, and replacement oil has, therefore, been added, possibly several times during the analysis cycle, then interpretation of the results can be difficult. Therefore, equations have been developed to take the addition of oil into account. In this connection one works with the concept of *wear rate* or *total wear rate*, where one attempts to compute the total quantity of a wear metal which has been generated since the last oil change for the system.

A number of computation equations have been developed (see References 2 and 3). One of the simplest looks like this:

$$C_{Tn} = C_n + \tfrac{1}{2} \sum_{i=1}^{n} [C_i + C_{i-1}] \times \Delta O_i / O$$

where:

C_n = the concentration measured in the last sample
C_i = the concentration in an arbitrary sample i
C_{i-1} = the concentration in the sample before i
ΔO_i = top-up oil added between between tests i and $i-1$

O	= total oil volume
n	= number of samples

By utilizing the concept *wear rate* one can obtain a clearer definition of abnormal wear for oil-consuming systems.

Limitations of the method

Users of the SOAP program claim that they find a very large proportion of the defects which could lead to a breakdown. But of course SOAP has its limitations. Thus the method provides no indication of:

- large particles, e.g. bearings can break down due to a few large particles.

- defects which occur quickly, e.g. due to the lack of lubricating oil or due to bearings which burn up.

- defects where no wear metals are formed, e.g. breakdowns due to metal fatigue.

Areas of application

The program is used in situations where breakdowns are catastrophic or expensive.

It is widely used in the military.

In the US it is used by the Air Force, Navy and the Army.

It is used by many civil aviation companies.

It is used for checking certain contractor equipment, and in England it is used for checking locomotives.

Training required/desired

No special rules have been established for training in this area, but as is apparent from the foregoing description, considerable knowledge is required and experience with the subject of measurement is necessary. Knowledge of both components and materials and knowledge of the mechanisms of wear is essential.

A large number of measurements must be made to determine the normal wear rate, and one must carefully adhere to the sampling schedule and keep a record of the amount of oil added to the lubricating system.

The laboratory must establish routines which ensure that the analysis instruments are always operating optimally, for even small changes in the wear metal content can significantly affect the interpretation of the results.

If all these prerequisites are fulfilled, then SOAP can provide the first indication of abnormal wear in many cases.

Finn Kristensen

References

1) *Joint oil analysis program.* Laboratory manual, NO.600-76-D-0596. Department of Defense, USA.

2) *Statistical analysis of wear metal concentration measurements in oil.* Karl Scheller, Auburn Univesity, Alabama, and Kent J. Eisentraut, Wright-Patterson AFB, Ohio.

3) *An investigation into the rate of contamination of the oil circuit of turbojet engine.* Lotan D., SNECMA document MFTM No. 10564.

Chapter 29

Strain gauge technology

NDE principle

Strain gauge measurement is a technique which is based upon the connection between the deformation of an object of measurement and changes in the electrical resistance of a conductor which is attached to it. The configuration of a typical strain gauge is shown in Figure 29.1.

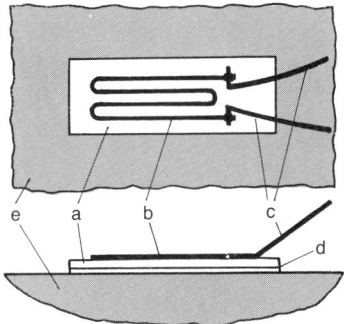

a) Supporting foil
b) Resistance wire
c) Connections
d) Adhesive joint
e) Object of measurement

Figure 29.1. The basic configuration of a strain gauge.

When an object of measurement is subjected to a load, it will be deformed. The cross sectional area of the electrical conductor in the strain gauge which is attached to the object will be altered, thus changing the ohmic resistance when deformations occur. Extremely small changes in ohmic resistance – and thus in the load on the object – can be recorded by means of a Wheatstone bridge. The measuring principle is illustrated in Figure 29.2.

It is possible to perform measurements on all types of materials which have known deformation data due to loads, e.g. metals, concrete and plastic materials.

Strain gauges are equally useful for the measurement of positive as well as negative changes in length (tensions as well as compressions).

Strain gauge technology

Development of the method

The strain gauge measurement technique was developed in the late 1930s by two American researchers, Simmons and Ruge. An ever increasing requirement for stronger and lighter constructions for use in the aviation industry made it necessary to be able to document the load conditions. Furthermore, using the strain gauge measurements it was possible to improve the theoretical models for complicated objects.

Today, this knowledge has spread to the development and maintenance departments of some production facilities.

The continuing increase in automated production on assembly lines makes it more and more attractive to establish warning systems based on strain gauge measurements in order to prevent untimely breakdowns of the expensive production equipment. Furthermore, the loss of production time can be an economic catastrophe which far exceeds the cost of an investment in preventive maintenance.

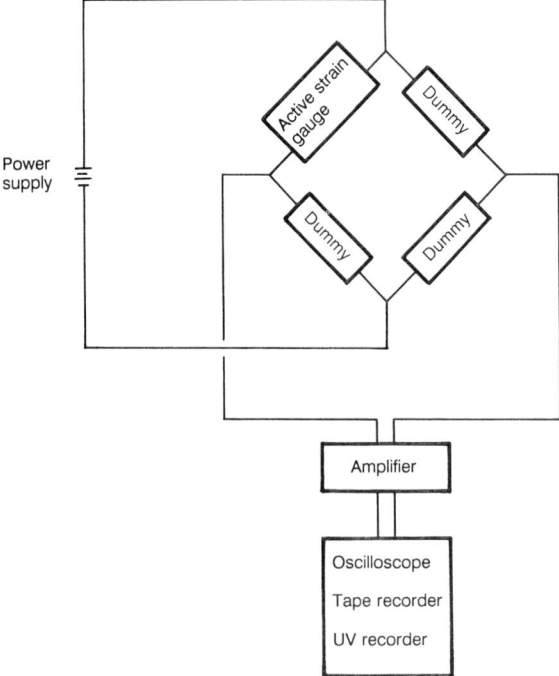

Figure 29.2. The measuring principle employed in the strain gauge is based on the detection of small changes in the resistance of an arm of a Wheatstone bridge.

Areas of application

Strain gauges are used in industry as transducers for weight and pressure measurements.

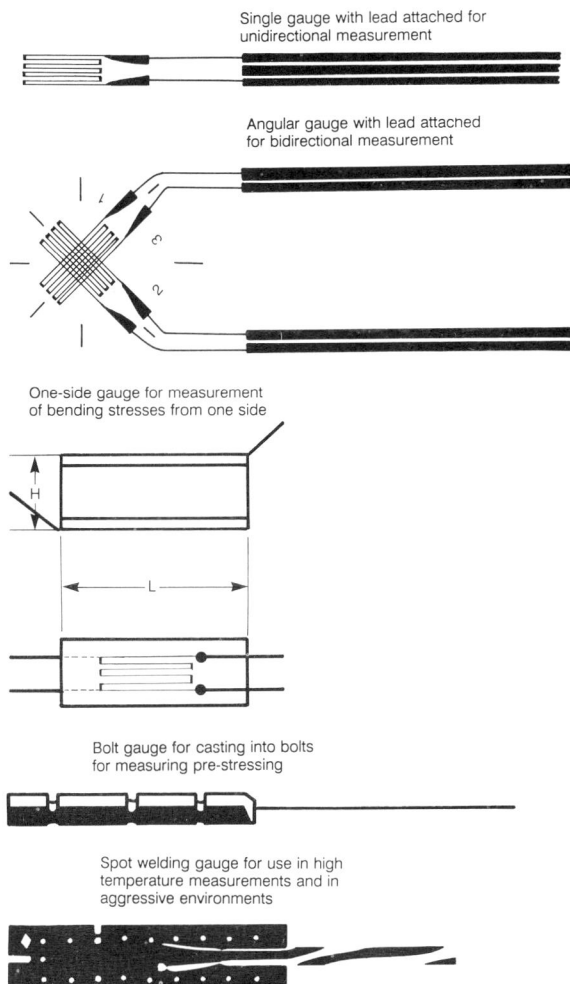

Figure 29.3. Various types of strain gauges are shown.

In addition to this industrial production of transducers based upon the principle in strain gauge measurements, the technology is a unique tool for the evaluation of strengths of materials, e.g. in connection with the analysis of failures, load conditions, residual lifetimes, vibrations, etc.

A designer may often have considerable knowledge of the design of a particular construction and the physical properties of the materials. On the other hand knowledge of the loadings are in many cases known from experience with similar constructions or estimated from the expected operating conditions.

When the strength of such constructions is to be evaluated in relation to the

Strain gauge technology

development of a new design, or when an existing design is to be evaluated for possible overloading (static or dynamic), the application of strain gauges can provide valuable information about stress conditions in individual components or in the entire structure.

Strain gauge measurements can be planned so that separate results are obtained for tension-compression, bending and torsion components. In more complicated structures measurements can be performed using special strain gauges to determine the principal directions of the stresses which are present.

Strain gauge signals can be converted and transmitted by means of a radio system so that, for example, stresses in a rotating system can be measured.

Strain gauges are manufactured in many different shapes each with its particular application. The measurement length of a strain gauge can vary from less than 1 mm to several hundred millimetres. Strain gauges are generally available for performing measurements in the temperature interval from 0°C-80°C. Certain special types can be used up to very high temperatures (400°C) e.g. for use when measuring components in power plants. Examples of some strain gauge types are shown in Figure 29.3.

When the conditions of stress are measured by means of the strain-gauge method, these results can be evaluated in relation to known, permissible stresses in the particular material, such as specified in data information sheets for the material, standards, fatigue diagrams, etc.

The accuracy which can be achieved when performing strain gauge measurements is very high. Under ideal conditions deformations on the order of one millionth of a mm per mm measurement length can be measured. This corresponds to a material stress of 0.21 MPa (0.21 N/mm^2). This is more than adequate for ordinary strength evaluations.

If it is necessary to measure residual internal stresses, as for example from welding, casting or tightening materials in a construction, then a special strain gauge measurement technology must be employed.

The strain gauge is mounted on the object of measurement. Then a small (1.6 mm dia.) hole about 2 mm deep is bored in the object. When the hole is bored, internal forces (residual stresses) are released. This change in the conditions of stress is recorded by the strain gauge. In this manner a measure of the magnitude of the internal forces is obtained.

The accuracy of this form of measurement is somewhat less than for ordinary strain gauge measurements. If comparative evaluations are to be performed of objects before and after hardening, machining, etc., then this measurement technique can be particularly useful.

Practical applications

In the following some abbreviated descriptions are provided concerning actual or suggested applications of strain gauge measurements.

Example 1:
- A 40 year old coal crane is due for renovation of winches, bogies and lifting mechanisms. But what is the residual lifetime of the steel construction?

    Calculations and material documentation is no longer available.

    The solution of the problem is that strain gauge measurements are carried out at those points in the structure which a structural analysis indicates to be the most severely loaded. The strain gauge measurements are carried out on the crane during normal operation. The results of the measurements are used to evaluate whether measurements have been performed at the correct locations. Next the data is used to evaluate the total "lifetime" of the structure (i.e. the maximum permissible number of loadings). The result is evaluated in relation to the load history of the crane, then the residual lifetime with an appropriate margin of safety can be established.

    The project can then be evaluated from an economic viewpoint. Should one invest in renovation of the crane, or is it economically attractive to replace it?

Example 2:
- Cracking is observed in the load carrying frames of some piece of transport equipment (bus, forklift, truck, train, ship, etc.). The cracking occurs in a butt weld near a flange junction. The broken section is cut out and it is determined that there is fatigue crack growth from the root side of an incompletely penetrated weld.

    Would the butt weld have sufficient fatigue strength if a complete penetration were performed?

    The forces in the frame system can not be computed directly, for they are caused by shocks which are not well-defined.

    A strain gauge measurement on the repaired butt weld is performed during normal operation. The results are evaluated in relation to fatigue data for the weld joint in question. The answer in this specific case: Yes, a completely penetrated butt weld has sufficient fatigue strength under the given test conditions.

Example 3:
- A steel smokestack exhibits resonant oscillations (transverse oscillations) when the wind strikes it from a particular direction at a certain speed. This is due to the fact that the damping ability of the chimney is inadequate.

    What effect will the mounting of a vibration damper have on the chimney?

A strain gauge measurement is carried out on the chimney without the vibration damper. The chimney is caused to oscillate by means of a thin wire rope (with catch-line) attached to the lifting fittings of the chimney. The steel wire is attached via a break-cord to a tractor or winch which can exert a pull. When the break-cord breaks, the chimney will begin an oscillation which is slowly damped. This is recorded by means of the strain gauge measurements.

A corresponding experiment is performed after the vibration damper has been mounted, and the data is compared.

The damping ratio was evaluated in this specific situation to be three times less than the minimum requirement specified by the building safety code before mounting of the vibration damper. On the other hand the damping ability of the chimney was twice as good as the minimum requirement specified by the code after the vibration damper was mounted.

The measurements were carried out while carefully checking (by means of the strain gauges) for the risk of overloading the chimney and while taking appropriate safety precautions to prevent damage or injury due to the falling wire rope.

Example 4:
- A motor console exhibits crack formation in a weld at the mounting bolts. It is observed that there are some vibrations in the motor during operation.

A strain gauge measurement is undertaken which by means of a frequency analysis reveals the resonant frequency to be 1 kHz (1000 oscillations per second) at a certain rotational frequency. Such a high frequency oscillation is evaluated to be due to torsional vibrations in a massive axle. Because the oscillations occur at a particular rotational frequency, they must be due to a repeated loading change which exactly corresponds to the torsional characteristic frequency of the massive axle.

The defect is expected to be located in the bearings of the axle system or in the transmission gears.

Disassembly of the axle system reveals a defect in a gear tooth in the transmission gears. The symptom was cracking in a weld. The cause was a bad tooth!

Training required/desired

Strain gauge measurements only provide information about the stresses at those points where the strain gauges have been placed. It is, therefore, of critical importance that the positions are selected by individuals with insight regarding the strength of materials.

The selection of the measurement points should be undertaken on the construction itself. There are often minor inconsistencies between old construction drawings and the actual piece of equipment, particularly with respect to small details.

It is generally very valuable to be able to see the data on an oscilloscope or chart recorder while the system is in operation. Particularly interesting operating conditions can then be repeated for a more detailed evaluation.

If one has the necessary insight regarding the strength of materials, then strain gauge measurement techniques can be learned by studying the literature, from suppliers of strain gauges and by performing small, practical experiments.

In England the British Society for Strain Measurements can provide a course of instruction on strain gauge measurements. It is advisable to invest in this training if strain gauge measurements are often performed in a company.

If strain gauge measurements are not performed often, training local personnel is probably too expensive, and it would, therefore, be wise to consult with firms or institutes which have sufficient expertise in this area.

Lars Tofte Johansen

References

1) *Non-destructive testing.* The Danish Welding Institute, Publication 72.07, 1988.

2) *Strain measurements*, Brüel & Kjær, 1975.

3) *Strain gauge technology*, A.L. Window and G.S. Holister, Applied Science Publishers Ltd., 1982.

Chapter 30

Stroboscopy

NDE principle

Stroboscopy is a term applied to a monitoring and measuring technique where a rotating or swinging object can be inspected in operation by illuminating it with flashes of light, where the frequency is equal to a multiple of the rotational frequency of the object, so that it appears that the object is standing still. If the flash frequency is increased or decreased in relation to the rotational frequency of the object, one can make it appear as if the object moves slowly backwards or forwards.

Development of the method

The method is based upon the physical principle that the human eye perceives the motion of a body in relation to the frequency of the light which illuminates it.

The first stroboscopes which produced something resembling this phenomenon were quite primitive and quite difficult to adjust to the desired frequency. However, the evolution of technology, particularly in electronics, has made the development of the modern stroboscope with digital readout and a wide range of applications possible. Some of these are discussed in the following.

Areas of application

The stroboscopic effect is used to check rotating or oscillating objects in many different situations, of which the following:

- monitoring direction of rotation and speed

- checking for cracks and breaks

are among the most important.

Practical applications

Checking rotation
The use of stroboscopy for checking the direction of rotation and the rotational frequency, etc. can be performed by directing a stroboscopic light source at an oscillating, sliding or rotating part, and then adjusting the flash frequency of the stroboscope so that the object appears to stand still. The frequency and direction

can then be read on the scale or the display of the stroboscope (see Figures 30.1, 30.2 and 30.3).

Figure 30.1.

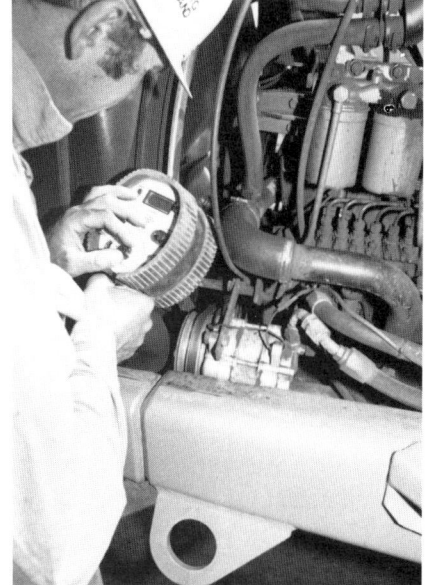

Figure 30.2.

Checking cracks and breaks
Using the stroboscope one can also examine rotating machine parts for breaks or cracks without having to stop the machine or the system. The light from the stroboscope is directed at the rotating part, the flash frequency is adjusted as described above, and the object is inspected visually for possible cracks or breaks.

Figure 30.3.

Condition monitoring

Condition monitoring of clutch plates and clutch bolts can be performed by "freezing" the clutch parts by adjusting the stroboscope until the clutch appears to be standing still. Then one can measure the separation of the clutch plates, for example by using an optical measuring instrument. But remember, the clutch is *not standing still*, and it must, therefore, absolutely not be touched during the check.

Checking of clutch bolts is performed by adjusting the instrument so that the clutch plates appear to move backwards and forwards, allowing the visual examination of possible wear and/or checking for missing bolts. This "on-line" method is a real time-saver, for when the machines are not operating it is possible to concentrate on repairing defects discovered previously (see Figures 30.4 and 30.5).

Figure 30.4.   Figure 30.5.

Training required/desired

The use of the stroboscope does not require actual training, but a brief period of instruction in its construction and practical application must, however, be recommended to assure satisfactory results from the beginning.

Personal safety
In connection with the instruction provided one should place considerable emphasis on the safety aspects, for the stroboscope is used for checking relatively fast-moving parts and gives the illusion that they are standing still while they are actually moving at normal speed.

Kaj Tranholm Olesen

Chapter 31

Test coupons

NDE principle

The basic principle is the determination of material loss by *measuring weight loss*. Weighed *test coupons* of the material to be studied are exposed to the operating environment. After an appropriate period, normally one to three months, the test coupons are removed. Any corrosion products are removed by immersion in an appropriate acid, then the cleaned plates are dried and weighed. On the basis of the weight loss the average corrosion rate is computed according to the formula:

$S = 3.65 \, V/(A \, \varrho \, d)$,   where

S = corrosion rate in mm/year
V = weight loss in mg
A = surface area in cm^2
ϱ = density of the metal
d = exposure time in days

The method thus measures the average corrosion rate over a period and tells nothing about whether or not there have been large variations in the corrosion rate during this period.

Development of the method

The test coupons, which can be rectangular or circular, are made of the same alloys as those which one wishes to examine. Because the weight loss normally must be determined on a laboratory balance, the weight should not exceed 200 g. In order to achieve the maximum possible sensitivity, the area/volume ratio should be as large as possible without running the risk that corrosion penetrates through the test coupons during the exposure. Beyond these considerations the shape can be determined by the accessibility of the areas where the test coupons are to be inserted.

When placing the test coupons some type of holder is required. This can be in the shape of a rack which can hold many coupons, if conditions permit, or it can be a holder for only one plate. Various types of holders with test coupons are shown in Figures 31.1-31.3. Figure 31.1b shows a simple mounting of a 10 x 100 mm coupon using a T-fitting. Figure 31.2 shows a rack with room for a large number of coupons. Figure 31.3 shows coupons in a pipeline, which via various mounting forms can be placed at different levels in the pipe. Notice that the holders are twisted 90° compared to the normal mounting orientation.

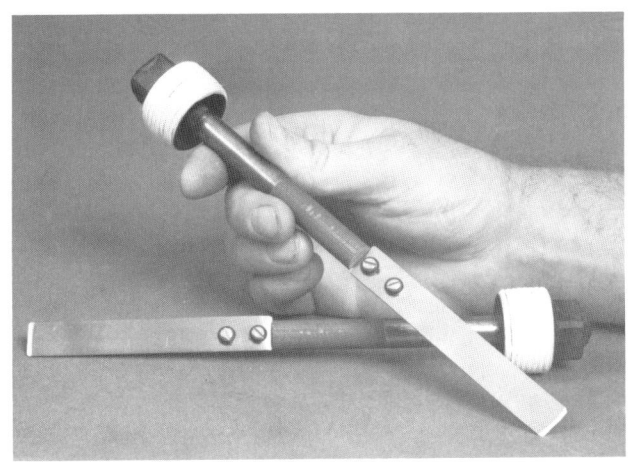
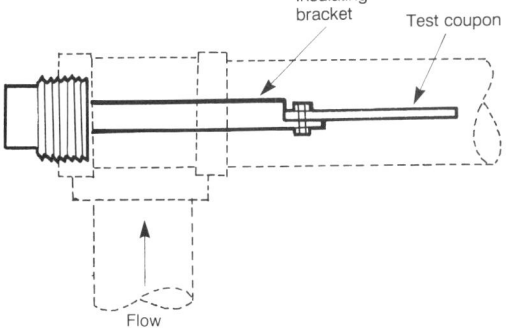

Figure 31.1. Simple mounting of test coupons. (a) Photo of coupons and holder ready for mounting. (b) Coupon inserted via a T-fitting.

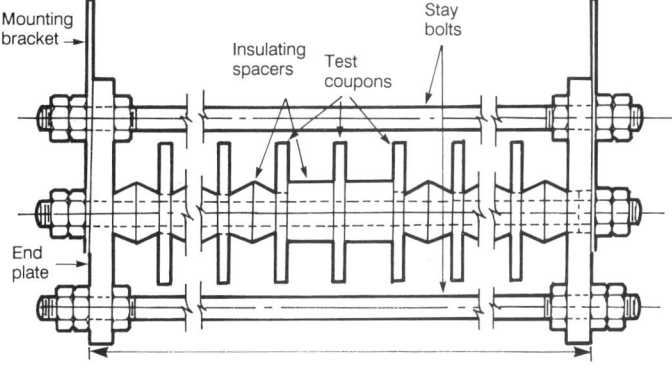

Figure 31.2. Rack with several coupons.

The test coupons are mounted via so-called access fittings which both permit mounting in systems with pressures up to 400 bar and also – using the necessary utility tool, a so-called retriever – permit the insertion and removal of the coupons while the system is under pressure.

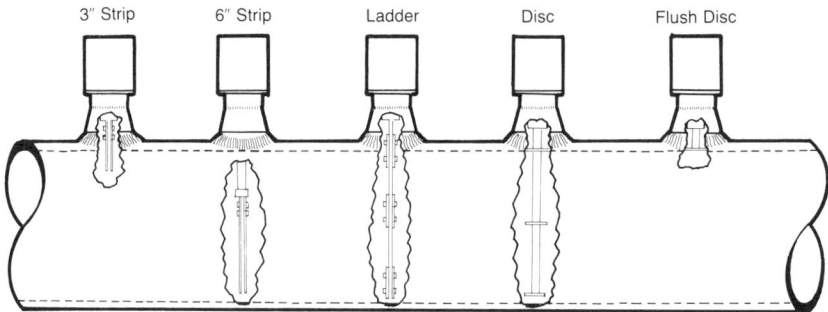

Figure 31.3. Coupons mounted in a pipeline via access fittings (Cosasco).

To avoid undesired galvanic effects, the test coupons should be mounted electrically insulated from the mounting screws, the holder and the system.

The placement of the test coupons is important. In order to relate data from the weight loss measurements to the actual parts of the system it is important the inserted test coupons "see" the same conditions – temperature, flow, etc. – as the system parts. It can also be necessary to take heat transfer conditions which can have an effect upon the rate of corrosion into account.

Areas of application

Weight measurements provide – in principle – correct results for the corrosion depth only when completely uniform corrosion occurs. In the case of strongly localized corrosion (pitting) the computed corrosion rate is substantially lower than the real corrosion penetration in the pits. In this case supplementary information can be obtained by measuring the depth of the pits by means of a measuring glass with a narrow tip which can go down into the pits.

Because the method is based upon real loss of material, it can be used for both corrosion and erosion measurements.

Because of the test coupons' relatively slow response, they can only record slow changes in the corrosion rate. They are, therefore, unsuitable for measurement of changes in corrosion due to transient changes in operating conditions.

Test coupons can be used to test new materials under operating conditions before changing the material used in the system.

Furthermore, test coupons are useful tools for checking the reliability of more sensitive but indirect methods such as ER-probe (electrical resistance) and LPR-probe (polarization resistance) measurements. In order to confirm reliability, the average corrosion rate of both the test coupons and the other methods should agree when compared over a longer period of time, e.g. one month.

Practical applications

Test coupons are widely used for the monitoring of corrosion rates in open water cooling systems. Corrosion protection is carried out by adding inhibitors which can be optimized by adding that quantity of inhibitor which is adequate to keep the corrosion rate sufficiently low. The inhibitor dosage can thus be adjusted monthly based upon the results from the test coupons.

Test coupons are often used as documentation in connection with the acid cleaning of boilers and heating systems to demonstrate that the process has not caused excessive corrosion in the system. The test coupons are present in the system during the entire cleaning process, and after the work has been performed a check is made to see that too much corrosion has not occurred, for example <2 mg/cm^2 per 24 hours. In this case the corrosion rate is so great that it is possible to obtain a weight loss result after only a few hours.

Training required/desired

The NDE operator must have access to and be able to handle a laboratory balance. Furthermore, he or she must also have knowledge of which cleaning processes can effectively remove the corrosion products formed without removing the underlying metal.

Ebbe Rislund

References

1) *Preparation, cleaning and evaluating corrosion test specimens.* ASTM G1-81 Standard Practice.

2) *Conducting corrosion coupon tests in plant equipment.* ASTM G4-84 Standard Method.

3) *Preparation and installation of corrosion coupons and interpretation of test data in oil production practice.* NACE standard RP-07-75 Recommended Practice.

Chapter 32

Thermography

NDE principle

Thermography makes use of the infrared spectral region. The shortest wavelengths lie near the wavelength limit which corresponds to the visual perception of deep red. The long wavelength end of the infrared spectrum overlaps the millimetre region of the microwave wavelengths.

The infrared spectral band is usually divided into four sub-bands, the limits of which also overlap. They include:

- the near infrared 0.75-3 μm
- the short wave infrared 3-6 μm
- the long wave infrared 6-15 μm
- the far infrared 15-1000 μm

Development of the method

At present the following major types of thermographic instruments have been developed, referred to informally as IR instruments (IR=infrared).

Point measurement instruments (radiometers, bolometers, pyrometers) are pistol-shaped units with widely ranging capabilities from the simplest type with a galvanometer display to microprocessor controlled units with input options for the emission factor and the ambient temperature. These units provide minimum and maximum temperatures as well as average temperature with one decimal on measurements and continuous storage in memory if desired. The emission factor depends upon which material is measured, and the factor can be entered into the apparatus. Table 32.1 provides a number of these emission factors.

Simple imaging instruments (heat-sensitive telescopes) are characterized by low weight. They are available with gas cooling and with electrical cooling. They are all portable. Figure 32.1 shows an example of a heat-sensitive telescope.

Line scanners are larger units which, as the name suggests, can provide profiles of an object, etc.

Table 32.1
Emission factors of various standard materials.

METALS	Temperature (°C)	Emission factor
Aluminium		
polished plate	100	0.05
anodized plate	100	0.55
Brass		
highly polished	100	0.03
oxidized	100	0.61
Copper		
polished	100	0.05
highly oxidized	20	0.78
Gold		
highly polished	100	0.02
Iron		
"polished" cast iron	40	0.21
oxidized cast iron	100	0.64
very rusty plate	20	0.69
Magnesium		
polished	20	0.07
Nickel		
electrogalvanized, polished	20	0.05
oxidized	200	0.37
Silver		
polished	100	0.03
Stainless steel		
(18-8)	20	0.16
oxidized	60	0.85
Steel		
polished	100	0.07
oxidized at 800 °C	200	0.79
Tin		
Commercial tin plate	100	0.07

Table 32.1 (continued)
Emission factors of various standard materials.

OTHER MATERIALS	Temperature (°C)	Emission factor
Diverse materials		
bricks (ordinary red)	20	0.93
Carbon:		
candle soot	20	0.95
graphite (uneven surface)	20	0.98
concrete	20	0.92
glass (polished sheet)	20	0.94
Lacquers		
white	100	0.92
flat black	100	0.97
Lubricating oil (thin film on nickel)		
nickel base only	20	0.05
0.025 mm thick film	20	0.27
0.051 mm thick film	20	0.46
0.125 mm thick film	20	0.72
thick coating	20	0.82
Oil-based paint		
average of 16 colours	100	0.94
Paper – white quality	20	0.93
Plastic with rough surface	20	0.91
Sand	20	0.90
Human skin	32	0.98
Dirt		
dry	20	0.92
water saturated	20	0.95
Water		
distilled	20	0.96
ice, smooth	−10	0.96
frost crystals	−10	0.98
snow	−10	0.85
Wood		
Oak planks	20	0.90

Figure 32.1. Heat-sensitive telescope.

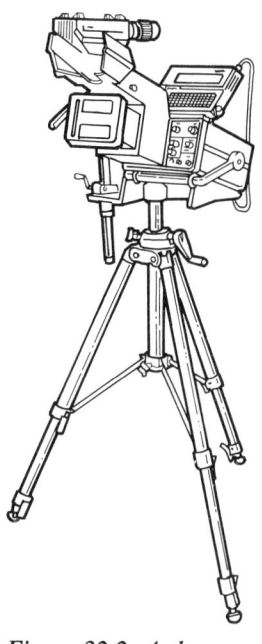

Thermography with imaging and measuring instruments. Systems which all can be supplied with options for colour rendition, data handling, direct temperature read-out, photography, video output, etc. See Figures 32.2-32.7c.

Figure 32.2. A thermographic system with "Hunter Husky" calculator with automatic real-time display of object temperature (Agema Thermovision© Type 870).

Thermography

Figure 32.3a. An example of a modern electrocooled scanner.

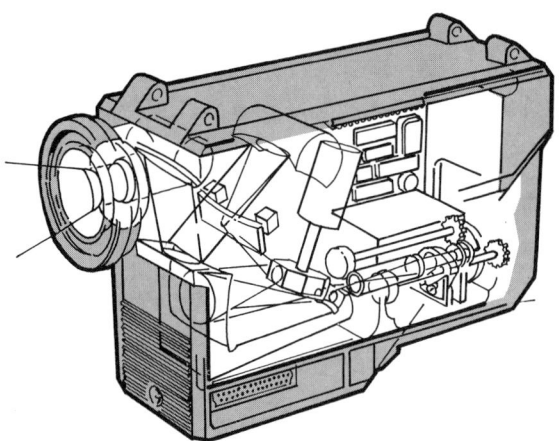

Figure 32.3b. The electromagnetic radiation from the object of measurement is focused via an infrared lens onto an oscillating mirror driven by a DC motor. The rays from the oscillating mirror are focused via three other fixed mirrors to a horizontal prism which rotates at 1800 rpm.

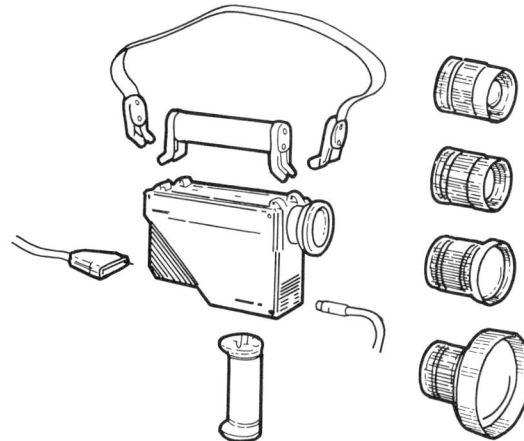

Figure 32.4. Exploded view of the scanner part. Note the portability options: a handle mounted under the scanner, a carrying handle above and a carrying strap if that is more convenient. Furthermore, optional lenses: close-up, wide-angle and telephoto.

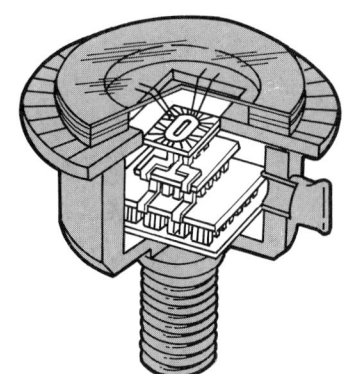

Figure 32.5. The detector is the heart of every IR scanner. An example of a modern electrically cooled (Thermovision© Type 870) device is shown here. The Sprite detector operates at a temperature of -70 °C. This is achieved by using an encapsulated, three-stage thermoelectric cooler (a Peltier element). The Sprite detector is sensitive to radiation between 3 and 5 μm.

Areas of application

Thermal measurements are used for routine maintenance of station machine units to prevent electrical faults, to locate defective materials and to optimize the efficiency of a process in other ways.

The prevention of unexpected break-downs is the aim of the technique. These same inspections, which are justified from the point of view of production advantages, can also offer energy savings. Many other problems are often regarded as maintenance problems but are in fact energy problems.

The extent to which emphasis is placed on these problems on the basis of energy or production efficiency depends upon the priorities set by management.

For example most industrial plants are highly motivated to keep the production line in operation at all costs. However, primary application of thermography for energy savings will usually give similar advantages. For example, an overheated electrical component localized and replaced by a new one today can easily prevent a breakdown tomorrow, but it is also true that an overheated component represents wasted energy. In addition there is the risk that production may be interrupted.

Thermal measurements are used as a means to localize and to quantify energy losses. When this information has been received, management can use these data to set standards for energy savings. These are normally based on cost-benefit parameters in connection with the desire to optimize energy use.

Many components are designed with a view to optimizing their function and not necessarily their energy efficiency. As a consequence many components today serve their function well but are perhaps enormously inefficient with respect to energy use. Thermal inspection is an ideal technique for the analysis of energy losses in existing components. Not only can the type of energy loss be identified, but after corrective action is taken thermography can be used to check how effective the corrective action has been.

When components have been designed and built correctly, thermography can routinely identify components which due to age or environmental conditions use too much energy and pose a threat of production break-down. This routine can thus be part of a *preventive maintenance programme*: "An ounce of prevention is better than a pound of cure."

Although many components are correctly designed, poor material quality can cause their energy use to be excessive. For example poorly applied insulation will often turn out to be the cause of big expenses during operation due to the energy losses.

Areas of practical application are summarized in the following:

Electrical power industry and electronics
Electrical power plants, transformer stations, line inspections, electronic and electrical component inspection.

District heating
Installations – all components used for heat distribution.

Building construction
Refrigeration installations, building insulation, hidden pipe installations, floor heating installations, doors, windows, garage doors.

Industry
Production monitoring, quality control, process control, ovens, boilers, ball bearings, cooling towers, motors, ventilators, piping installations, reactors, low and high temperature measurements, factory chimneys, steam lines, vessel corrosion.

Practical applications

Electrical power industry and electronics
Within the electrical industry thermography is used to check for preliminary cracking, loosened bolts, wear, and corroded or environmentally damaged components. These defects can be localized during operation and at a safe distance, because they radiate more heat than undamaged components. Of course loading will produce heat, but the balance in a series of components will readily reveal if one or more of them is out of balance. The necessary repair will then be able to be performed in a timely manner, before the damage gets worse and perhaps gives rise to a break-down and the subsequent economic consequences.

District heating
In the area of heating technology thermography is used to check for leaks at all stages, even under the surface of the earth, in concrete shafts, etc., whereby expensive excavations and traffic disturbances can be avoided.

Furthermore, thermography is often used in conjunction with other NDE techniques such as ultrasonics and radiography to check the piping systems in ovens. Thermography is well suited here to reveal external and internal coatings, corroded regions, etc.

Building construction
In the area of building technology thermography is used for identification of defects in hidden piping installations and for checking for moisture, insulation, cold bridges, etc. In connection with the above applications accessories such as those shown in Figure 32.8 can be used.

Industry
In other areas of industry thermography is used among other things for checking junction boxes where defects can be revealed before machines are interrupted from service. Defects can also be revealed due to excessive pressure in bearing races and pressure rollers, and thermography is also used for certain forms of production monitoring where thermal aberrations in products characterize their quality.

Medical applications
For completeness it should be mentioned that thermography has also found application in the medical and veterinary sector. The method is used, for example, for examination of deep venothrombosis, peripheral circulatory disturbances, breast cancer, migraine and for the examination of the appendix and tumors.

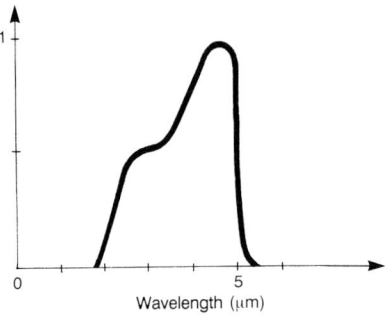

Figure 32.6. Optical coating. Thermovision© Type 870 uses a broad-band coating which enhances the relative response in the spectral region from 2-4 μm, an advantage where spectral filters are required, e.g. with petrochemical or laser appliations.

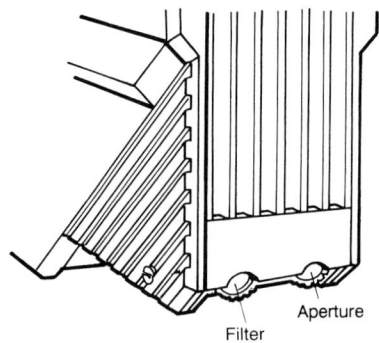

Figure 32.7a. Filter selection. On the back of the scanner there is a filter wheel which can be adjusted to three areas (apertures) marked 0, 1 and 2. The apertures are placed so that object temperatures in the region from -20 °C to +500 °C can be measured without using filters. Object temperatures up to 1500 °C can be measured using filters. The smallest aperture (2) is used for the highest temperature which can be measured.

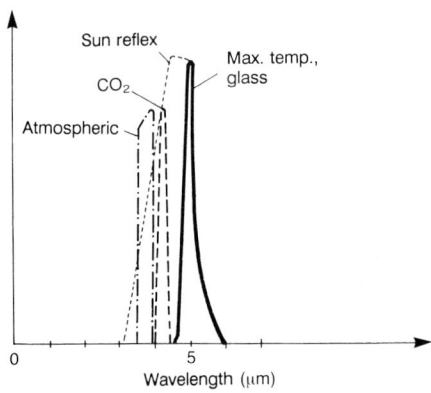

Figure 32.7b. Filters for IR systems where the "peak" power is utilized.

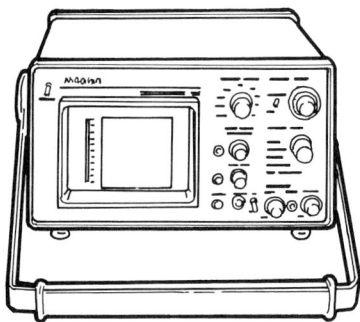

Figure 32.7c. The operating panel of a Thermovision© Type 870.

Training required/desired

The use of thermographic equipment generally requires only a brief introduction of a few hours' duration. This instruction is usually provided in connection with the delivery of the equipment.

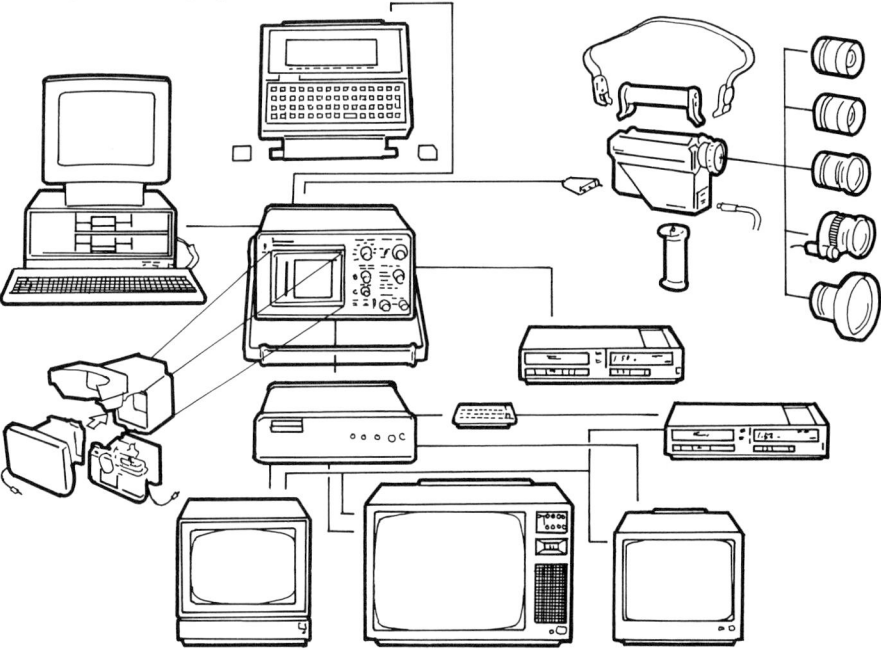

Figure 32.8. Examples of the accessories which can be attached to an Agema Thermovision© Type R870.

Thermography

Interpretation of the results, however, places greater requirements on experience with the system and insight regarding its operation. It is, therefore, recommended that to achieve the best possible results, the operator should participate in a short course dealing with the interpretation of results. Such courses are provided by the equipment manufacturers as required within the areas: electricity, building construction and process systems.

Thomas Dresler
Henning Pamperin

References

1) *Manual of remote sensing*, L.W. Bowden, American Society of Photogrammetry, Falls Church, Virginia, USA, 1975.

2) *Infrared detectors*, R.D. Hudson Jr. and J.W. Hudson, Halsted Press, New York, 1975.

3) *Infrared system engineering*, R.D. Hudson Jr., Wiley-Interscience, New York, 1969.

4) *Fundamentals of optics*, F.A. Jenkins and E.H. White, McGraw-Hill Book Co. Inc., New York, 1975.

5) *Elements of infrared technology: generation, transmission and detection*, P.W. Kruse, L.D. McGlaughlin and R.B. McQuistan, John Wiley & Sons Inc., New York, 1962.

6) *Thermal imaging systems*, M.J. Lloyd, Plenum Press Inc., New York, 1975.

7) *Electro-optical systems analysis*, K. Seyrafi, Electro-Optical Research Company, Los Angeles, California, 1973.

8) *Infrared radiation*, I. Simon, The Commission on College Physics, D. Van Nostrand Co. Inc., New Jersey, 1966.

9) *The detection and measurement of infrared radiation*, T.A. Smith, F.E. Jones and R.P.Chasmar, Oxford University Press, London, 1968.

10) *Practical applications of infrared techniques*, R. Vanzetti, Wiley-Interscience, New York, 1972.

11) *Infrared radiation*, A. Vasko, SNTL, Prague, 1968.

12) **The infrared handbook**, W.L. Wolfe and G. Zissis, Office of Naval Research, Washington, D.C., and Environmental Research Institute of Michigan, Ann Arbor, Michigan.

Chapter 33

Ultrasonic leak detection

NDE principle

If a component is subjected to an internal overpressure due to a gas, audible noise will occur when the gas flows out through leaks above a certain size. It is therefore often possible to detect leaks by listening for them. Unfortunately the human ear is not very sensitive to the sound which occurs due to small leaks.

One can, therefore, improve the method by using a listening apparatus which can record low noise levels and also detect the high frequencies (40-50 kHz) which occur due to turbulent flow.

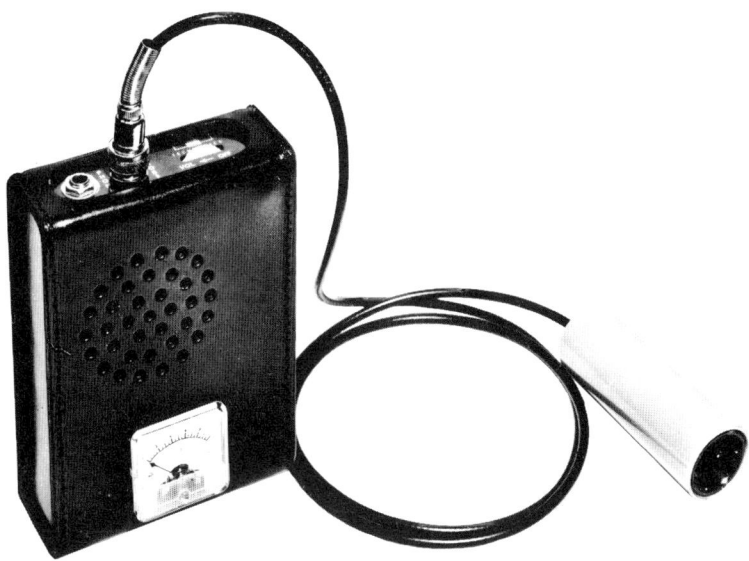

Figure 33.1. Listening apparatus with directional microphone.

The listening apparatus is equipped with a narrowly directionally sensitive microphone with an amplifier and a filter which can eliminate lower frequencies so that "ordinary" audible noise does not affect the equipment. The signals detected are converted by the apparatus to audible sound over a loudspeaker or in earphones. The equipment may also be equipped with a galvanometer whose reading varies with the input signal. The apparatus is supplied with a 9-volt battery, it is simple and solidly built which means low investment and maintenance costs.

Figure 33.1 shows a listening device with a built-in speaker, galvanometer and directional microphone.

Sensitivity
The sensitivity of the listening device is defined as the minimum leak rate which can be detected. Because the leak rate and thus also the noise from the leak depends upon the pressure difference between the two sides of the leak, the absolute pressure and temperature and the distance to the leak as well as upon the size and shape of the opening, there is no simple relationship between a given leakage (leak size) and the measured leak rate.

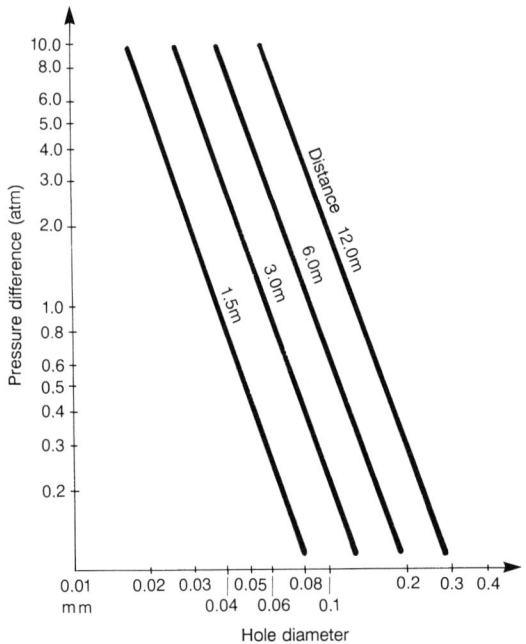

Figure 33.2. The maximum distance between the leak and the microphone of the listening apparatus for a given hole size and pressure difference.

Figure 33.2 gives an idea of the sizes of the smallest holes which can be detected depending upon the pressure difference between the two sides of the leak (in atmospheres) and the distance of the microphone from the leak (in metres).

Calibration
The sensitivity of the listening apparatus can be checked and adjusted by testing the equipment with a well-defined reference defect leaking at a given rate and at the desired distance from the microphone.

Advantages

- The listening apparatus can be used regardless of the type of gas.
- The apparatus can be used with both pressurized and vacuum systems.
- Great sensitivity can be achieved with this technique when used correctly.
- A leak can be detected quickly and at great distances.
- The method is simple and user-friendly.
- The method can be automated.
- Investment and maintenance costs are modest.

Limitations

The equipment is sensitive to high frequency background noise, which in some circumstances may make the examination impossible.

The method is not sensitive enough for certain monitoring tasks.

Development of the method

The listening method discussed here as applied to the detection of leaks in materials is relatively new.

Even though a number of unclarified questions remain in connection with the mechanisms which cause ultrasonic noise, the method has already become quite widely used and recognized.

Areas of application

The listening equipment is particularly useful in connection with leak testing in air pressure systems. Most air pressure systems operate at a pressure of up to about 6 bars. Any leak will, therefore, normally generate a loud noise which is easy to localize using such a listening device.

Another important area of application is the testing of piping systems for oxygen, acetylene and other gases such as are used in aircraft, welding workshops and hospitals.

Electrical discharges in high-voltage lines generate noise with the same high frequency as gas escaping through a leak and can thus be detected easily with the listening apparatus.

Instead of a microphone the listening apparatus can be provided with a contact probe for recording noise in materials. With this technique it is possible to monitor flows in pipes and tanks. A fault-free flow can be recorded, or incorrectly operating valves can be localized. This technique can also be used to localize leaks in buried pipes.

By using a noise generator, e.g. in connection with leak testing of a container (where very strict air-tightness requirements are not in force) it may not even be necessary to pressurize the component.

Figure 33.3. Leak testing of a piping system with the listening apparatus.

The noise generator emits high frequency noise, and the noise will penetrate even the smallest openings and be detected by the microphone of the listening apparatus which is placed on the opposite side of the container wall from the noise generator.

The above technique can also be used to check for leaks in doors, windows, ventilation channels, etc.

Practical applications

As noted in the previous section the listening apparatus can be used for many purposes. Figure 33.3 shows an example of leak testing of a piping system.

Training required/desired

Because the method is simple and user-friendly the operator can be readily trained in the use of the listening apparatus.

Hardy Hansen

References

1) *The ultrasonic leak detector: What every engineer should know.* Gerald L. Anderson. **Materials Evaluation, 42,** October, 1984.

2) *Information Lechsuch - Gerät für Gase.* Nekton ApS, Classen & Co.

3) *Leak detection with ultrasonics.* Andrex Radiation Products A/S, Bycosin AB.

4) *Non-destructive testing handbook, volume 1: Leak testing,* ASNT-ASM. Robert C. McMaster.

Chapter 34

Ultrasonics

NDE principle

Ultrasonic examination by means of the pulse-echo method can be described in its simplest form as a type of material radar. One transmits a very short, high-frequency pulse of mechanical oscillations into a material. The pulse has the character of sound oscillations, but it cannot be heard because of its high frequency and is, therefore, termed ultrasonic.

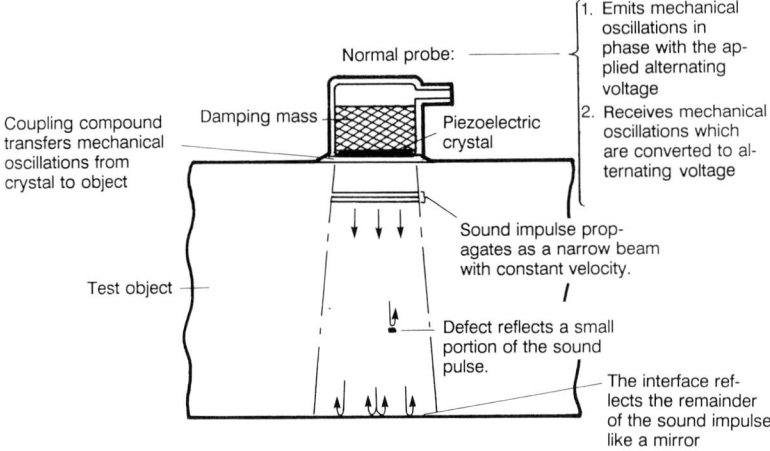

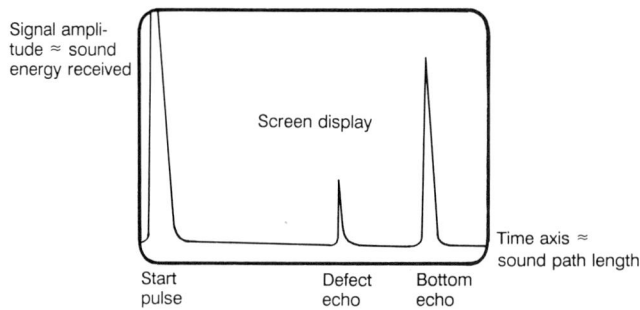

Figure 34.1. Principle of ultrasonic examination.

The pulses propagate in the material in a very narrow beam until they strike an interface such as the opposite surface of the test object or an internal defect. The pulses are entirely or partly reflected back to the transmitter, which now functions as a receiver. See Figure 34.1. The time interval between the transmission and the reception of a pulse depends upon the distance traversed, for the propagation speed is constant for a particular type of material. If one sends a pulse normal to the surface of a plate and measures the transmission time, one can compute the thickness of the plate. This is the principle on which the ultrasonic thickness gauge is based.

If one wants to localize internal defects in an object, one must know the starting point of the sound pulse and the direction in which it has travelled as well as the distance traversed. Larger defects are mapped by moving the probe or transducer which transmits and receives the sound pulses across the surface of the object.

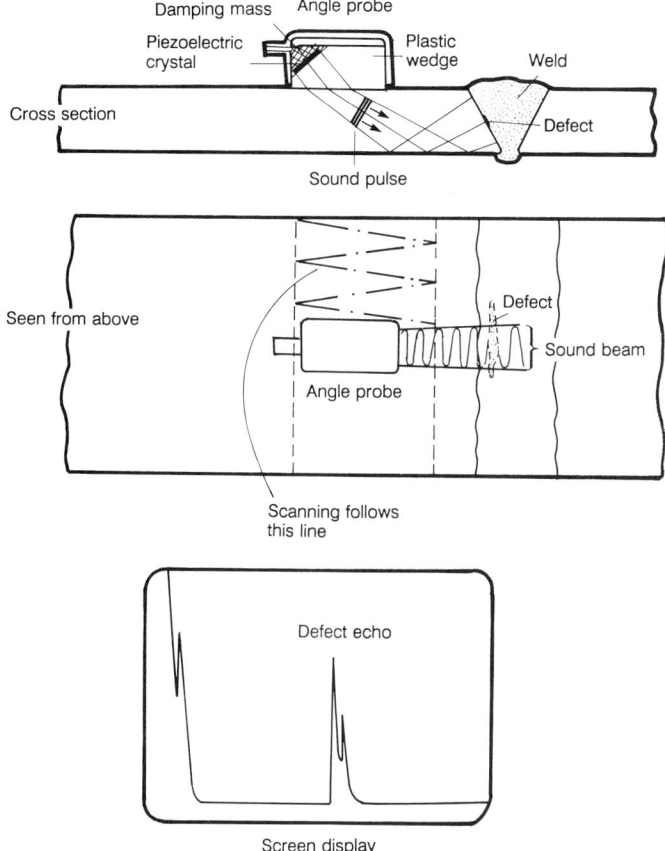

Figure 34.2. Ultrasonic examination of a weld with an angle probe. Cross section (above) and top view (middle). The lower picture shows the appearance of the screen image.

This search pattern is called a *scanning* and often follows a certain pattern, see Figure 34.2.

The emission is repeated so many times per second that the cathode ray tube of the apparatus displays an apparently stationary image. On the other hand there must be sufficient time between pulses that the reflected pulses, which can sometimes be reflected back and forth many times in objects with smooth surfaces and low attenuation, will have had time to dissipate before the next pulse is emitted.

The system is called the pulse-echo method and can be used both for the determination of the dimensions of a material when the speed of sound propagation is known and vice-versa for determination of the speed of sound in the material when the thickness of the material is known.

Development of the method

Examination by means of sound waves has been in use for many years in the form of simply listening to the ringing of an object (a coin, cup or wheel tyre) which is emitted when it is caused to oscillate by means of an appropriate blow.

During World War II electronics was developed dramatically, and – among other things – radar was developed.

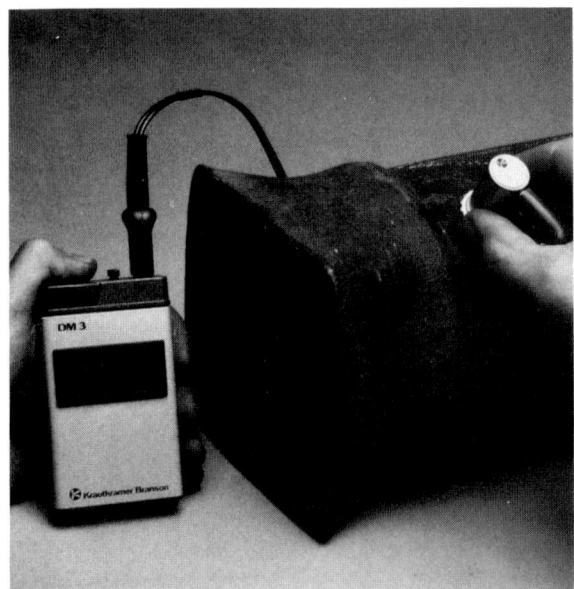

Figure 34.3. Thickness measurement in the range 1-100 mm in steel (Das Echo no. 32, p. 45).

Because it had also been discovered that slices of certain crystals, e.g. quartz, underwent mechanical oscillations when an alternating current was applied to the surface, it was possible to produce the first ultrasonic devices which operated by means of the pulse-echo method. Completely satisfactory equipment which worked according to this principle came onto the market in the early 1950s. Since that time they have been developed a great deal and applied to a variety of special purposes, but the basic principle behind their operation is still the same.

Today a large number of very small battery-powered thickness gauges are available which can be used to measure the thickness of metallic objects between 1 and 100 mm thick or more. See Figure 34.3.

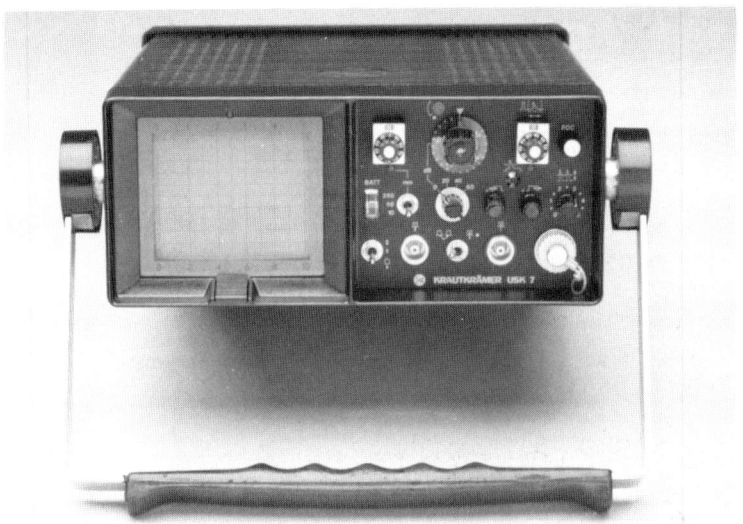

Figure 34.4. Lightweight ultrasonic equipment for manual examination of welds and materials (Das Echo no. 31, p. 49).

Slightly larger, battery-driven equipment for manual examination of welds and materials of all types which can be penetrated by ultrasonic waves (see Figure 34.4) are available.

Furthermore, a wide range of probes have been developed (see Figure 34.5) which are adapted for the various materials and geometries which are to be examined.

There are also large, mains power supplied devices which are used for automatic examination of mass-produced items and parts such as plates, pipes and sections. Such equipment is often equipped with several probes at once, and does not, as a rule, require constant observation of the display screen image during use, see Figure 34.6.

Figure 34.5. Standard probes for various purposes (Krautkrämer).

Figure 34.6. Multi-channel ultrasonic equipment for automatic examination (Das Echo no. 30).

Finally, there is ultrasonic equipment with automatic registration of all measurements both with respect to echo size as well as reflector location in an object, e.g. the P-scan system developed by the Danish Welding Institute, see Figure 34.7. This rather expensive equipment is used particularly at nuclear power stations where – in connection with completely automated scanning systems for the probes, *automatic scanners* – it can substantially reduce the time which the service personnel need to occupy areas with a high level of radioactivity.

Figure 34.7. P-scan recording ultrasonic equipment (Danish Welding Institute) connected to a fully automated weld scanner.

The P-scan system can also be used advantageously for a number of ordinary examinations because the automatic recording of all important signals significantly enhances the reliability and reproducibility of the method. See also Chapter 23 on **P-scan**.

Areas of application

Ultrasonic examination is used mostly today in the steel industry but also to a lesser extent for examining concrete (and for medical examinations).

The most important applications in the steel industry are for checking the quality of welds (see Figure 34.8), examination of rolled (Figure 34.9), extruded and forged (Figure 34.10) items for production defects, examination of cast items for casting defects and for checking dimensions (Figure 34.11).

Ultrasonic testing is particularly used in connection with the condition monitoring of systems in operation for thickness measurement, whereby reductions of wall thicknesses due to erosion (mechanical wear) or corrosion (chemical wear) can be ascertained (see Figure 34.12).

Figure 34.8. Ultrasonic examination of welds on pipes (Krautkrämer).

Figure 34.9. Ultrasonic examination of rolled plates for laminations (Das Echo no. 29, p. 54).

Ultrasonic examination can also be used for condition monitoring to measure the changes in structure which can occur in certain materials due to the action of high temperatures and hydrogen, for example, which can penetrate the material. In this connection measurements of attenuation coefficients for high frequencies (e.g. 10 MHz) may be performed, whereby the degree of hydrogenation is revealed by increased attenuation due to microcracks (see Figures 34.13 and 14). Or larger cracks within the material may have developed, which can be demonstrated directly by reflection of the sound waves. In both cases it is of great importance for the evaluation of the results of the examination that the initial condition of the new component is known so that one can determine the degree to which the condition has changed during the period of operation.

Ultrasonic examination is also well suited to the identification of cracks in connecting bolts in flange junctions, clutches and machines, and in stay bolts in diesel engines and presses. Here it is particularly advantageous that the examination can generally be carried out without removing the bolt, as it only requires access to one end surface of the bolt.

Figure 34.10. Examination of forged turbine blades (Das Echo no. 32, p. 33).

Figure 34.11. Examination of cast wheels (Das Echo no. 32, p. 33).

Figure 34.12. Thickness measurement of piping systems at an oil refinery (Das Echo no. 32, p. 45).

Figure 34.13. Hydrogenated material in a boiler pipe. A thick, laminated coating of magnetite is seen on the water side, and behind this the steel has become so brittle that a fracture has occurred (Materials Evaluation no. 43, p. 1164).

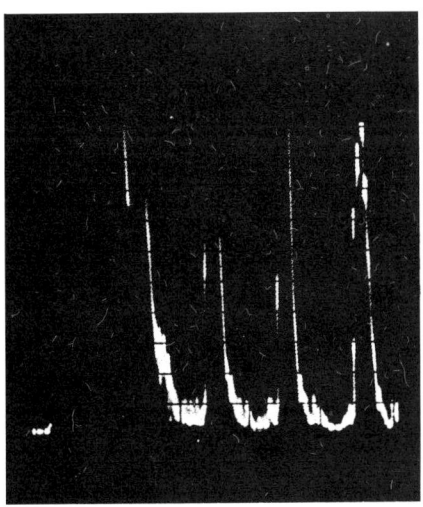 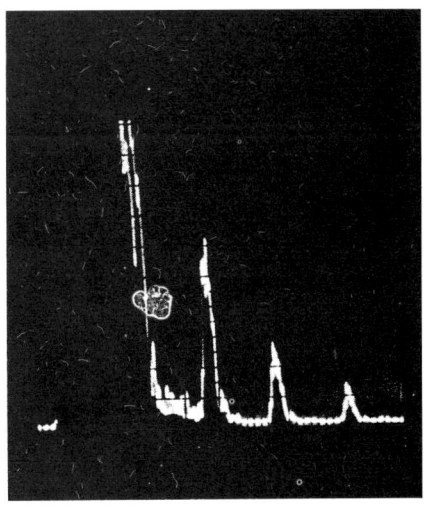

Figure 34.14. Ultrasound indications from undamaged (left) and hydrogenated (right) piping walls. The latter shows strongly enhanced attenuation (Materials Evaluation no. 43, p. 1166).

It is worth noting that the cracks are often due to incorrect tightening of the bolt so that the pre-tension is either too low or too high. This incorrect pre-tension is due to the use of torque wrenches, but also here ultrasonic examination provides the possibility of achieving much more accurate tension. The method is in fact accurate enough to allow the measurement of the small physical elongation which occurs when a bolt is tightened (see Figure 34.15). When the elongation is known, one can compute the tension in the bolt using the elastic modulus of the material. Incidentally, the measurement of the increase in length using ultrasonics is facilitated by the fact that the speed of sound in the bolt under tension is slightly lower than in the unloaded condition. Therefore, the pulse takes a little more time to move through it. Because the bolt is also slightly elongated due to tension and sound moves more slowly as well, the ultrasonic measurement indicates an apparent elongation which is 3-5 times as large as the real physical elongation. One must of course make corrections for this effect.

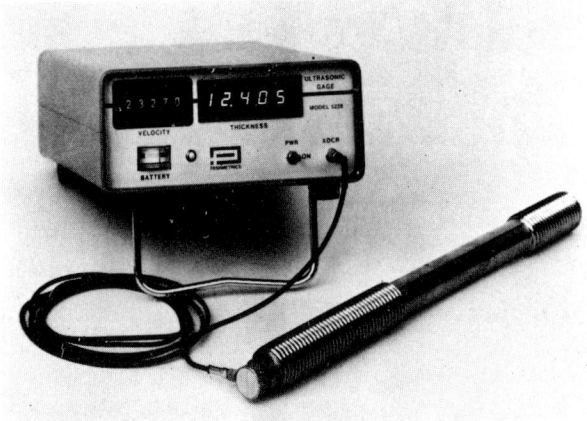

Figure 34.15. Ultrasonic instrument for the measurement of the elongation of bolts (Jan Houlberg).

Ultrasonic examination is also used to check the bonding between two different materials, e.g. white metal on steel in bearing liners, tin and silver brazed to copper in braze metal joints (see Figure 34.16), adhesive bonding in honeycomb panels of aluminium and plastic for aircraft constructions (see Figure 34.17), fibreglass reinforced plastic to steel in windmill blades.

Examinations of this sort are carried out in connection with production monitoring as well as condition monitoring.

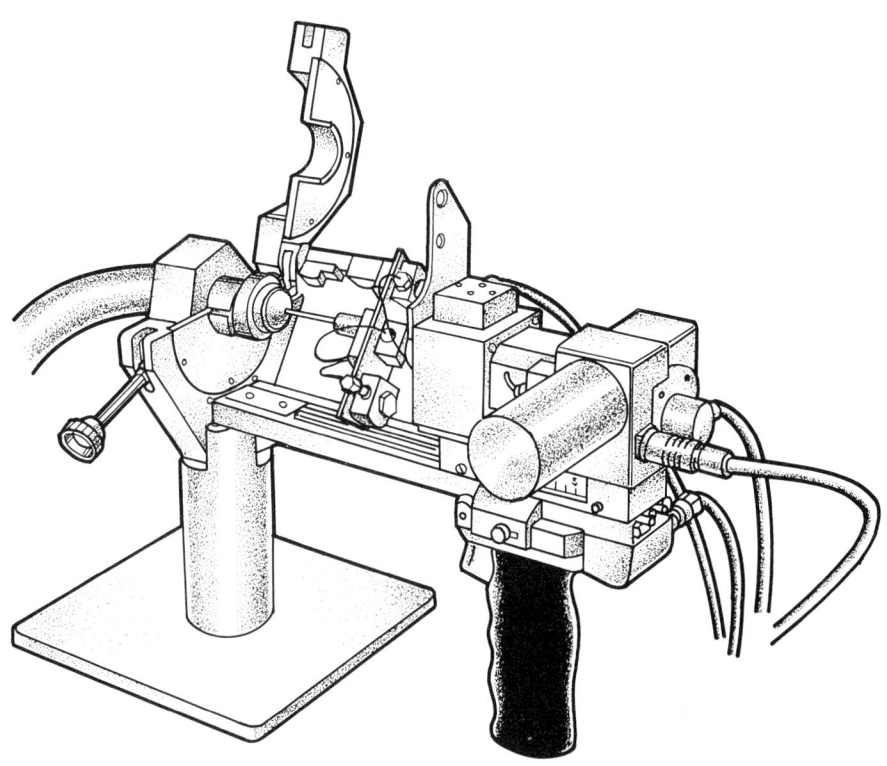

Figure 34.16. Automatic ultrasonic scanner for examination of brazed joints on pipe ends (Danish Welding Institute).

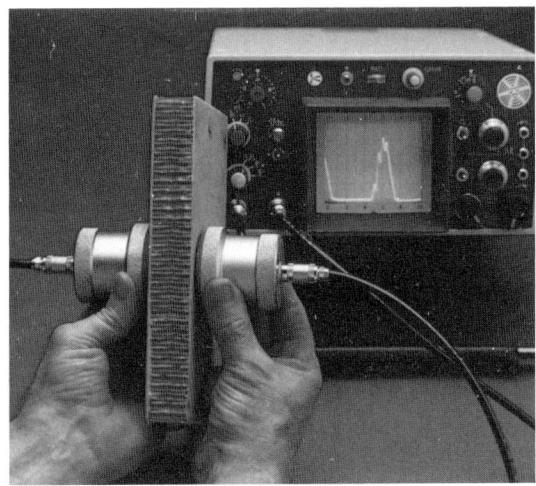

Figure 34.17. Examination of adhesive bonding in a honeycomb panel (Das Echo no. 32, p. 34).

Training required/desired

In spite of the dramatic evolution which has taken place with respect to apparatus, ultrasonic examinations place increasing demands on the training and practical experience of operating personnel. Even with thoroughly trained personnel the method is relatively uncertain when the size of the defect in a weld is to be measured, and the risk of overlooking smaller defects is quite significant.

Therefore, it became apparent at quite an early stage that it was necessary to train and to examine the personnel who would work with the method.

Similar training programs for personnel for the other NDT methods has also been introduced, and documentation for satisfactory completion in the form of a NDT certificate has become a requirement from many government agencies and clients, where the safety and environmental consequences of failures of the structures in question are very great.

nordtest

NDT LEVEL 2 DIPLOMA

Number: Test method: Category:

Expiry date:

This is to certify that

Address:

Date of birth: Personal code No.:

fulfils all requirements of NORDTEST DOC GEN 010 regarding training, experience and physical fitness and has successfully completed an examination in this method and its application for:

☐ Examination of all objects within the category

☐ As above plus objects specific to production/in-service inspection* of:

NORDTEST EXAMINATION CENTRE:

Address:

Date of issue:

(Date) (Signature NEC examiner)

(Date) (Signature external examiner)

*Delete as appropriate

nordtest

NDT LEVEL 3 CERTIFICATE

Number: (same as diploma) Category:

This is to certify that our employee:

Date of birth: Personal code No.:

is QUALIFIED for and AUTHORIZED to carry out and direct nondestructive testing activities with the methods shown below.

NORDTEST ACCEPTED TEST-CENTRE:

Adress:

(Date (Signature NAT)

NATIONAL STEERING COMMITTE ACKNOWLEDGEMENT

This is to certify that we have recieved the following documentation:

- Satisfactory qualifications of the holder
- Consent that the holder will strictly follow the rules of NORDTEST DOC GEN 010 for responsibilities and duties of certified level 3 personnel

The documentation has been reviewed and accepted

NATIONAL STEERING COMMITTEE in: (Country) (Signature NSC)

Endorsements

Date of	Radiography	Ultrasonics	Magnetic particles	Liquid penetrant	Eddy current	Leak detection
Issue						
Expiry						
Sign NSC						
Prolong						
Sign NAT						

Figure 34.18. Nordtest diploma and certificate as documentation for the qualifications and recognition of NDT personnel, particularly in connection with production monitoring. The training course is, however, required for condition monitoring of pressurized systems.

The designation NDT is used here because the training is particularly directed at production control. It should, however, be noted that the instruction is also required in connection with the condition monitoring of systems under pressure.

In Denmark it has been decided to accept a joint Scandinavian system for the examination and certification of NDT personnel called the *Nordtest system*. It has operated since 1978 and is based upon three levels of qualification termed levels 1, 2 and 3.

The fundamental idea of this system is that one should acquire a certain degree of practical experience and theoretical knowledge at each level and that one be tested at a national examination centre. After passing the examination, the candidate receives a diploma, see Figure 34.18. Then the industrial company where the candidate is employed can issue a certificate which can be regarded as a sort of driver's licence for the candidate and at the same time a guarantee from the firm that it stands behind the measurements which result from the employee's examination activities, see Figure 34.18.

Aksel Feddersen

References

1) *Werkstoffprüfung mit Ultraschall.* (Materials testing with ultrasonics.) J.H. Krautkrämer, Springer Verlag, Berlin 1961.

2) *Handbook on the ultrasonic examination of welds.* Doc 115/11W 527-76 from the International Institute of Welding.

Chapter 35

Vibration monitoring

NDE principle

All rotating machines vibrate. Put your hand on any rotating machine, and you will be able to feel the machine's vibrations. These vibrations contain valuable information about the condition of the machine. By analysing the signals one can in many cases predict whether a machine is about to develop a fault. One can thereby avoid break-downs, for a possible defect can be corrected during the next scheduled pause in production.

Figure 35.1. All rotating machines vibrate.

For simple machines this vibration monitoring can be accomplished using very simple and inexpensive instruments of the sort shown in Figure 35.2. For example, with such an instrument one can identify bearing defects (how is described later).

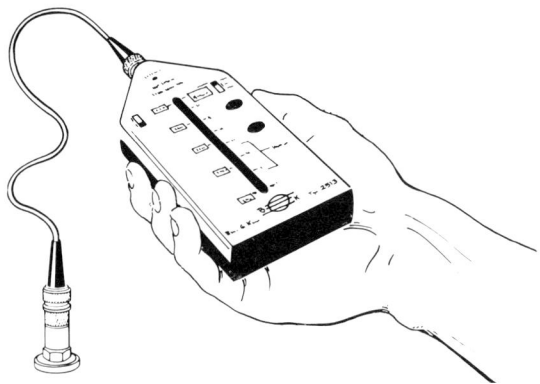

Figure 35.2. A simple, inexpensive instrument for vibration monitoring.

In the case of more complicated machinery more advanced methods must be used. Figure 35.3 shows a type 2515 instrument from Brüel & Kjær which is a combined fault detection analyser and a diagnostic analyser (FFT-analyser).

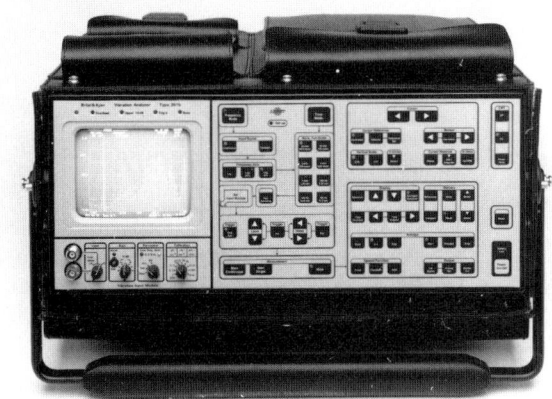

Figure 35.3. FFT analyser for combined fault detection and diagnosis.

For extremely expensive equipment or where break-downs must be avoided for (almost) any price, monitoring equipment can be permanently attached to the machines in question, and the analysis can be carried out from a common control room using a tabletop computer, such as an IBM compatible PC, to carry out the automatic analysis and to alarm service personnel when the possibility of a machine defect has been detected. Figure 35.4 shows the basic setup of such a system.

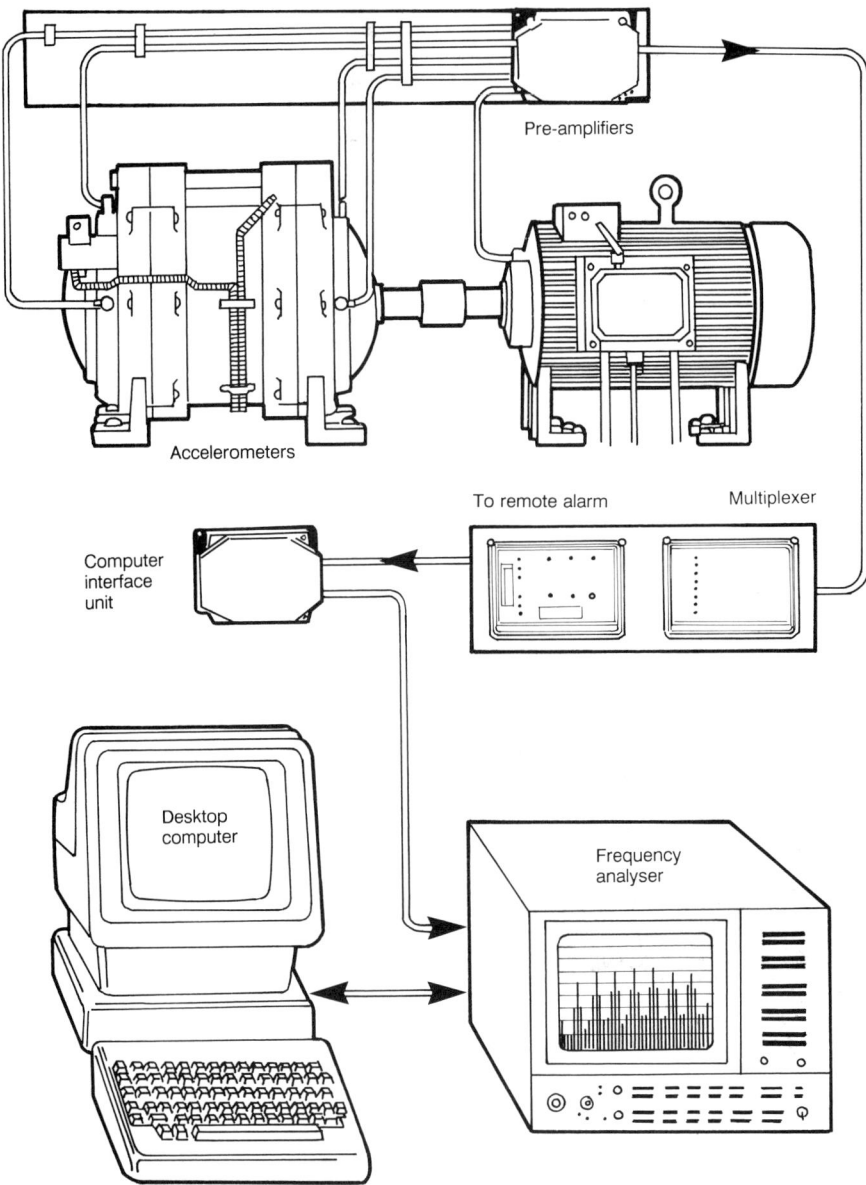

Figure 35.4. Monitoring equipment permanently attached to expensive machinery can provide constant monitoring and alarm services.

Economics

As mentioned earlier the main objective is to avoid unplanned interruptions of service due to machine defects, for they can be discovered at such an early stage that repairs can be planned and carried out during a pre-scheduled outage in operations.

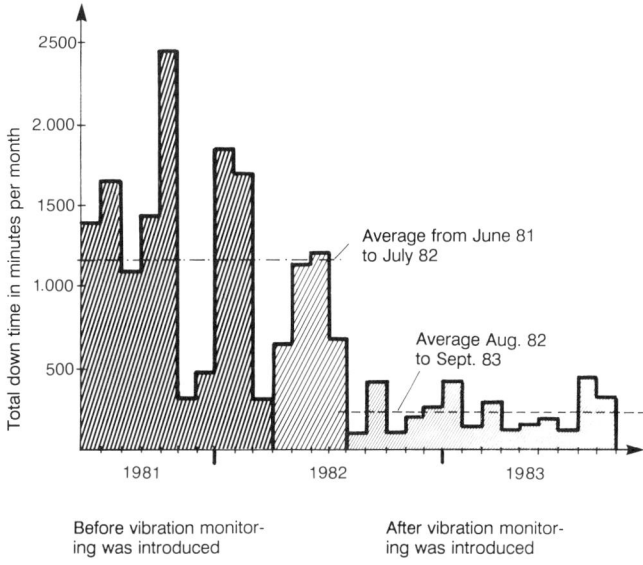

Figure 35.5. The total down time is reduced in this case by 81% due to the introduction of vibration monitoring.

There are examples of situations where bearing defects (in both journal bearings as well as roller bearings) have been detected several months before replacement was definitely required.

Figure 35.5 shows lost production time in operating minutes per month at a Canadian paper mill. During 1984 the firm introduced a maintenance system based upon vibration analysis and thereby reduced unplanned interruptions in service by over 80%.

Other examples show that the investment in vibration monitoring equipment is often paid back within the first year of operation.

Development of the method

Utilizing the vibration signal from a rotating machine to evaluate the condition of a machine is a technique which is just as old as rotating machines themselves.

At a very early stage operators learned to press the blade of a screwdriver against the bearing housing of the machine and the shaft against the ear to transmit the vibration signal on to "vibration analysis" in one's brain.

Through the years this method proved to be so effective that the electronics industry decided to develop instruments which could perform the same function, but reproducibly, i.e. independently of the memory or state of mind of an individual.

The first instruments which appeared on the market were vibration measuring devices which indicated the RMS value of the speed in the vibration signal over the entire usable frequency region, which at the time was from about 10 Hz to about 1000 Hz. The measurements were based upon electromagnetic transducers which, just as a phonograph needle, provide a voltage which is proportional to the speed of the sensitive portion of the transducer.

Later the accelerometer was introduced, i.e. a transducer which provides a voltage which is proportional to the acceleration to which it is exposed. At the same time the electronic integrating circuit appeared, making the measurement of oscillation velocity or displacement possible by means of accelerometers.

These early measurement techniques, which today are widely used, turned out to be very useful. Trained operators can, using small, hand-held instruments, often detect future bearing defects or imbalances before other parameters such as temperature or visible vibrations indicate any abnormality. Using this method, however, often means that repair can be carried out the same day that the fault was detected.

It was recognized at an early stage that there was a need to utilize all of the information which was present in the vibration signal and that more advanced methods were required. If only a simple frequency analysis had been carried out, many weeks warning would often be achieved, and the repair could be included in a longer term maintenance plan.

The desire to achieve better frequency analysis led, therefore, naturally to the development of the fast frequency analysers – the so-called FFT (Fast Fourier Transform) analysers.

The first FFT analysers were, however, both too big and too complicated to use in industry, and furthermore, they were very sensitive to water, dust and shocks. They were in fact not particularly suitable to an industrial environment. In spite of this they were used anyway for a decade or so.

On the basis of these facts the battery-powered FFT analyser, such as the one shown in Figure 35.3, was developed.

Areas of application

Figure 35.6 shows a typical spectrum, i.e. a graphical display of how the energy in the vibration signal is distributed over the measured frequency band, in this case from 0 to 10 kHz.

The basic idea behind frequency analysis is that the vibrations at various frequencies can be related to various sources in the machine and provide, therefore, an indication of where the condition of the machine has changed since the last analysis was performed.

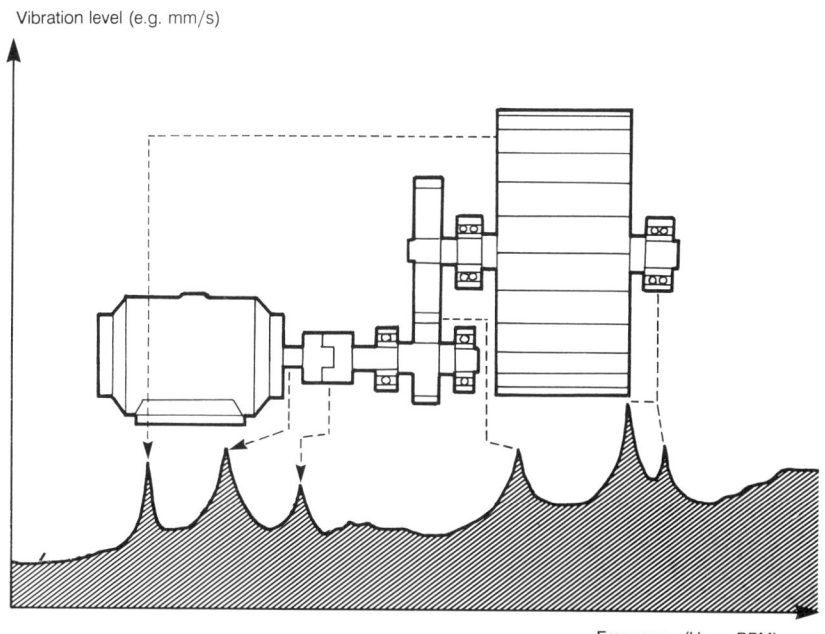

Figure 35.6. Frequency analysis – a machine monitoring tool.

Low frequency region

We define the *low frequency components* as the "peaks" which are observed in the spectrum near the rotational frequency (revolutions per second) of the shaft and up to three to five times this value.

When we refer to vibrations in the following, we mean vibrations measured radially (i.e. normal to the rotational axis of the shaft) unless otherwise specified.

The vibrations at the rotation frequency are often due to imbalance or misalignment.

Vibration monitoring

At the second harmonic of the rotation frequency (i.e. twice the rotation frequency) we find components which also can be due to misalignment but can also be caused by a static deflection of the shaft. We will see later how phase measurements can help to distinguish these various types of defects from one another.

In connection with axial measurements (measured parallel to the shaft) at a pressure bearing, a strong enhancement of the second harmonic of the rotation frequency will also indicate misalignment.

By comparing previous measurements with newer ones, one can thus determine – among other things – if the machine in question has become more imbalanced or out of alignment.

Journal bearings also give low frequency signals if defect conditions arise. The main problem here is hydrodynamic instability in the oscillating system consisting of the shaft, oil film, and bearing housing.

The phenomenon is called *oil whirl* and consists of vibrations which arise because the centre of gravity of the shaft rotates within the tolerance which there always is in a journal bearing at a speed which is often just under half the rotational speed, typically between 40% and 50% of the rotational speed.

Hysteresis whirl is another self-generating phenomenon. It consists of strong vibrations which arise at a critical speed when the rotor speed approaches the critical speed in an attempt to pass it. This can be due – among other things – to small mechanical motions in the mountings of the axle system. These small slippages then excite the characteristic frequency of the system: the critical speed.

A third type of vibrations in rotating machines with flexible rotors is induced by variations in gas or steam flow. It has a different background but occurs at the same frequency as hysteresis oscillations.

The last type of defect we will examine here is mechanically loose assemblies.

Often loose assemblies, e.g. in bearing housings, generate vibrations below the rotation frequency and between the rotational frequency and its harmonics, often referred to as the "half harmonics", "one-and-a-half harmonics", etc.

The phenomenon can be seen in Figure 35.7 where the bearing bolts had not been properly tightened after a repair.

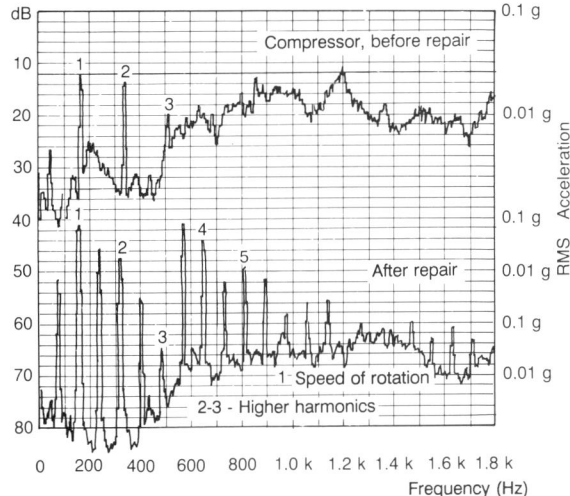

Figure 35.7. Half-harmonics occur due to mechanically loose components.

Middle frequency region

At higher frequencies in the spectrum we find components which are due to the teeth in the gearbox. The gear meshing frequency is defined as the rotational speed of the shaft times the number of teeth on the gear. The region near the meshing frequency and the lower harmonics of this frequency are referred to in this connection as the *middle frequency region*.

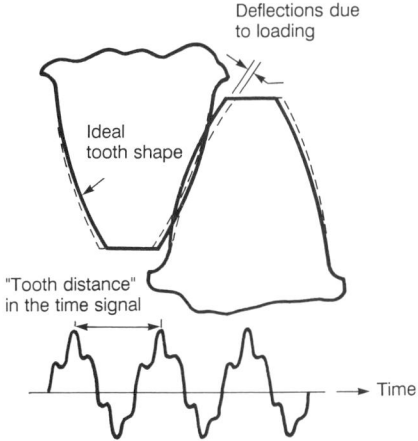

Figure 35.8. Typical signal from a set of gears.

Vibration monitoring 347

A new and properly functioning gearbox will have a high level of vibration at the meshing frequency, but not just here.

Due to the loading of the teeth, the teeth which are meshing will bend slightly, and they will bend differently according to whether one or more sets of teeth are meshing.

The bending will also vary, because the torque transferred differs depending upon whether the meeting point between two teeth is close to the root of the drive wheel tooth or near its top.

The vibration signal will, therefore, in principle appear as in Figure 35.8, i.e. we will not only observe the meshing frequency but also its higher harmonics as shown in Figure 35.9.

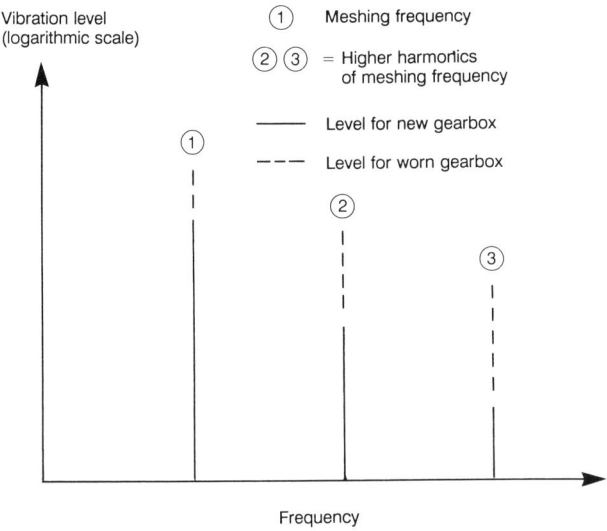

Figure 35.9. Frequency spectrum due to the meshing of gears.

Furthermore, new gearboxes will often show relatively strong vibrations at yet another frequency often called *the ghost component* because the vibrations are a remnant from the "previous generation" of gears. The ghost component is a strong vibration level at a frequency which corresponds to the number of teeth present on the index gear when the gear in question was cut. It arises due to geometric imperfections in this index wheel and is not a sign of defects in the gearbox.

In contrast to the meshing frequency "the ghost component" will not vary in strength due to varying loads. It will become weaker as time passes due to wear.

When the gearbox is worn, as shown in Figure 35.10, the teeth are worn at the spots where a sliding motion occurs between the two meshing gears and along the entire surface of the tooth *except* precisely on the pitch point. Here only a pure rolling motion occurs and the wear will, therefore, normally be minimal.

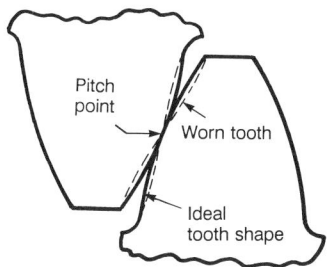

Figure 35.10. *The effect of wear on the meshing of gears.*

This means that a worn gearbox will show an increased level of vibration at the second harmonic of the meshing frequency, and probably also at higher harmonics, but not at the meshing frequency itself. This is illustrated in Figure 35.9.

But we are still only considering a correctly operating but worn gearbox.

If, on the other hand, the gearbox acquires a significant defect, e.g. due to a cracked tooth, there is nothing wrong with the operation of the gearbox, but after a period of operation the defective tooth may fall off, possibly falling into the meshing of the gears and destroying the entire gearbox.

This cracked tooth will, however, bend more when loaded and will, therefore, emit a different and larger vibration signal once each revolution.

The vibration signal can thus be regarded as a signal from an operational but perhaps worn gearbox with a signal which can be represented as a small pulse once each revolution superimposed upon it. See Figure 35.11.

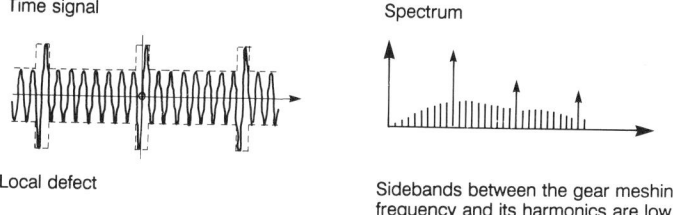

Figure 35.11. *The effect of a beginning defect upon the vibration signal.*

It can be demonstrated mathematically that a series of pulses which occur with a constant period, as shown in Figure 35.12, will produce a spectrum identical to the spectrum from a single impulse but separated into lines, and that these lines will be separated by a distance corresponding to the impulse frequency.

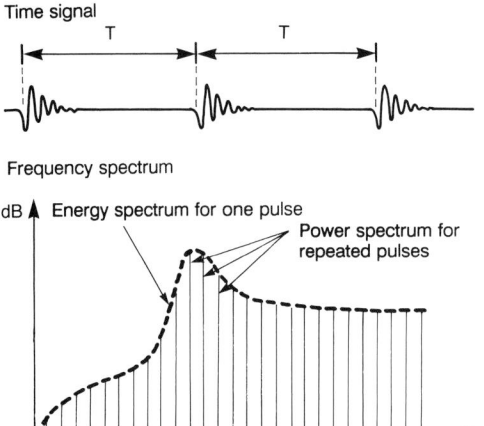

Figure 35.12. Frequency analysis of repeated pulses.

This means that a defect under development in a gearbox where the defective wheel rotates 24 times per second (i.e. at 24 Hz) will appear as a line spectrum with lines 24 Hz apart. In other words a defect will *not* appear as an increased vibration level at the meshing frequency but as a practically global enhancement of the background level or "the grass" between the high components at the meshing frequency and its harmonics.

Figure 35.11 is shown with a logarithmic amplitude scale, i.e. the vibration level at the meshing frequency and its harmonics can be 10 to 100 times stronger than the signal which indicates that something is about to go wrong. Therefore, a measurement without frequency analysis as described in the introduction (e.g. between 10 and 10,000 Hz) would not detect this defect.

The only way to find these incipient defects is to perform some type of frequency analysis and compare the levels between the gear meshing frequency and its harmonics from previous measurements with the most recent ones. How such a comparison of spectra can be automated will be described later.

As the signal "spreads out" from the original impulse to an enhanced vibration level over a broader region, the vibration levels around the meshing frequency and its harmonics increase and close in upon them.

Figure 35.13 shows a schematic illustration of the spectrum due to a more serious defect in a gearbox.

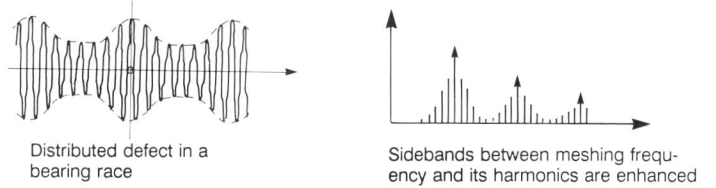

Distributed defect in a bearing race

Sidebands between meshing frequency and its harmonics are enhanced

Figure 35.13. The effect of a widespread defect on a vibration signal.

Figure 35.14 shows measurements from a gearbox, first in the original, new condition and later in damaged condition.

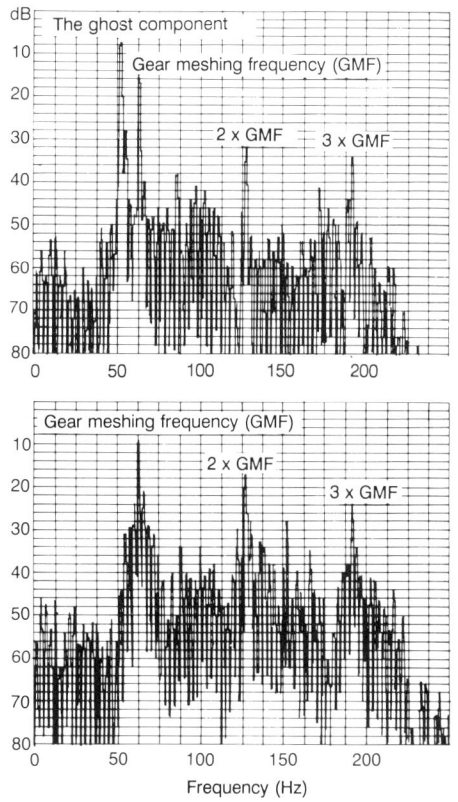

Figure 35.14. The signal from a gearbox. (a) In new condition. (b) Defects appearing.

(In order to achieve clarity in these spectra a *time averaging technique* is used, a method which will be described later and which does not matter for the conclusion drawn here.)

Vibration monitoring 351

The spectra display the following symptoms:

- Sign of ordinary wear, for the levels at the second and third harmonics of the meshing frequency have increased considerably.

- Sign of ordinary wear because the "ghost component" appears to have disappeared.

- Sign of an incipient local defect, for strong increases occur in the level between the meshing frequency and its harmonics.

Thus it will often be possible to evaluate the condition of a gearbox during operation.

The high frequency region

The upper end of the frequency spectrum on the original illustration, Figure 35.6, has not yet been discussed. The frequency components we find here are in this connection referred to as the *high frequency components*.

We will typically find signals from ball and roller bearings up here in the frequency region from 2 kHz to 20 kHz, depending upon the rigidity of the mechanical construction.

An incipient defect in a ball or roller bearing will typically be a small crack or a little corrosion pit in either the inner or the outer bearing race or in one of the balls.

This crack will cause a little bump or shock each time a ball or roller passes. Because the crack is very small at first, these small shocks will be very brief and sharp like tiny hammer blows.

These small hammer blows will excite the bearing race which will, therefore, vibrate for a brief moment at its own resonant frequency until the oscillation dies out due to the damping of the mechanical structure. The situation can be compared with small hammer blows on a bell. No matter how hard or how rapidly the bell is struck, the bell will always ring with the same tone, corresponding to the resonant frequency of the bell.

The mechanical structure acts in this manner like a mechanical amplifier, and it is precisely this phenomenon which is used for the detection of bearing defects.

If the defect is on the stationary ring, the crack will always be excited by balls with the same pressure, whereby a row of pulses will look like those in Figure 35.15 (left).

If on the other hand the defect is on the rotating ring, the balls which excite the crack will have the greatest pressure at the "bottom" of the race and lesser pressure when the crack is "uppermost" in the race. The row of pulses will then appear as shown on the right in Figure 35.15. This "amplitude modulation" is important when *envelope curve analysis* technique is used. This technique is discussed later on.

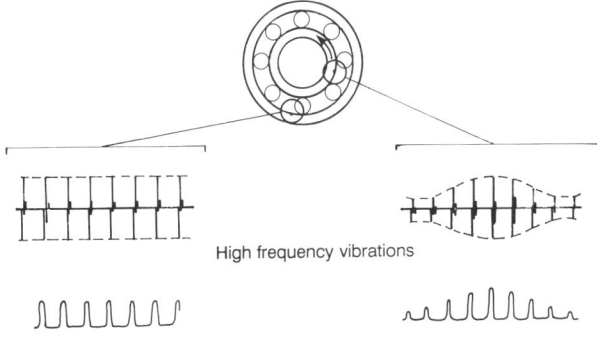

Figure 35.15. Defects in a bearing race.

Because these cracks are very small (they can often not even be seen with the naked eye) the pulses which they cause will be very small and sharp and, therefore, contain energy from low to often very high frequencies.

Because the lower portion of the spectrum is dominated by much more energetic signals, namely the signals described earlier for the rotation frequency region and the gearbox frequencies, an incipient bearing race defect must be detected in the high frequency region where the spectrum does not contain other signals.

This fact combined with the fact that the small pulses excite mechanical resonances which also are quite high in frequency make it possible to detect these defects by monitoring the vibration level near one of the resonant frequencies of the mechanical structure.

Figure 35.16 provides an unusually clear example of this phenomenon because the energy from the defective bearing race is so large that it also appears in the lower and middle frequency ranges.

It is, however, clear that a little peak is present between 9.6 and 11.3 kHz precisely corresponding to a mechanical resonance and this is clearly an early sign of a bearing defect.

It is extremely easy to ascertain where these mechanical resonances are located. While the machine is at rest it is struck with a hammer, and the vibrations which arise from the machine are analysed in an FFT analyser. A battery-powered analyser can be brought almost everywhere, and this *impact test* can be carried out in just a few minutes.

Vibration monitoring

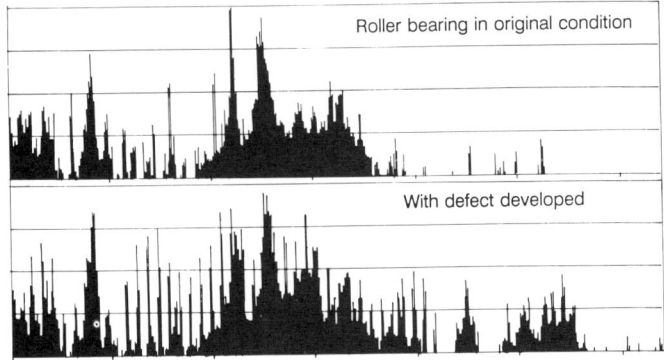

Figure 35.16. (a) Spectrum for roller bearings races in original condition. (b) Spectrum for roller bearing races after flaws have developed.

Figure 35.17 shows that there is more than one resonant frequency and that in principle it does not matter which is used, for a bearing race defect will probably be visible at all points, but with the same amplitude. (Where bearing races run alone in simpler machines – where other shock-like signals such as cavitation, pulses from pressure lubrication, etc. do not occur – other, simpler methods can be used, e.g. rms/peak factor measurement, with an instrument as shown in Figure 35.2.

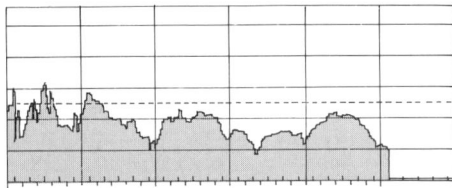

Impulse test on bearing support with motor not rotating - Frequency axis 0-20 kHz

Figure 35.17. Impulse test of a bearing race.

Summary

It has been described in the foregoing sections how defects are revealed in the frequency domain for the most common machine components.

We have seen how the low frequency region, defined as the region around the rotation frequency up to about the fourth to the sixth harmonics of this frequency contain information about imbalance, alignment defects, static shaft bending and instability in journal bearings.

We have seen how the high frequency region, defined as the region where structure resonances are dominant often starting between 1 and 5 kHz and extending up to 10 or 20 kHz, sometimes even higher, contain information about incipient defects in roller and ball bearings.

We have seen how the middle frequency region, defined as the region lying between the two other regions, contains information about the state of wear of a gearbox as well as incipient defects.

Fault detection

In previous sections it has been described how the vibration signal contains valuable information about the condition of simple machine components such as ball, roller and journal bearings, gearboxes, and general defects such as imbalance, alignment errors, static bending and mechanically loose components.

This information can be used for monitoring machines with the goal of predicting breakdowns long before they otherwise would manifest themselves. This makes it possible to plan repairs well in advance and to integrate them into the production plan whereby loss of production due to machine stoppages can to a large degree be avoided. See Figure 35.5.

In order to avoid lengthy and complicated signal analysis in the daily work routine, the following requirements must be fulfilled by a monitoring system:

- The system must detect most defects as quickly as possible.

- The system must give as few false alarms as possible.

- The system must provide sufficient information to allow a reasonable guess to be made about *what* is about to go wrong and not just *that* something is about to go wrong. This will give management the option of deciding whether a more thorough analysis should be undertaken, possibly requisition specialists, order spare parts, etc.

- Error detection must be simple enough to use that a specially trained operator can carry it out. The daily routine work of vibration monitoring should not be performed by highly trained personnel.

Broad-band measurement

For many years efforts have been made to locate defects by comparing the vibration level measured in the broad-band from 10 Hz to 1000 Hz using previous measurements on the same machine or with publicly available standards such as the international ISO 2372, the British BS 4675 and the German VDI 2056 as shown in Figure 35.18.

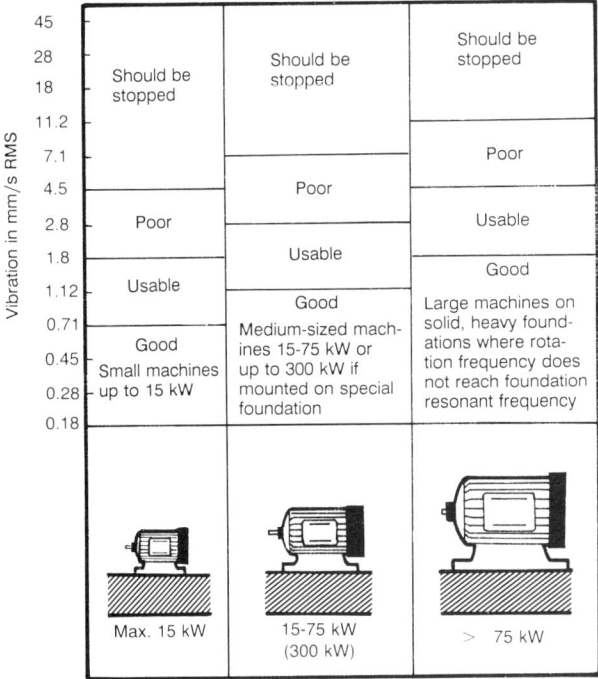

Figure 35.18. Vibration evaluation chart from the German VDI 2056 standard.

The idea behind the standards is that comparable machines, e.g. electric motors, on the same order of size, e.g. from 15 kW to 75 kW, will have the same vibration level and can, therefore, be compared. There are, however, a number of fundamental difficulties with this method.

Early in this section on **Areas of application** it was described how very energetic signals due to imbalance will almost always be present. Since they are the strongest, one will not be able to detect defects which manifest themselves at other frequencies but at lower levels using a broad-band measurement. These signals will be overwhelmed by the strong signals.

Another problem which arises when using a standard is that there is a big variation from machine to machine in how much the signal is damped between its source (e.g. in a bearing race) and the measurement point (e.g. on the surface of the bearing housing).

Figure 35.19 shows the results of an experiment. It is apparent that the same machine type of the same order of size will experience attenuation between the bearing and the selected measuring point which can vary by a factor of 1000. That is, on two machines *with the same defect* one would in theory be able to measure a 1000 times higher vibration level on one compared to the other. Therefore, the limits specified by the standards are quite unreliable.

For these reasons we must conclude that fault detection (defect detection) must be undertaken by comparing measurements with earlier data in order to detect *changes* in the condition of the machine of interest. Measurements from other machines can *not* provide any indication of the condition of a machine.

A broad-band measurement using portable as well as permanently installed equipment will only provide an indication of those defects which will yield a strong vibration signal (e.g. imbalance or misalignment).

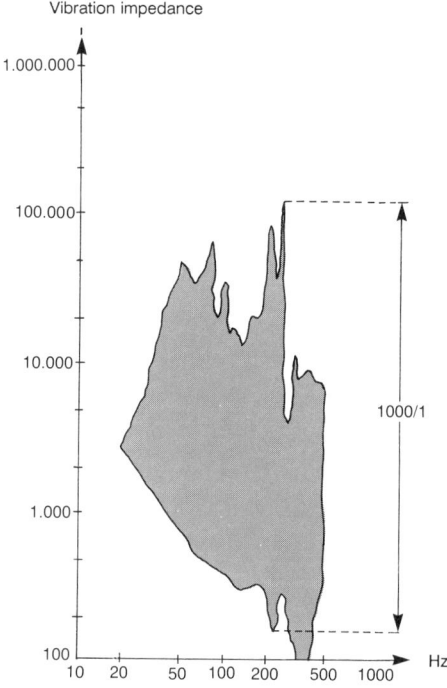

Figure 35.19. *Within this region lies the attenuation measured on 14 different ethylene compressors. Thus the absolute vibration level cannot be used as a criterion. (AMSE paper no. 71, Vibration 96, September 8-10, 1971, Toronto, Canada.)*

In the case of bearing or gearbox defects this method will provide a very late warning, and it is possible that temperature sensors or other process parameters will detect the defect well in advance.

Therefore, in order to achieve the earliest possible defect detection, some form of frequency analysis must be undertaken.

1. Early discovery of a defect provides longest possible advance warning about necessary repair

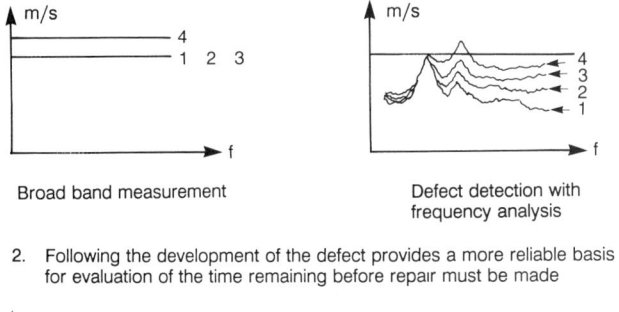

Broad band measurement   Defect detection with frequency analysis

2. Following the development of the defect provides a more reliable basis for evaluation of the time remaining before repair must be made

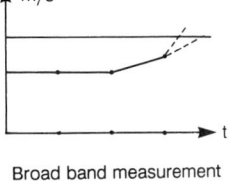 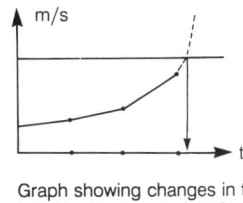

Broad band measurement   Graph showing changes in the individual frequency bands

Figure 35.20. Frequency analysis used for the detection of defects.

Frequency analysis

Figure 35.20 shows the principle involved in frequency analysis. If a defect develops and manifests itself at a frequency different from the one at which the highest level of the spectrum lies, it will not be discovered using a broad-band measurement before it turns up near this highest peak. The first three measurements performed using the broad-band technique will not indicate any increase (the vertical axis is logarithmic). Only after the fourth measurement will there be any indication that something is about to go wrong.

Using frequency analysis on the other hand one will already be able to see that something is changing after the first measurement.

FFT analysers

As indicated in the foregoing one must use some form of spectrum comparison to achieve the earliest possible recognition of a defect in a machine, and it is therefore tempting to use the "save and compare" functions which are available on the most common single channel FFT analysers. Here, however, two problems are encountered.

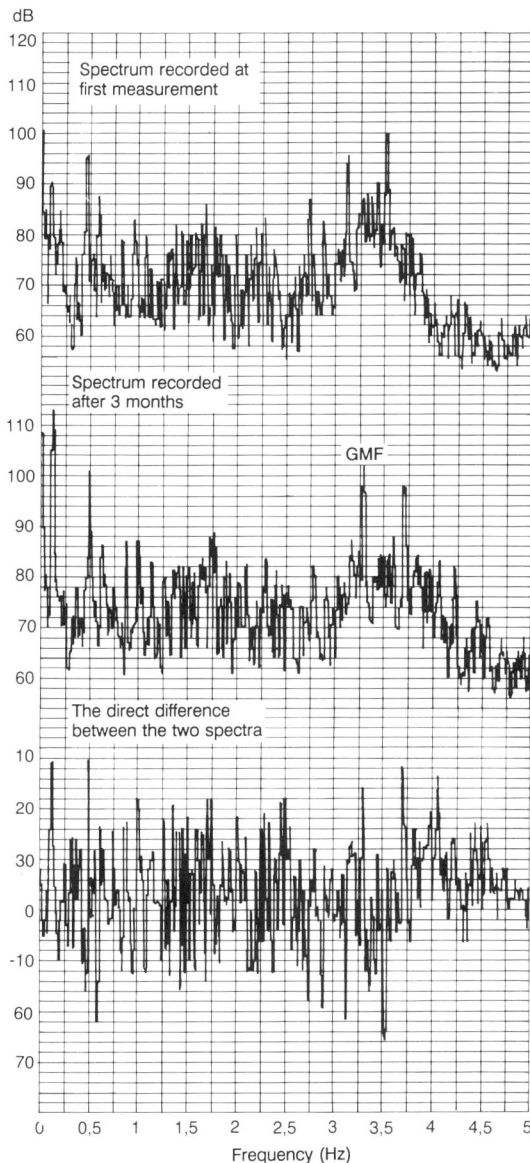

Figure 35.21. The direct difference between two narrow-band spectra, such as the 400 lines in this case, cannot be used to detect defects. Even small random variations give large differences due to the excessive frequency resolution.

For the first it is apparent that the frequency resolution is often much too high. Figure 35.21 shows a spectrum from a gearbox. Even very small changes in the signal will make a big difference in the individual lines and thereby in the spectrum as a whole. In the example a difference of ±30 dB or a change of about 30 times the original level is observed. It seems clear that if one must detect defects with small changes in the level between the gear meshing frequency and its harmonics, one cannot work with *random* excursions of 30 dB.

The next problem is that if there is a change in the rotation speed, then all the rotation dependent frequency components will change their position in the spectrum, whereby a direct comparison cannot be made. The meshing frequency in the first measurement is seen to be about 3 kHz and in the second to be about 3.2 kHz. Clearly these two spectra cannot be "subtracted from one another" to see the differences and thereby to see whether defects are under development.

The solution is to reduce the data, i.e. to use all the information available to create a synthesized spectrum which can:

- suppress random changes which have no background in the condition of the machine, and

- compensate for changes in speed.

Spectrum synthesis

Figure 35.22 shows such a synthesized spectrum which is formed by adding the energy present in a number of the narrow frequency bands from the FFT analysis.

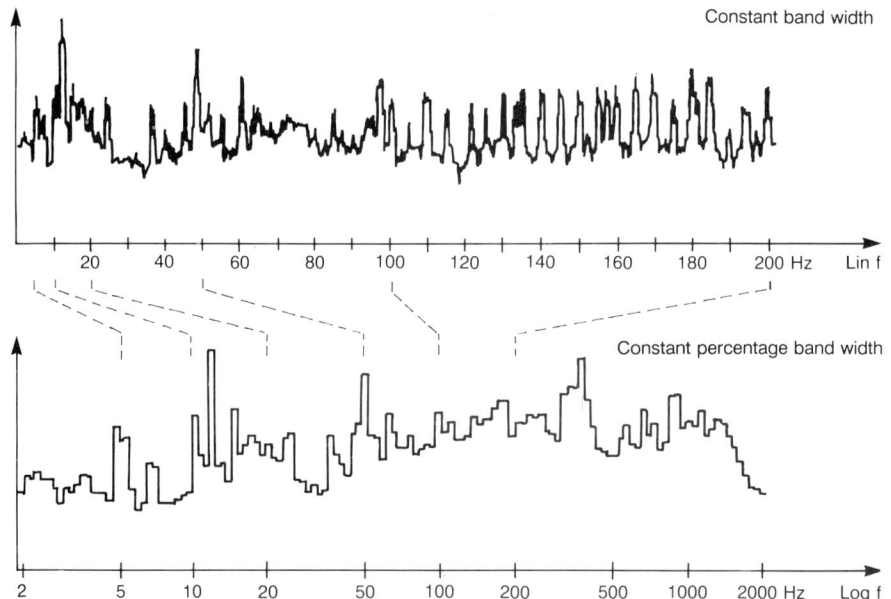

Figure 35.22. The effect of the synthesis of a 6% spectrum with a logarithmic frequency axis.

This new spectrum must have bands which are so wide that they encompass the random variations between various measurements at the same measuring point (of the same machine).

Thus octave analysis (i.e. frequency analysis with filters which extend from one frequency to twice the frequency) has been used for many years by the naval authorities of many countries, and the method has long ago demonstrated its superiority to broad-band measurements. But it does not have adequate resolution for a gearbox where an incipient defect will appear as a higher level of vibration between the gear meshing frequency and its harmonics.

In order to achieve this the resolution must be sufficiently large that there are at least three bands in each octave, referred to as one-third octave analysis, or analysis with filters with 23% band-width. In this fashion one can ensure that at least one of these bands does *not* contain the energetic gear meshing frequency or its second harmonic.

However, we have one problem remaining. We imagine a gearbox with the gear meshing frequency right on the edge of one of the bands which we have now synthesized. If a minor speed increase occurs before the next measurement, then this high component will overlap into the next band and, therefore, cause an alarm due to the level increase even though there is no defect. Therefore, we must "spread out" the synthesized spectrum so that each band has the maximum value of its own value and its two neighbours.

Reference masks

Figure 35.23 shows the result of this "spreading" on a segment of the spectrum shown in Figure 35.22. It will correspond to cutting the spectrum out in cardboard, placing it on a table, shaking sand over it and then pushing the spectrum one bandwidth to either side. The lines which the piled up sand now shows form our new reference. The originally synthesized spectrum (Figure 35.22) is called the *reference spectrum*, and the spread out spectrum (Figure 35.23) is called the *reference mask*.

The next time we measure and synthesize a spectrum no band in the newly synthesized spectrum may exceed the lines in the mask from the first reference measurement. If this happens, it is a sign that a defect is perhaps developing. This small change has, however, increased our effective band-width three times from the original one-third octave to one octave.

As mentioned earlier this is not sufficient, so the band-width of the originally synthesized spectrum must be further decreased. In practice 4-6% spectra are most often used as reference spectra. Figure 35.24 shows a portable analyser, Brüel & Kjær type 2515, which uses a 6% band-width in the original synthesis. The synthesis is performed automatically as the measurement is performed, and the reference spectra are stored in the analyser for later comparison.

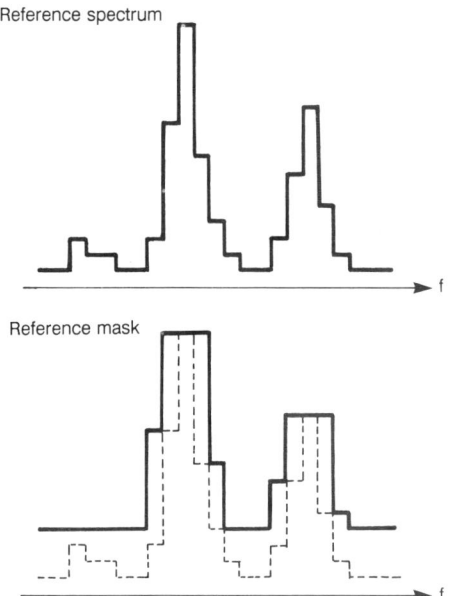

Figure 35.23. Sometimes a minimum level is also superimposed, below which no alarm is to be issued. The minimum level is determined among other things by the quality of the tape recorder (if one is used).

Figure 35.24. Water and dust proof analyser in use in an industrial environment.

Speed compensation

The last problem which we must compensate for is changing speed.

If the speed of a machine changes by more than an amount corresponding to the tolerance which is inherent in the selected band-width, then the energetic vibration components will lie at frequencies outside the limits of the mask and thereby indicate that a defect has arisen even though the reason is just a change in the speed of rotation.

Because the spectrum which is synthesized is made in such a way that the new spectrum has a constant percentage band-width and is displayed with a logarithmic frequency scale, a change in speed will correspond to the same number of bands in the spectrum at all frequencies.

If it is assumed that the synthesis has been performed with a 6% band-width, and the speed is increased by 10%, this corresponds to shifting the entire spectrum slightly less than two band-widths to the right.

Due to the logarithmic frequency scale we thus have a relatively simple means of compensating for the speed changes which would not be possible if the FFT's linear scale had been retained.

There are, however, certain problems connected with this procedure. Not all components in the frequency domain will change with rotational speed. We have already mentioned structure resonances which provide us with an indication of incipient roller and ball bearing race defects. These are, however, relatively wide, often up to an octave, so even using a logarithmic scale they will rarely cause problems for moderate speed changes. If in fact it is a matter of significant changes in speed, one must be aware of this possibility of false defect indications.

Also other components will be independent of the rotational speed of the machine and thereby very clearly visible in the spectrum. Here it can be mentioned that the vibration of electric motors is at 100 Hz, twice the line frequency. It is due to the contraction of the iron because of the electric field, and it is often the highest component in electrical motors which drive simple components such as centrifugal pumps. But this component does not contain any information about the mechanical condition of the motor.

If we work with fault detection using speed-compensated spectral comparison as described above, the risk of false alarms will always be present due to these fixed components which should not have been speed-compensated but are present with the rest of the spectrum.

If the resolution of the spectrum is sufficiently high, these components can be quickly identified and thus expenses in connection with further analysis of the machine in question can be avoided. But if the components lie quite close to other relevant components, the 50 (or 60) Hz line frequency will be close to the rotational frequency just below 50 (or 60) Hz in a two pole induction motor, then more thorough and significantly more time-consuming analysis must be undertaken. In the case of the electric motor an FFT analysis with zoom (whereby the frequency resolution becomes 5 to 10 times finer) near the 50 (or 60) Hz will clearly reveal which is which.

Figure 35.25 shows a synthesized spectrum recorded with a 6% band-width synthesis from a bearing on a paper machine (the solid line) along with the reference mask (the dashed line).

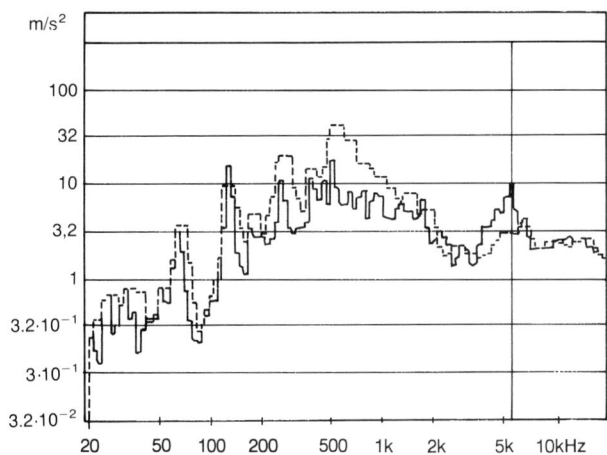

Figure 35.25. The reference mask (dashed line) and a measurement which shows an imminent bearing race defect in a paper machine.

It can be seen how a beginning bearing race defect manifests itself in the high frequency region. Notice also that the total vibration level has fallen so that a broad-band measurement would show an improved bearing race condition, while the opposite is in fact the case.

Testing procedure

Portable, water- and dust-proof, battery-powered FFT analysers are available on the market today with this synthesis built-in. Figure 35.24 shows such a device. A specially trained worker follows a route each day where he or she records the synthesized spectra on the machines in operation and compares them with the reference masks which are stored in the memory of the analyser.

Using the cursor, one can indicate the rotational speed of the machine, whereby the analyser will automatically carry out the speed compensation.

If desired the new spectra can be stored and brought back to the shop where they can be transferred to a desk-top computer such as an IBM-PC or a Hewlett

Packard series 200. If at any time a fault is suspected, the old spectrum can be used to develop trend curves to monitor developments.

When the results for the day have been read into the computer it will produce a list including:

- Measured points.

- Points which should have been measured during the day in question but which were not for some reason, e.g. because a machine was not in operation.

- Points which have shown such large mask overlaps that additional analysis should be performed.

Finally the computer will take the reference spectra which are to be used the next day and read them into the analyser's memory in the order in which the operator must use them. All of these operations can be performed automatically.

Such a monitoring program, when it becomes routine, can be performed at a rate corresponding to 10-20 measuring points per hour and should, if experience from the system of interest does not indicate otherwise, be arranged so that one returns to each measuring point about once per month.

Faulty diagnosis

When a defect is recognized by means of the spectral comparison with speed compensation as described in the foregoing section, yet another analysis must be carried out.

First, someone from the maintenance department with some experience with the diagnosis of machine defects must review the synthesized spectra which led to the error report to be sure that one of the factors indicated earlier which can cause false alarms is not the culprit.

Next, the frequency with which the excursion has taken place provides an indication of what might be the cause of the defect:

- In the low frequency region we find, among other things, signs of imbalance, poor alignment, bent shafts and instability in journal bearings.

- In the middle frequency range we find, among other things, information about the state of wear of gearboxes; possible gear tooth defects such as eccentricity, alignment defects in the gearbox itself and uneven gear wheels will also turn up here.

- In the high frequency region we find the first signs of defects in ball and roller bearings and indications of other phenomena which result in a series of small, sharp pulses.

On this basis, an FFT analysis of the signal will often provide the diagnosis as indicated earlier in this chapter.

If this is insufficient, then a number of other methods of analysis can be brought into play, as described below.

Phase measurements

The use of phase measurements at the frequency of rotation is often valuable.

As shown in Figure 35.26 a static imbalance will produce radial vibrations which are in phase, while a dynamic imbalance as shown in Figure 35.27 will also produce radial vibrations, but out of phase.

An externally forced rocking motion which, among other things, can arise due to alignment errors, Figure 35.28, will give both radial and axial vibrations and both out of phase.

A shaft with static deformation, i.e. a bent shaft, will similarly give both radial and axial vibrations, see Figure 35.29, where the axial vibrations will be out of phase, and the radial vibrations will be in phase. These defects will most often give higher harmonics of the rotation frequency. The imbalance itself, which only has energy at the rotation frequency, will often show higher harmonics due to non-linearity in the mechanical system, and, therefore, the presence of higher harmonics cannot be used as an indicator that a possible defect is not an imbalance.

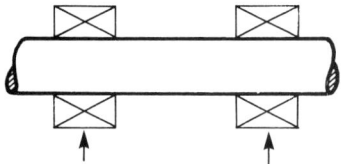

Figure 35.26. Static imbalance. The vibrations on the two bearing mounts at the rotational frequency are in phase.

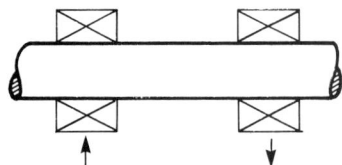

Figure 35.27. Dynamic imbalance. The vibrations on the two bearing mounts at the rotational frequency are out of phase.

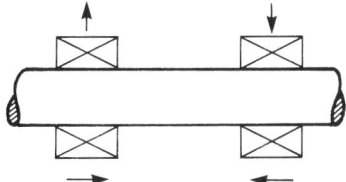

Figure 35.28. Forced rocking motion. Both the axial and radial vibration at the rotational frequency are out of phase.

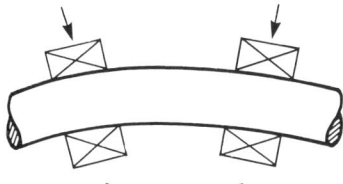

Figure 35.29. Static bending of the shaft. Both axial and radial vibrations are at the rotational frequency. The axial are in phase, the radial are out of phase.

Note here that a two-channel FFT device will give the correct phase if the phase is measured with triggering in one of the two channels and the vibration signal in the other. A single-channel FFT analyser must, on the other hand, be specially designed in order to provide a phase measurement which can be used for this purpose. The internally computed *FFT phase* cannot be used directly. If it is not clear from the operating manual of the FFT analyser that it has amplitude and phase interpolation, then it is uncertain whether it can be used as described here. In any case consult with your supplier.

Cepstrum

Cepstrum is a method of analysis for the identification of families of harmonics and side-bands and their relative strengths. Simply expressed, the cepstrum which is used for machine monitoring is a frequency analysis of a frequency analysis. Just as a series of pulses in time will become a line in the frequency spectrum at the pulse frequency and at higher harmonics, thus also a series of equidistant lines in the *spectrum* will become a line in the *cepstrum* and higher harmonics in the cepstrum.

The scale in the cepstrum is 1/Hz or seconds, and the position of the first component will correspond to the distance between the individual bands in the spectrum's series of harmonics or side-bands.

Figures 35.30 and 35.31 illustrate the use of cepstrum.

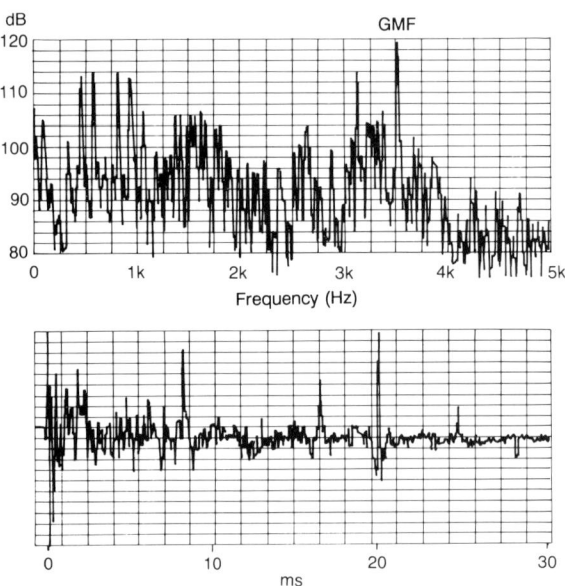

Figure 35.30. Spectrum containing many components. The cepstrum shows that there are two strong families of side-bands and/or harmonics at 120.7 Hz and at 49.8 Hz.

That the spectrum of Figure 35.30, which was recorded from a defective gearbox, contains a series of harmonics and/or side-bands seems clear, but *which* it is almost impossible to see.

The cepstrum in Figure 35.30, on the other hand, clearly shows that the spectrum contains two series of harmonics or side-bands, namely a series with a fundamental frequency of 49.8 Hz and another series with a fundamental frequency equal to 120.7 Hz.

Figure 35.31 shows how one can separate these harmonics by computing the cepstrum of a part of the frequency spectrum. The cepstrum of Figure 35.31 has been calculated for that portion of the spectrum shown in the upper part of the figure, and the cepstrum shows that only the family with a mutual separation of 49.8 Hz remains.

Thus we now know that the series of harmonics with a mutual separation of 120.7 Hz lies in the lowest portion of the spectrum where it probably indicates imbalance or bad alignment. On the other hand that portion of the harmonics which has a mutual separation of 49.8 Hz lies in the upper portion and, therefore, indicates side-bands near the gear meshing frequency, i.e. signs of initial gear tooth defects.

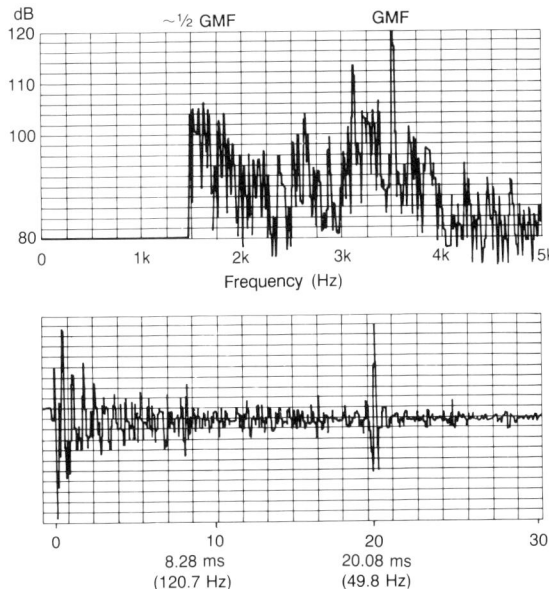

Figure 35.31. If the cepstrum is computed based only on the signal near the gear meshing frequency, one of the families of harmonics disappears. See the conclusion in the text.

The cepstrum is highly insensitive to phase differences in the original signal. Particularly in gearboxes both amplitude and frequency modulation can arise due to the varying opposing moment of the machine being driven. Depending upon the mutual phase relationship of the modulating signals, one can obtain widely differing spectra, but the cepstrum will still show the two frequencies clearly.

The relative amplitude in the cepstrum gives the relative strengths of the harmonic families or side-bands.

Finally, the cepstrum is highly insensitive to differences in attenuation in the mechanical structure; cepstra computed from signals from two different ends of a gearbox will often be very similar.

At present certain portable FFT analysers for machine monitoring have built-in cepstrum computation, as that shown in Figure 35.24 (or Figure 35.3). Thus one can obtain the cepstrum of the portion of the spectrum which is marked by the cursor on the screen by just pressing a button.

By performing a zoom analysis around the gear meshing frequency and then selecting cepstrum, the cepstrum will indicate which side-bands lie around this frequency. This is a very valuable tool for the diagnosis of gearboxes.

Vibration monitoring

Envelope detection

It has been shown that an incipient fault in a rolling-element bearing will create a series of sharp pulses, but pulses with very little energy content, see Figure 35.15.

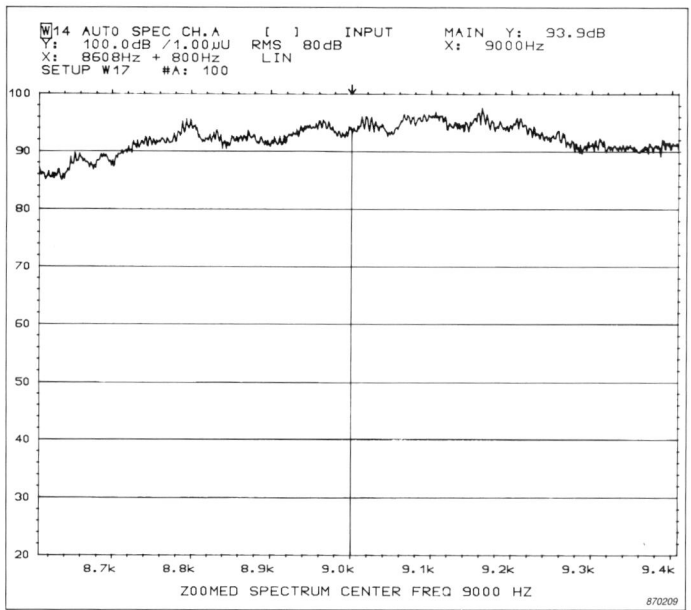

Figure 35.32. A zoom spectrum illustrating smearing due to changes in the rotational speed.

These faults are detected by monitoring the increase in vibration level at structural resonances, but the analysis needed to establish that it is a faulty bearing and not a signal originating from for example a forced lubrication system or another spiky signal cannot be made by simple FFT analysis.

The structural resonance frequency where such a fault shows up will be in the kHz range, while the rotational speed could be as low as between 1 to 2 Hz, which could be less than the resolution to find the harmonics of the original series of pulses amplified by the mechanical structure. This is viable provided the signal measured from the machine is absolutely stable. In real life, however, this is seldom true. Due to changes in rotational speed, the heavily zoomed spectrum will be smeared to such a degree that the line spectrum from the original pulses will disappear, see Figure 35.32.

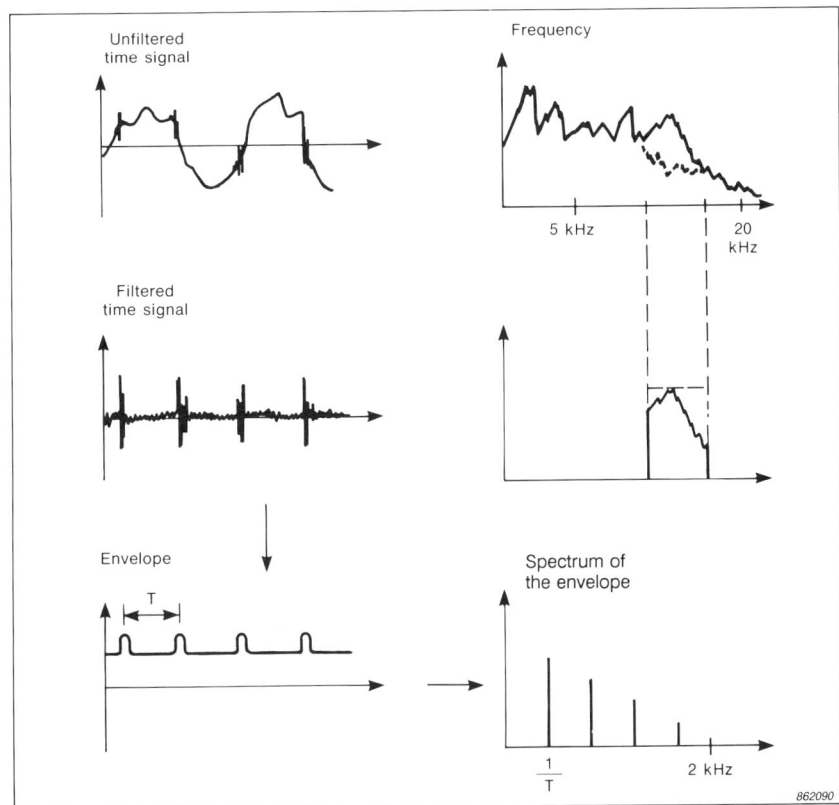

Figure 35.33. The principle of envelope detection.

An easy way to overcome this is to use analogue envelope detection, as follows:

The time signal is filtered approximately around the frequency region where the increase is detected, see Figure 35.33. This leaves the high frequency signal which we know contains the pulse-excited vibration of the bearing housing without most of the contaminating signals.

This signal is then rectified and low-pass-filtered at a frequency approximately one-half the bandwidth of the bandpass filter. The signal now looks somewhat like the original pulses from the bearing, but what is most significance is that we have thus recreated the pulse frequency. By analysing this signal in the FFT analyser, the pulse frequency can be determined exactly.

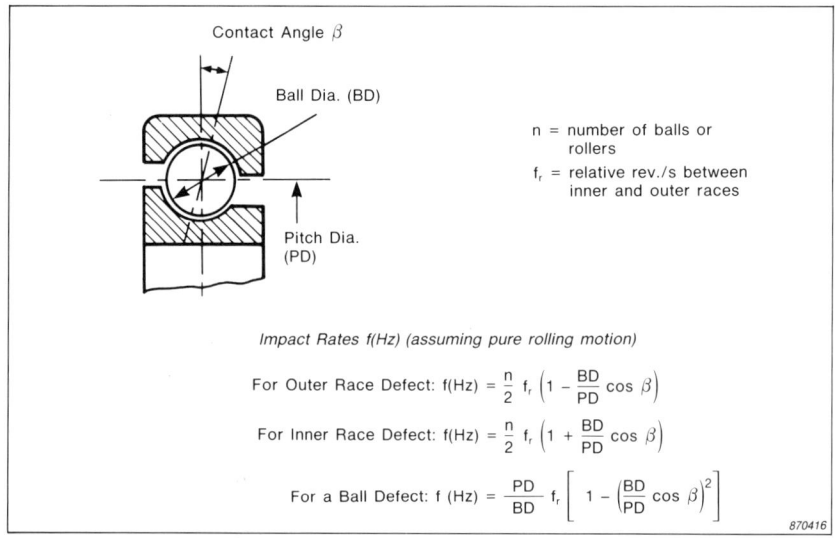

Figure 35.34. Determination of pulse frequency from physical dimensions of the bearing and the number of balls or rollers.

Since the impulse rate can be calculated, see Figure 35.34, for an inner race fault, an outer race fault or a roller fault, then for each known bearing, the source of the fault can be pinpointed. Note that the real frequency will be slightly lower than the calculated frequency due to sliding.

If the fault is on the rotating race, it is sometimes possible to see the amplitude modulation from the varying load on the crack. This modulation effect will turn up as sidebands around the lines corresponding to the pulse rate, spaced at rotational speed, see Figure 35.35.

Synchronous time averaging

When performing ordinary FFT analysis, a series of FFT computations are often performed, and the spectrum which is shown on the screen is computed as an average of e.g. 64 individual spectra. This is done to reduce the influence of random noise on the spectrum; the spectrum is smoothed out so that it is easier to read.

When performing synchronous time averaging, first a time signal is created by recording several time signals and finding their average and then afterwards carrying out a single FFT computation.

Figure 35.36 shows the principle. This can be used, for example, for the analysis of gearboxes which have shown excursions in the frequency comparison and where the signal can easily become complicated due to differing rotational speeds of two or more shafts.

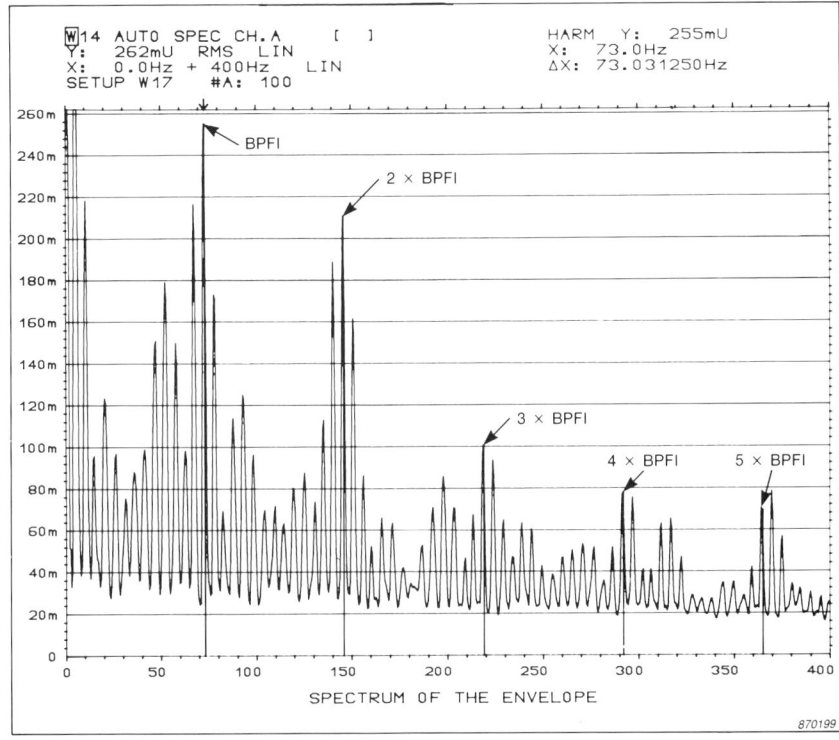

Figure 35.35. FFT analysis of the envelope. Amplitude modulation is detected by sidebands around the ball passing frequency and its harmonics.

In such cases it can happen that signals which are *not in phase* from one recording to the next will be reduced in the final FFT spectrum, while signals which *are in phase* from one recording to the next will be enhanced.

The shaft which is to be analysed is provided with an optical trigger, whereby the signals are recorded synchronously with the rotation of this shaft.

In other words: the signals which are synchronous with the trigger pulse will remain in the spectrum, while the signals which are not synchronous with it will be weakened and finally disappear from the spectrum.

Using synchronous time averaging one can isolate that part of the vibration signal which is caused by one of the shafts of the gearbox and weaken those which arise from other components. Synchronous time averaging can be regarded as a kind of "acoustic magnifying glass", whereby one can focus upon the machine part one wishes to examine.

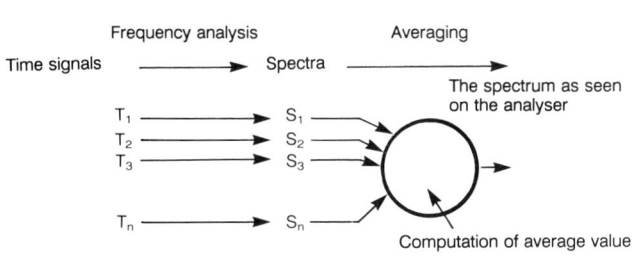
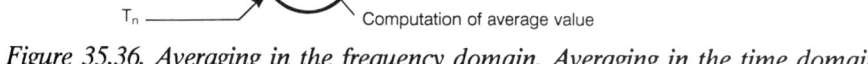

Figure 35.36. Averaging in the frequency domain. Averaging in the time domain.

If the gearbox contains a middle shaft which is not accessible and one cannot, therefore, obtain a trigger pulse from it, then one can make use of a so-called "trigger multiplier" to multiply the trigger signal with a given factor corresponding to the number of teeth on the toothwheels and thus "focus" on a shaft which is not physically accessible. Figure 35.37 shows an example of such a frequency multiplier (Brüel & Kjær type 5898).

This method is also applicable where simple frequency averaging cannot be applied, e.g. with piston engines.

Defect development

These days it is quite a widespread practice to record certain process parameters over time, and thereby to get an indication of when the process has changed in an inappropriate manner. The parameter may be oil pressure, exhaust gas temperature, coolant flow, etc. This technique is also applicable in connection with the vibration monitoring of machines.

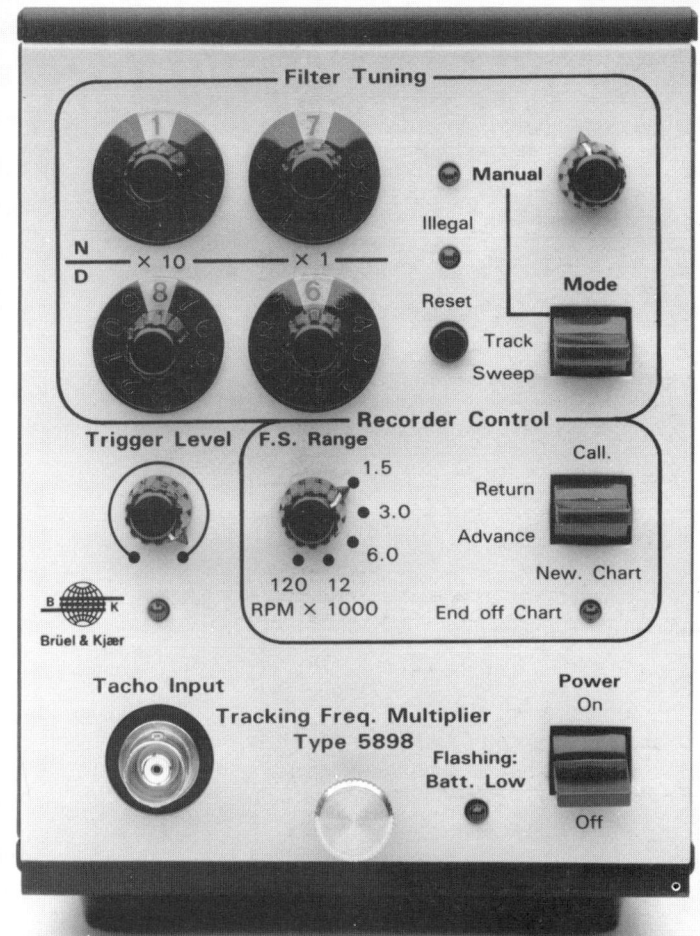

Figure 35.37. Frequency multiplier for advanced diagnosis of gearboxes etc.

With routine daily monitoring one will obtain a series of spectra without any significant changes. Should some excursion beyond the limits of the reference mask suddenly be recorded in one or more bands, it may be due to an imminent defect such as in a gearbox or a roller bearing race.

The defect will perhaps develop slowly, and corresponding to this the vibration level in one or more bands will slowly increase.

A graphical representation of the vibration level in these frequency bands (often called a trend curve) is an important tool for providing an estimate of the time remaining until a repair must be initiated.

Vibration monitoring

Figure 35.38 shows how a defect develops over the course of several months. At the maximum permissible vibration limit, which the responsible manager has established, the operator can consider whether an unplanned interruption of operations should be recommended to management.

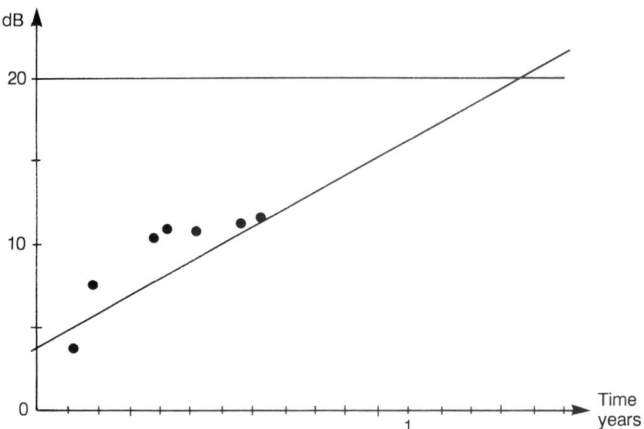

Figure 35.38. Trend curve analysis. The frequency band in which a change in the condition is observed is used for the trend curve analysis. The energy in the rest of the signal is not included in the calculation.

However, far from all defects develop in this fashion. For instance, journal bearing instability is hardly likely to develop slowly and smoothly, so that the trend curve of the vibration level identified for it cannot provide any indication of the remaining safe operating time.

Trending can thus be a useful tool, provided the diagnosis has already been made, and it is possible to conclude that the defect in question is a type which will develop in *an even, predictable manner*, provided we know with respect to this type of fault that the vibration level which is associated with the diagnosed defect will develop in a parallel manner.

Trend curve analysis can be carried out on a tabletop computer which, as described earlier, carries out the administration of reference spectra, masks and newly recorded spectra.

The trend curve in Figure 35.38 was carried out on a computer which simultaneously computes the remaining safe operating time on the basis of the indicated maximum vibration level.

These programs should work interactively so that the operator or the engineer can perform trial and error experiments with the data and observe on the screen which trend curve seems to provide the best agreement with the measured values.

Maximum values

The choice of maximum permissible levels of vibration is always difficult, and it is natural to utilize the experience which the operator gains through working with the system.

If no values based on prior experience are available, then the recommendations from VDI can be used. They specify that:

- an increase in the level of vibration with a factor 2.5, or as sometimes specified 8 dB, is an indication that the vibration spectrum and other related process parameters should be investigated,

- an increase of a factor 10, i.e. 20 dB, should – if no other parameters provide a reliable indication to the contrary – cause the machinery to be shut down.

These values are, as mentioned above, reasonably safe to use until a basis of experience with the machinery in question has been established.

The most important thing to remember is, however: undertaking a trend analysis without prior diagnosis is like driving a car blindfolded. You might manage, or you might be on your way to a catastrophe, *but you won't know it.*

Martin Rørbye Angelo

References

Introductory literature
1) *Measurement of vibrations: Machine health monitoring,* Brochures from Brüel & Kjær.

2) Reports produced by Ødegård Danneskjold-Samsøe for the National Technology Council.

Signal analysis
3) *Frequency analysis,* Brüel & Kjær.

4) *Shock and vibration measurements,* Brüel & Kjær.

5) *Shock and vibration handbook,* Cyril M. Harris and Charles E. Crede, McGraw-Hill Book Company, New York, 1987.

Machine analysis and maintenance theory
6) *Machinery analysis and monitoring,* John S. Mitchell, Pennwell Publishing Company, Tulsa, Oklahoma, USA, 1981.

7) *Practical machinery management for process plants*, Heinz P. Bloch, Gulf Publisher, 1988.
   Vol 1: *Improving machinery reliability.*
   Vol 2: *Machinery failure analysis and troubleshooting.*
   Vol 3: *Machinery component maintenance and repair.*
   Vol 4: *Major process equipment maintenance and repair.*

Chapter 36
Visual inspection

NDE principle

In principle *visual inspection* only includes inspection by means of eyesight. However, in practice the term visual inspection has a somewhat broader meaning. In any inspection one or more of the senses (hearing, sight, feeling, smell and taste) will be used as primary "NDE methods".

In this section the term "visual inspection" is to be understood in the broad sense. In connection with visual inspection a number of simple, auxiliary tools are used. These will be dealt with later.

Development of the method

The development of the method visual inspection is primarily a matter of sharpened attention to details. An inspection becomes more thorough as the user gains experience with the system or machine installation. Through the acquired knowledge the senses are put to better use.

This experience together with the right auxiliary tools and systematic inspection can yield a good record of observations which can be applied in the decision-making process.

Areas of application

Visual inspection can be applied within almost all areas of preventive maintenance work. There will, however, be methods of inspection where the use of advanced measuring equipment will give a more detailed description of the state.

Inspection can be performed while the machine/system is running, as well as during stoppage.

Visual inspection can be immediately applied in a number of areas, such as:

- inspection of cleaning
- checking for corrosion, erosion and deformities
- checking for ruptures, cracks and wear
- monitoring of manometers, pressostats and temperatures
- monitoring of oil level, greasing and greasing apparatus
- monitoring of the operational condition of systems or machines

It is important to be aware of the advantage of the human senses as tools of inspection as they are able to register deviations from a desired or normal condition.

Therefore, it is important to have references when performing visual inspection. In many cases it would be ideal to have an intact sample on hand when inspection is performed, but that is of course rarely possible. Instead it is necessary to know what the object looked like when new and how it looked at previous inspections.

The original drawings and notes from earlier inspections can be used to this end as can photos taken at the inspections.

The experience of the inspector is of course of great importance, but even the experienced inspector can benefit from a check-list, especially when inspecting larger and more complicated machines or systems.

The quality of visual inspection will depend on the inspector's experience and systematic approach as well as upon his or her imagination and thorough understanding of relevant degradation processes.

Auxiliary tools
The following auxiliary tools can be used in connection with the visual inspection:

1. Original drawings
2. Notes from earlier inspections
3. Notebook and pencil
4. Lamp
5. Scraper and steel brush
6. Chalk or similar item for markings
7. Mirrors
8. Magnifying glass
9. Straight steel ruler, square and level
10. Vernier, micrometer and measuring tape
11. Calipers and measuring compass
12. Telescope
13. Crane
14. Weights
15. Magnets
16. Grinding tools
17. Camera
18. Check-list

In this list some of the most necessary auxiliary tools have been mentioned, but others may be needed, depending on the inspection job in question.

On the basis of careful planning the experienced inspector can determine which tools are necessary for a given inspection. However, the inspection job should never become routine work so that imagination no longer is part of the process.

Below are brief comments on the auxiliary tools listed above.

Original drawings
The importance of good knowledge of the original drawings and notes made during earlier inspections has already been mentioned. The applicability of notes from previous inspections is very dependent on the systematic approach and care with which the notes originally were made.

Notebook and pencil
A notebook and pencil are indispensable for any inspector, especially if several inspections will be made before there is time to make out the final inspection report. It is important to allocate time for the necessary notations while the impressions from the inspection are still fresh.

Lamp
It is obvious that good illumination is required for a successful visual inspection. A hand held lamp is indispensable. Special lamps such as those fastened to the forehead can be chosen for special jobs. The hand held lamp can be used for revealing faults by shining the light parallel to the surface, whereby irregularities and corroded areas will stand out like a lunar landscape.

Scraper, steel brush and chalk
It is often necessary to inspect equipment which has not been sufficiently cleaned. In order to make a preliminary inspection before further cleaning is carried out tools such as a scraper, a steel brush and a pick hammer can be useful. Corrosion products and the like hiding the surface and preventing visual inspection can be removed with these tools.

The inspector should be aware that by knocking on the material the sound can often reveal areas lacking in material thickness or areas with other defects.

If the inspector suspects such areas, these should be marked with chalk, and a more careful inspection can be performed later, e.g. using ultrasonics.

Mirrors
When inspecting areas not easily accessible a mirror can be of great help. Depending on the job it can be of any size from the small dentist's mirror which can be stuck into small openings to a much larger mirror which can make the job easier when making external inspection of pipelines placed near the ground or close to a wall.

Magnifying glass
A magnifying glass can be used for closer inspection of suspicious-looking areas just as it often will be a natural tool for inspections of precision instruments.

Straight steel ruler, square and level
These tools will normally be put to use before the actual measuring takes place. Sometimes custom-made shapes may have to be constructed in order to perform an inspection.

Vernier scale, micrometer, measuring tape, measuring compass and calipers
Simple measuring tools will often reveal insufficient material thickness or deformations. Depending on the desired accuracy, micrometer, vernier or measuring tape can be used. Indirect measuring by means of measuring compass or calipers may of course also be used, provided one takes into account the risk of errors that may result from indirect measurement. However, new types of compasses have been constructed with built-in measuring dials so that direct measurement can be performed.

When inspecting many similar objects, e.g. pipes in a heat exchanger, it may be a good idea to construct a measuring mandrel with maximum and minimum measurements so that sorting can be performed quickly. On the other hand, it will often be interesting to keep an eye on an advancing corrosion from one inspection to the next by means of accurate measurements, and in that case a tool that only separates the acceptable pipes from the unacceptable ones will be insufficient.

The human eye can only distinguish finer details when the distance to the object in less than 1 metre, so visual inspection is usually confined to objects that one can get close to.

Telescope
If it is impossible to get close enough, one can often obtain good results using a good telescope or binoculars.

Crane
When the objects of visual inspection are large and heavy, it may be necessary to use a crane or a jack in order to inspect all the surfaces.

Weights
In some instances, weighing the object to be inspected may reveal inadequate material thickness. In this case it is important to have careful references with which to compare the new observations. If the decision is made to weigh a large, cast valve in order to reveal inadequate material thickness or slag in the casting, this will only succeed if the exact weight of a flawless valve is known. When the aim is to decide the material composition of smaller objects, a simple identification of the density of the material may be helpful.

Magnets
In connection with an inspection there may be a desire to identify the individual materials that are part of a given piece of equipment. When one is dealing with metals, there are only a few which can be readily identified through visual inspection. Here a magnet can be used to distinguish the ordinary magnetic types of steel from the non-magnetic stainless steel types.

Spark patterns
The table below is taken from *APIs Guide for inspection of refinery equipment*, Chapter IV, page 34.

METAL	Volume of spark stream	Relative length of spark stream[1]	Colour of stream near wheel	Colour of streaks near end of stream	Quantity of spurts	Nature of spurts
1. Wrought iron	Large	160	Straw	White	Very few	Forked
2. Machine steel (AISI 1020)	Large	175	White	White	Few	Forked
3. Carbon tool steel	Moderately large	140	White	White	Very many	Fine, repeating
4. Gray cast iron	Small	60	Red	Straw	Many	Fine, repeating
5. White cast iron	Very small	50	Red	Straw	Few	Fine, repeating
6. Annealed malleable iron	Moderate	75	Red	Straw	Many	Fine, repeating
7. High speed steel (18-4-1)	Small	150	Red	Straw	Extremely Few	Forked
8. Austenitic manganese steel	Moderately large	110	White	White	Many	Fine, repeating
9. Stainless steel (type 410)	Moderate	125	Straw	White	Moderate	Forked
10. Tungsten-chromium die steel	Small	90	Red	Straw[2]	Many	Fine,[2] repeating
11. Nitrided Nitralloy	Large, curved	140	White	White	Moderate	Forked
12. Stellite	Very small	25	Orange	Orange	None	
13. Cemented tungsten carbide	Unusually small	5	Lt. orange	Lt. orange	None	
14. Nickel	Very small[3]	25	Orange	Orange	None	
15. Copper, brass aluminium	None				None	

1) The figures are only relative. Actual spark streams depend on grindstone, pressure etc.
2) Bluish-white spurts
3) Wavy streaks

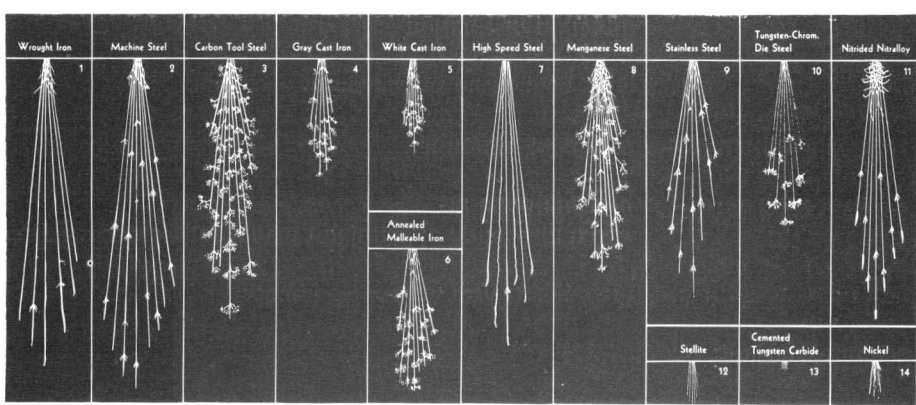

Spark patterns which can be used as a guide to assist in the identification of various metals.

Visual inspection 383

One method is to grind the material and to identify it from the resulting spark pattern, as indicated on the previous page. It should be added that this method requires a good deal of experience, and in any case it will be a great advantage to have some known reference blocks on hand that can be used for comparison.

Material determination by chemical means
A crude quantitative and qualitative material determination can be performed using chemicals, usually acids or mixtures of these. Chemicals are applied to the object to be examined, and from the reactions and possible colouring the type of material is determined.

Camera
In cases when it is impossible to supply observations with concrete measurements, a photograph may be of great value when trying – perhaps years later – to evaluate the development. Many misunderstandings can be avoided if a picture is available as additional documentation, for instance when deciding on repairs to be made.

Documentation
If visual inspection reveals irregularities or changes, the next step is to decide whether repairs are due now or if they can wait until the next inspection, or whether further inspection using more refined NDE methods is called for.

In order to carry out such an evaluation it is necessary to know the history of the equipment. Systematic and well-documented notes from both recent and earlier inspections will be of critical importance when making decisions. When notes are made after an inspection they should of course be both systematic and well-documented. In some cases years will go by before the same piece of equipment is examined again, and then it is important that notes taken previously are clear and understandable.

Nomenclature
In order to describe phenomena that are difficult to measure it is advantageous once and for all to establish a clear and unambiguous nomenclature. It is a good idea to work with tables of the various types of errors that can be expected in a given job and to separate each into appropriate subtasks. This makes it relatively easy for a future reader to make out what various comments mean.

Standard forms/sketches of the installation
Of course it is also important to be able to indicate the location of an observed fault unambiguously. To this end it is advisable to produce standard forms or measuring drafts of the various pieces of equipment which one is responsible for. Each piece should be given a number and name plus perhaps an indication of measuring points so that misunderstandings are avoided.

These standard forms or measuring drafts are a very handy auxiliary tool when passing on results and also later on, should one wish to analyse the results as historical material.

Practical applications

Hydraulic systems

Check these points with the system stopped:

- that the oil is clean and clear

- that the system is clean and dry

Check these points with the system in operation: listen for unusual noises from the motor, pump and piping system, and check:

- that the oil is clean and clear (must not foam) and that the temperature is correct

- that the system is intact (no leaks) and that the filter indicators are OK

- for possible draining of condensation

Visual inspection

Belt pulley

Check these points with the system stopped:

- missing belts and screening

- belt tracks (belts must not work into the bottom of tracks)

- that the belts are of equal length (maximum difference equal to half the thickness of the belt)

- that the belt pulleys are flushed

- belt tension by pushing the belt downward by hand (downward movement must not exceed 10% of the centre distance between the pulleys)

- the condition of the belts; possible cracks

Check these points with the system in operation:

- belt noise, missing belts and screening

Electric motor

Check these points with the system stopped:

- motor mountings

Check these points with the system in operation:

- listen for unusual noise
- temperature of bearings
- working temperature of motor (max. 60 °C)
- motor for possible vibrations

Visual inspection

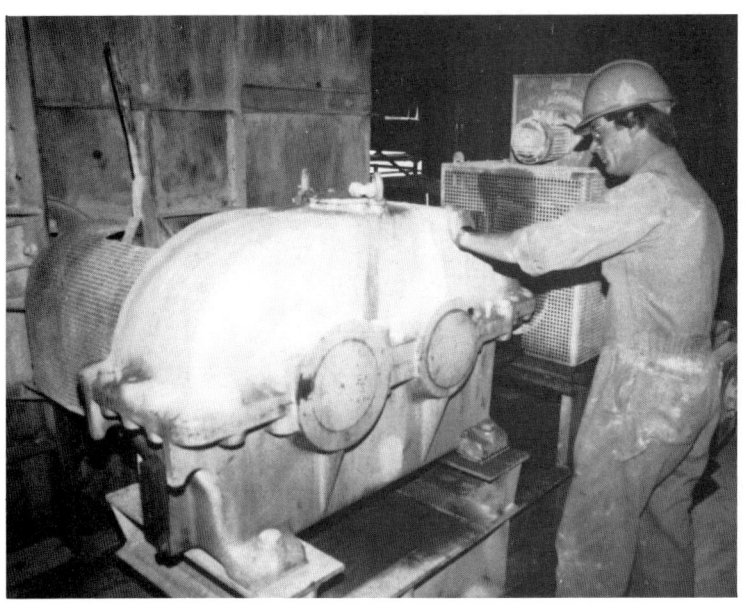

Gearboxes

Check these points with the system stopped:

- oil level and spilled oil

- gearbox mountings

Check these points with the system in operation: Listen for unusual noises from bearings and gear wheel. Check also:

- working temperature of gears (max. 60 °C)

- oil level and spilled oil

- gearbox mountings

- gear vibration

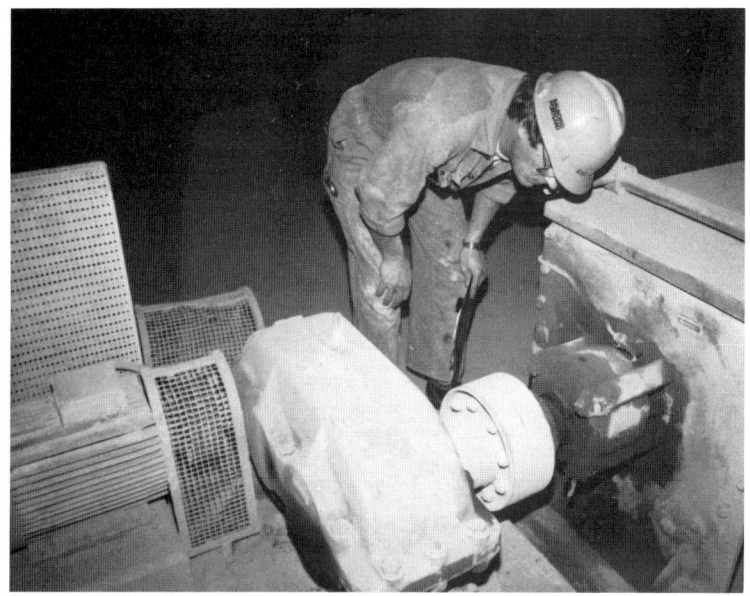

Clutches

Check these points with the system stopped:

- that the clutch plates are secured to the axle
- that the clutch plates are 2-4 mm apart
- that the clutch axles are parallel and precisely aligned
- for warping (e.g. using measuring dial with magnetic support) max. 0.1 mm out of true for fast-moving clutch

Check these points with the system in operation:

- listen for abnormal noises from the clutch
- that the mountings are OK

Visual inspection

Cooling ventilator

Check these points with the system stopped:

- visual inspection of ventilator blades for wear, cracks and possible build-up
- ventilator mounting

Check these points with the system in operation:

- direction of rotation
- for vibrations (may need balancing)
- listen for unusual noises from the ventilator

Compressor *Check these points with the system in operation:* Listen for unusual noises from motor. Check also the following:

- temperature of bearings
- working temperature of motor (max. 60 °C)
- motor for possible vibrations
- fastening of motor and possible screening

Read manometer for

- Terminal pressure max. working pressure:	bar
- Oil pressure normal:	bar
- Filter minimum:	bar
- Filter max. filter indicator:	mbar
- Water temperature intake normally:	°C

Report any irregularities to the supervisor in charge.

Visual inspection

Transportation worm

Check these points with the system stopped:

- fastening of trough
- worm thread (by opening all inspection hatches)

Check these points with the system in operation:

- listen for unusual noise from middle bearings and worm threads
- bearing at pulling end for temperature (max. 60 °C)
- gasket housing at pulling end for leakages
- gasket housing and bearing opposite the pulling end

Conveyor belt

Check these points with the system stopped:

- visual inspection of rubber belt for holes and cracks plus scraper for wear
- alignment of the belt on rollers
- state and fastening of all rollers (upper, lower, tightening, driving and turning rollers)
- bearings
- the supports, lubricating pipes and lubricating fittings
- for build-up of dirt

Visual inspection

Chain drive

Check these points with the system stopped:

- for wear of chainwheel teeth

- for wear on roll chain

- tightness of chain by pushing the chain down by hand (downward bending must not exceed 10% of the centre distance between the chain wheels)

- that the chain wheels are in alignment and that they are not loose on the axles

- that screening is all right.

Water pump

Check these points with the system in operation: **Listen for unusual noises from engine and bearings. Check also:**

- temperature of engine (max. 60 °C)

- gasket housing and pipes for leakages

Welds

Visual inspection of welds is the easiest and also the most important kind of inspection of the state of welded joints.

When performing visual inspection, the first checking should take place without prior cleaning of the weld for rust and dirt because it is often easier to discover any crack formation when the seam has not been cleaned. After this initial inspection a more thorough cleaning of the surface of the weld should be carried out.

During the subsequent visual inspection one should be aware that many irregularities in the weld can look like cracks.

It is recommended that one check the cleaned welds with dye penetrant or even better using the magnet particle method and ultrasonic examination and perhaps even radiography.

Kaj Tranholm Olesen
Anders Korsbæk

References

APIs guide for inspection of refinery equipment, Chapter IV, Inspection Tools, 1st ed. 1960, American Petroleum Institute, Division of Refining, 1271 Avenue of the Americas, New York 20, NY, USA.

Chapter 37
X-ray crawlers

NDE principle

The radiographic principle has been dealt with in Chapter 26: **Radiography**, so only the crawler principle itself will be treated here. An *X-ray crawler* is in principle circumferential beam X-ray equipment mounted on wheels. The X-ray crawler is used mostly when one wishes to perform radiographic inspection of a large number of similarly welded circular welds during a construction project.

By way of introduction it should be pointed out that isotope crawlers based on Ir-192 have been used extensively for the same purpose. Today X-ray crawlers can be built for pipe dimensions just as small as the ones for which isotope crawlers were intended (6" pipes). Also, due to greater lack of film resolution and weak contrast the latter provides poorer picture definition when examining relatively moderate wall thicknesses. Consequently, the use of isotope crawlers is diminishing.

Practical applications

The main application of the X-ray crawler is welding inspection during the construction of pipelines and district heating pipes but it is also used for later inspection of welded joints. Compared to the conventional double-wall single-picture technique the X-ray crawler has great advantages. The former usually requires much more time and the price is about five times as high due to the size of the equipment. Also, the image quality speaks in favour of the crawler as the circular beam technique has a penetrameter sensitivity of 1.5% compared to the sensitivity of the double wall technique of 1.8%. Finally, for reasons of radiation safety the crawler technique is of course preferable.

The X-ray crawler technique is included in this handbook as it can be used for certain tasks within the field of condition monitoring.

Development of the method

The X-ray crawler was originally developed for quality control of welded joints in long pipelines but it can of course be used in connection with subsequent off-stream control. One can distinguish between cable-equipped crawlers and crawlers carrying their own power supply.

Push-rod crawler

A one-tank circumferential beam X-ray device on wheels with a supply cable in a long, jointed rod constitutes the simplest imaginable X-ray crawler. The action radius for such a crawler is limited to about 40 metres, i.e. a maximum of three welds for the average length of pipe.

Cable operated X-ray crawler

A one-tank circumferential beam apparatus on wheels equipped with a motor can, by reeling out a cable or pulling a cable behind it, move about 100 metres into a pipe, the equivalent of about 7 circular welds. The action radius is of course highly dependent upon the thickness of the cable. However, many circular beam tube-heads of standard type can be driven at near maximum output from a cable containing only two conductors. By adding a third conductor a permanent magnet motor can be manoeuvred back and forth. The method is highly dependent on making changes in the X-ray apparatus and will thus require the authorization of radiological health authorities and of the manufacturer.

X-ray crawlers equipped with petrol apparatus

An X-ray crawler consisting of a one-tank circumferential beam tube-head on wheels pulled by an electric motor may be powered by a gasoline-electric aggregate source that is carried along and thus be made independent of power supply from outside sources. The capacity of such a crawler is very large, up to 8 hours' working time, but the apparatus requires a regular supply of oxygen, is not very stable and may induce vibrations which may blur the focal point and/or shorten the life span of the filament.

Battery-powered X-ray crawlers

A one-tank circumferential-beam apparatus on wheels with driver motor and equipped with a storage battery and an inverter can carry along sufficient electrical capacity to drive several kilometres into a pipeline and make a large number of exposures. The power supply from the battery is sufficiently stable to make the exposures uniform, and the crawler can even be equipped with a compensatory device which extends the exposure time as the battery is depleted.

The X-ray crawler should be constructed so that the centre of gravity lies as low as possible. The centre of gravity must at least be below the centre line of the pipeline. A typical Danish-built 10-12" 160 kV_p (kV_p = peak kilo-volts, refers to X-ray tube voltage) X-ray crawler is shown in Figure 37.1.

Controlling the crawlers

Both the petrol-driven and the battery-driven crawlers have no connection to the outside world. In order to control such crawlers the following methods are used:

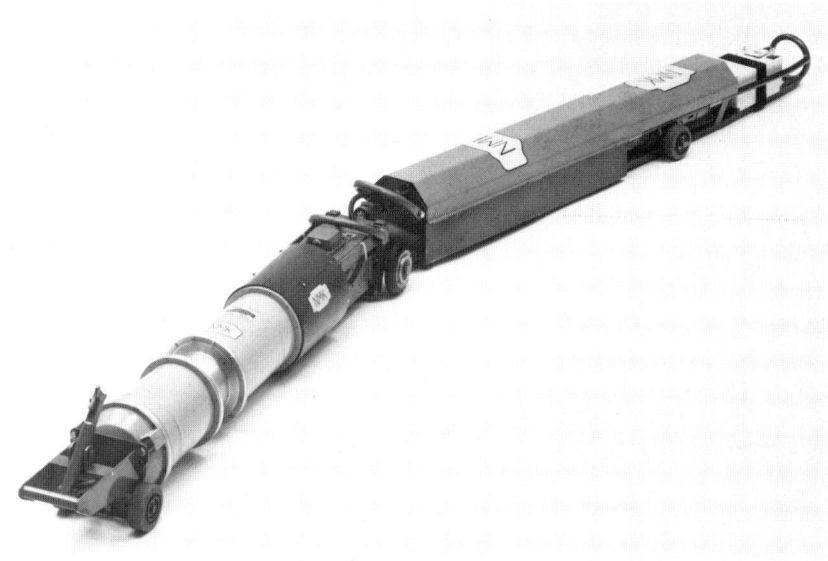

Figure 37.1. 160 kV_p NMK X-ray crawler for 10-12" pipelines.

Star wheel control

A "star wheel" (Figure 37.2) consisting of heavy rectangular sheet metal with slightly concave sides and mounted in a spring-loaded fork that can be turned is pulled along the inner surface of the pipe wall. When the star wheel meets the root of a weld seam, it will turn 90°. During this turn and after a short time delay (to avoid exposing the star wheel) a tripping mechanism activates the precise stopping of the crawler and the following exposure. After the finished exposure the crawler will automatically move on and expose the next weld.

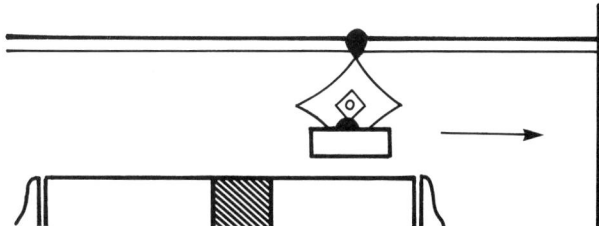

Figure 37.2. Star wheel during its motion along the interior wall of the pipeline.

In order to distinguish between circular and spiral welds two or more star wheels can be used. They will only trip if they are activated at exactly the same time. As a whole the star wheel is not a precise control mechanism as it will often trip due to rust or smaller defects on the inner surface of the pipe. Partly for this reason and partly to control the manoeuvring of the crawler it is necessary to be able to provide some control from the outside. This is done by means of a built-in sensor consisting of a geiger tube which can be activated through the pipe wall by a weak Cs-137 isotope. As the star wheel method has thus been supplemented by isotope control, the star wheel method is no longer often used but has been replaced by the pure isotope control.

Isotope control of X-ray crawlers

Using this method two or possibly three sensors in the shape of geiger tubes are built into the crawler at the proper distance from one another. Two or three sensors are enough to perform all necessary commands through the pipe wall such as "stop/expose", "go forward to next" or "go back".

Isotope-governed crawlers can also be equipped with star wheels but these then only serve the purpose of informing the operator of the position of the crawler with their scraping sounds and clicks at the welds.

Working procedure

The procedure during work with isotope-governed X-ray crawlers is as follows:

When the device is placed in the pipe, the focal point elevation is adjusted in such a manner that the focal point follows the centre line of the pipe and film is placed on the desired number of welds. The crawler is started and the controlling isotope is placed at the proper distance from the first weld. When the first geiger tube arrives below the controlling isotope, the speed of the crawler is halved, and when the second geiger tube is below the isotope, the crawler stops with the focal point exactly in the desired position, and exposure is performed in accordance with preset kV_p, mA and time. After the exposure the crawler is kept standing still below the controlling isotope. When the isotope is removed, the crawler will move backwards out of the pipe. However, if the geiger tube in front detects the removal of the isotope, the crawler will move forward. At the next weld it can then be stopped by the controlling isotope and the exposure procedure will be repeated.

The manoeuvring of the crawler inside the pipe can be monitored by means of the sound from the star wheel. During exposure the process can be followed by hand-held monitors (bleepers).

Dimensions of pipes and crawlers

The requirements of crawlers for small and large pipe diameters are conflicting. When working inside small pipes the size of the focal point is crucial due to the small film focus distance (f.f.d.), whereas it is necessary to take along considerable battery capacity when working inside pipes of large diameters. It has, therefore, been appropriate for the author to develop crawlers of six different sizes for production in Denmark. The crawlers range in size from 6" to 40" of which the larger can be adjusted to operate in a considerable diameter interval. Depending on size, the crawlers have been designed with power supplies delivering 140 to 225 (even 300) kV_p with anode currents from 1 to 4 mA. Typical exposure times are between 10 seconds and 1½ minutes. To ease driving around curves and to facilitate insertion and removal of the crawlers, they are designed with joints and the larger ones have been equipped with battery baskets for changing batteries (Figure 37.3).

Figure 37.3 "Stripped down" 200 kV$_p$ NMK X-ray crawler for 16-24" pipelines. From the left: star-wheel, single-tank circumferential beam apparatus, regulation transformer, gear motor, battery box with baskets, sensors 1 and 2, the control unit, and finally the inverter is just visible.

Practical applications

Nearly all circular welds on the Danish natural gas distribution network have been inspected using crawlers like the ones described above.

The film used has been Type D.5, and it has given full exposure without the use of fluorometallic screens complying with both the old 15 mA-minute rule as well as the new ISO-standards.

The typical number of exposures per recharging has been 60 to 100. As an example of a successful operation we can mention that on one day 296 films were exposed on a 20" pipe using a 200 kV$_p$ crawler and three sets of batteries.

Break-down

Even if a crawler is equipped with a bumper, it may get stuck if it meets piles of rubble or beams in the pipe. From pipelines of diameters of 20" and up the crawler can usually be freed by a person riding into the pipe lying on an inspection board (Figure 37.4). So-called retrievers are used to salvage crawlers from pipelines of diameters less than 20". The retriever is a strong carriage that is sent into the pipe. At the moment the crawler is hit, the retriever hooks on to the crawler, reverses direction and pulls the crawler out, possibly assisted by a wire.

Figure 37.4. Person on an "inspection board" operating in a 24" pipe.

Figure 37.5. Front end of the "retriever" just before collision.

Training required/desired

As an educational background corresponding to qualified technician in radiography is desirable, supplemented by specialized training in work with crawlers.

O.L.Høppermann

References

1) *Recommended practice for radiographic examination of welded joints – Part 3: Fusion welded joints in steel pipes up to 50 mm wall thickness.* International Standard 1106/3

2) *Radiographic examination of metallic materials by X-ray and gamma rays.* International Standard 5579.

Chapter 38

X-ray diffraction

NDE principle

Residual stresses

Residual stresses are stresses which are present within a material even though it is not acted upon by external forces. It is possible to measure the deformation of a material by means of X-ray diffraction and thus to compute the residual stress.

Diffraction

Most metals consist of many small crystals of metal. When such a material is irradiated by a beam of X-radiation, the radiation is reflected (diffracted) from the atomic planes in the crystal. Diffraction only occurs if the following relationship between the interatomic plane distance d, the radiation wavelength λ and the angle of incidence θ is fulfilled (see Figure 38.1).

$$\lambda = 2d \sin \theta \qquad \text{(Bragg's law)}$$

If λ and θ are known, one can compute the plane distance d.

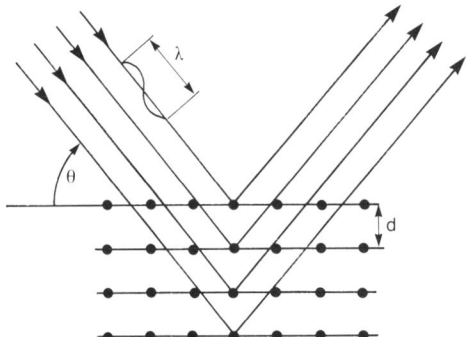

Figure 38.1. Diffraction can only occur if the wavelength λ, the interatomic distance d between planes and the angle of incidence θ fulfil the condition:
$\lambda = 2d \sin \theta$

If the intensity of the diffracted radiation is measured as a function of θ a curve like the one shown in Figure 38.2 is obtained.

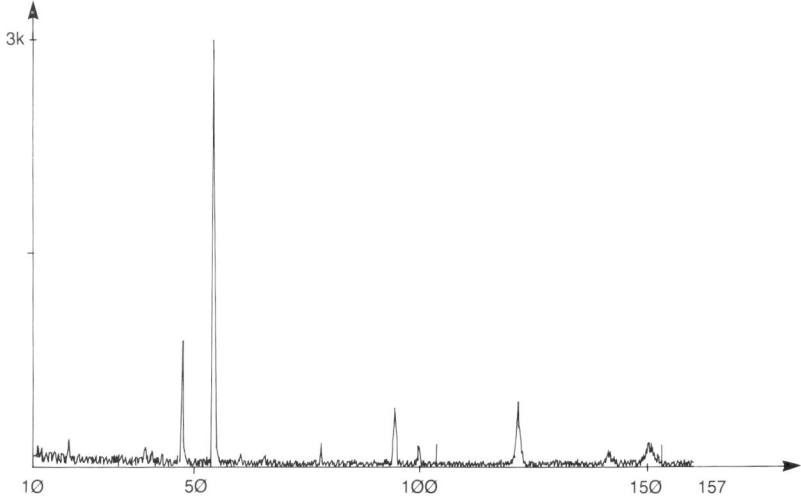

Figure 38.2. The intensity of the diffracted radiation as a function of the angle of diffraction 2θ. To measure residual stresses a peak with the highest possible angle 2θ is used.

When the material is subjected to a mechanical stress σ and thereby subjected to a strain ϵ, the lattice constant d and thus θ are changed.

$$\epsilon = \Delta d/d_o$$

where:
  Δd is the change in the interatomic planar distance.
  d_o is the interatomic planar distance in the unloaded condition.

By measuring how much the diffraction top has moved from its equilibrium position one can thus compute the strain. If one performs a measurement normal to a uniaxial stress, then the following expression is obtained:

$$\sigma = (-E/\nu)(\Delta d/d_o)$$

where E is the elasticity coefficient or Young's modulus and ν is the Poisson ratio (about 0.3 for most metals).

This method requires, however, precise knowledge of the interatomic planar distance in the unloaded condition d_o, and one must, therefore, use another method in practice (illustrated in Figure 38.3). For various angles ψ the position of the diffraction peak 2θ is measured. Due to considerations of measurement sensitivity long wavelength X-ray radiation is generally chosen, and diffraction then occurs from about the upper 10 μm of the material. In this case one can assume the presence of plane stress, i.e. no stress normal to the surface.

X-ray diffraction

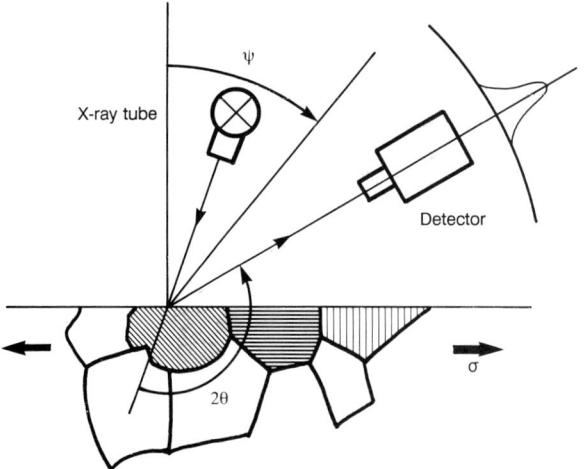

Figure 38.3. The principle for determining residual stresses with X-radiation is illustrated. If there are tensile stresses in the surface, the distance between the atoms in the direction parallel to the surface is increased, while they lie closer together in the direction normal to the surface. By measuring the distance in various directions ψ the stress can be computed.

Using Bragg's law one can obtain an expression for the stress in the surface (see Reference 1 for a more thorough treatment).

$$\sigma = K\, d(2\theta)/d(\sin^2\psi)$$

where $d(2\theta)/d(\sin^2\psi)$ is the slope of the graph of 2θ in degrees as a function of the expression $\sin^2\psi$.

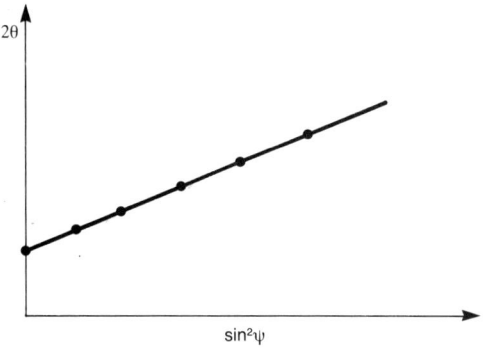

Figure 38.4. If the diffraction angle 2θ is displayed as a function of $\sin^2\psi$ a straight line is obtained. The slope of the line is proportional to the residual stress in the surface.

Residual stress constant

$K = -E \cot \theta \, (\pi/180)/[2(1 + \nu)]$ is called the *stress constant*. Because of the poor penetration of the X-radiation, only the stresses in the outermost surface layers are measured. If one is interested in the level of residual stresses in deeper layers, then one must remove material (e.g. by means of electropolishing) or use another method (strain gauges or neutron diffraction). These methods do not, however, have the same spatial resolution.

Development of the method

X-ray diffraction has been known and applied since von Laue and W.L. Bragg used the technique in 1912 to determine the crystal structure of NaCl, KCl, etc., and the method has been an important tool in the determination of micro-structures and the identification of unknown powder samples ever since.

X-ray diffraction equipment is usually large and heavy and, therefore, unsuitable for use in the field. The measurement of residual stresses has thus been limited to small samples and thus of limited practical interest.

Transportable, fully automatic equipment

However, transportable equipment has been developed which makes it possible to perform measurements on large objects. Such equipment is fully automated and completely dedicated to the measurement of residual stresses. A built-in computer performs the necessary data processing at once, and the result can be read directly as a stress value.

Other methods

Among the other methods for the measurement of residual stresses the most important – and the most commonly used – is strain gauge measurement. This method requires that a hole about 1 mm in diameter be drilled, and it can thus not be said to be entirely "non-destructive". On the other hand it can be used directly on most materials. Furthermore, there are other methods such as: neutron diffraction (which, however, requires access to a nuclear reactor and is therefore not transportable), ultrasonics and a magnetic technique which are under development.

Areas of application

Checking the heat treatment of welds

Residual stress measurements can be used to access the quality of welds and to determine whether or not subsequent heat treatment is required. A check of the effect of a stress-relieving heat treatment can also be performed using this method.

Material strength

It is important to know whether or not large residual stresses occur in a construction, for these must be added to the load to which the material is subjected during normal operation. In unfortunate circumstances it can mean a reduction in strength.

Surface treatments

The method is particularly well suited in connection with the characterization of surfaces and coatings. Most coatings are applied at high temperatures, and during the cooling process large residual stresses can occur because of differences in the thermal coefficients of expansion. Also surface treatments such as "shot peening" and laser hardening can cause residual stresses. Often these methods are used precisely to produce compressive stresses in the surface to avoid fatigue cracks and tension corrosion. With X-ray diffraction the effects of such treatments can be checked.

Direct measurement of applied stresses

Finally, one can of course use the method to directly measure the stress in prestressed constructions, etc.

Material requirements

X-ray diffraction can be used to measure the residual stress in most materials. The requirement which the material must fulfil is simply that reasonably strong diffraction will occur at the wavelength used with a large 2θ value. This is because a factor cot θ is present in the stress constant K. The quantity K should be as small as possible to make the measurements accurate, and therefore θ should be close to 90°. In order to achieve high intensity, long wavelength radiation is used, providing large angles of diffraction.

Surface requirements

In connection with the measurement itself there are certain requirements which the surface must fulfil. Due to the shallow penetration depth (c. 10 µm) the surface must not be too rough, for one will also obtain diffraction from stress-free protrusions in the surface and thereby introduce errors into the measurement (see Figure 38.5). The surface must, therefore, preferably be polished. This usually means that the surface must be treated before the measurement is carried out; and to remove the residual stresses which arise during polishing, the process is concluded with an electrolytic polishing process which removes material without acting on it mechanically.

Measurement accuracy

The accuracy of measurements is quite dependent upon the material, but for ordinary steel one can under normal conditions achieve accuracies within 5-15

MPa. A measurement can be carried out in less than 15 minutes if fully automated equipment is used. The measurement takes more time to perform on aluminium and stainless steel, and the measurements are not quite so precise.

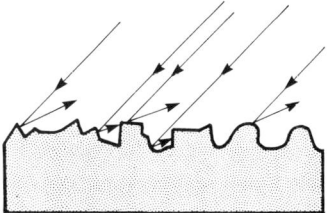

Figure 38.5. If the surface is too rough, at high angles of incidence to the surface one obtains mostly diffraction from stress-free protrusions on the surface.

Personal safety

Finally, there is of course the requirement for adequate shielding of the point of measurement during operation of the equipment to avoid exposure of the operator to the scattered X-radiation.

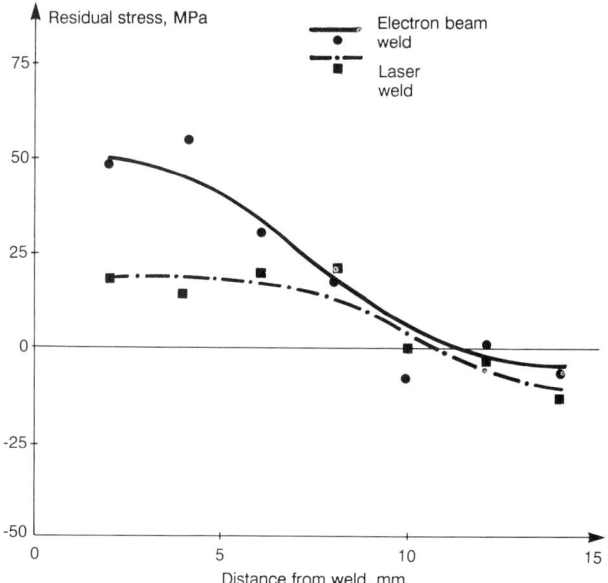

Figure 38.6. An example of a practical application. The graphs show the residual stress measured normal to two different types of welds as a function of the distance from the weld joint. The two types of welds are an electron-beam weld and a laser weld, respectively. Positive values indicate tensile stresses.

Practical applications

An example of a practical application is the checking of stresses around a weld. In Figure 38.6 the stress normal to two types of welds, *an electron beam weld* and *a laser weld*, in the same material (steel) is shown as a function of the distance from the weld.

This type of measurement can be used to determine whether or not it is necessary to apply a heat treatment to the weld. In this example there are tensile stresses in the surface. This can be dangerous if the weld is to be subjected to varying loads, which lead to fatigue.

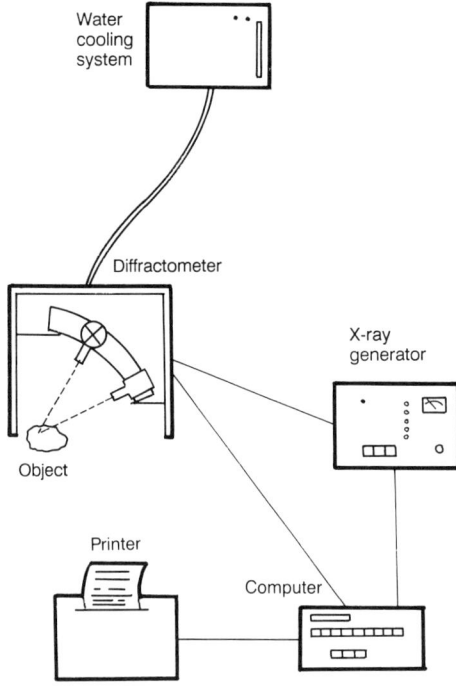

Figure 38.7. An overview diagram of an automatic setup for the measurement of residual stresses. The computer controls the diffractometer and receives and processes the data. The results and the treated data are printed out on the printer, which can be built into the computer.

Training required/desired

The operator of fully automatic equipment should be trained in the metallographic sample preparation skills (grinding, polishing and electropolishing) and should also have fundamental knowledge of X-radiation and the principles behind the measuring technique. It is also important that the user understand the decisive importance which the condition of the surface has upon the measurement. Line-up of the equipment is also very important to the accuracy of the measurement, and this the user should be aware of. Even though the equipment contains a computer for handling the data, it is not necessary for the user to have special knowledge of computers or programming.

Thomas Fich Pedersen

References

1) *Elements of X-ray diffraction.* B.D. Cullity, 2nd edn., Addison-Wesley Publishing Company Inc., New York, 1978.

2) *Residual stress. Measurement by diffraction and interpretation.* I.C. Noyan and J.B. Cohen, Springer Verlag, New York, 1987.

Index

Abrasive wear, 96
Accelerometer, 344
Acoustic cross correlation, 9-16
Acoustic emission, 17-23
Activatable tracers, 131
AE measurements, 17
AE sensors, 22
Agema thermovision, 313,319
Air renewal measurements, 134
Aircraft motor, 292
Aircraft parts, 50
Ampère, 44
Analytical ferrography, 91
Annular rings, 216
Arago, 44
ASNT, 59
ASTM, 21
ASTM E 109-63, 186
ASTM E 125-63, 186
ASTM E 138-63, 186
ASTM E 140-84, 114
ASTM E 432-71, 158
ASTM E 1009-84, 65
Atomic absorption spectrometry, 289
Automatic scanners, 242
Auxiliary electrode, 159
Auxiliary tools, 380
Aviation industry, 296

Baird Spectromobile, 67
Balconies, 271
Ball test, 136
BAM, 216
Banana, measuring tool, 24
Beams, measurement of, 271
Belt pulley, inspection of, 386
Boiler defects, 283
Boilers, 77
Bolometers, 310
Bolts, elongation of, 336
Bonding, ultrasonic examination, 336
Borescopes, stiff, 72
Bragg, W.L., 407
Bragg's law, 404
Brazed pipe joints, 337
Breaks, checking for, 303
Bremsstralung, 263
Brinell
   hardness unit, 103
   measurement, 104
British Gas Corporation, 198
British society for strain measurement, 301
Broad-band measurement, 355
Brüel & Kjær, 341

BS, 59
BS 3636, 158
BS 6072, 185
BS 4397, 185
BS 4416, 39
BS 4675, 355
BS M.35, 185
Building construction, 317
Bundesanstalt für Materialprüfung, 216

Calipers, 382
Cast wheels, examination, 334
CAT-scanning, 142
CCD camera, 231
Central Research Laboratories, 214
Cepstrum, 367
Chain drive, inspection of, 394
Chalk-alcohol method, 33
Charge coupled device, 232
Circumferential magnetization, 174
Clutch plates, examination, 304
Clutches, inspection of, 389
Coating thickness, 24-31
Coating thickness, radioisotope checking, 138
Coatings, testing for pinholes, 254
Collimated photon scattering, 140
Component condition monitoring, 196
Compression checking, soil, 137
Compressor, inspection of, 391
Computer assisted tomography, 142
Computerized tomography, 213
Concrete floors, 271
Concrete reinforcing rods, 270
Condition monitoring, 1,19,195,304
Conveyor belt, inspection of, 393
Cooling ventilator, inspection of, 390
Corrosion
   damage, P-scan measurement, 247
   measurement, 50,163,306
   protection, 163
   rate, 82
   rate, with LPR probe, 159
   testing, 124
Corrosion Centre, Danish, 169
Coulomb, 44
Coupons, test, 307
Cracking, 299
Cracks
   checking for, 303
   detection, 50,280
   indications, 180
   type determination, 279
   ultrasonic examination, 333
Crane, example, 299

Index 413

Cranes, 52
Crankshaft, 187
Creep, 280
Cross correlation function, 13
Cross correlation method, 10
Cross reference table, 4-5
CT method, 202,214
Cylinders, 77

Danish Isotope Centre, 141
Danish Maintenance Society, v
Danish Welding Institute, 200
Decibel scale, 224
Defect development, 374
Demagnetization, 174
Diffraction, 404
Digitization, 232
DIN, 59
DIN 54.150, 284
DIN 50150, 114
Directional microphone, 321
Distance measurement, 144
District heating, 317
Dorns, 193
Double exposure HI, 210
DR ferrograph, 90
Dye penetrant examination, 32-41
Dye penetrant method, 151
Dye penetrants, 35
Dry powder, 178
Dry powder developers, 36
Dynamic imbalance, 366

EDAX, 70
Eddy currents, 27,42
Eddy current testing, 42-59,202
Electrical power industry, 317
Electric motor, inspection of, 387
Electrocooled scanner, 314
Electrolytic polishing, 276
Electron beam weld, 410
Electron microscopy, 93
Electronic halogen leak detection, 155
Elevators, 52
Emission factors, table, 311
Emission spectrometers, transportable, 67
Emission spectroscopy, 60-71,288
Encircling probes, 46,49
Endoscope, video, 75
Endoscopy, 72-80
Envelope curve analysis, 353
Envelope detection, 370
Equotip hardness test, 109
ER probe, 81-87
Ernst type STE, 104
Erosion damage, P-scan measurement, 247
Erosion measurement, 163
EWGAE, 21

Fast fourier transform, 344
Fatigue, detection with PA, 220
Fault detection, 355
Faulty diagnosis, 365
Feeler gauges, 193
Ferritic welding, 50
Ferrograms, 90
Ferrograph, direct reading, 89
Ferrography, 88-102
Ferrography analysis, 88
FFT analyser, 341,344,358
FFT analysis, 373
FFT phase, 367
Fibrescopes, 72
Fibrescopes, flexible, 73
Flatness, measurement with laser, 146
Flow rates, measurement of, 129,131
Fluorescence, 66
Fluorescent
  magnetic liquid, 185
  magnetic powder, 179
  penetrants, 35
Foam generators, leak localization, 152
Fork lift truck arm, 186
Förster, 45
Frequency analysis, 350,358
Frequency multiplier, 375
Frozen fringe HI, 210

Gabor, Dennis, 211
Gamma radiation, 128
Gamma ray examination, 263
Gamma sources, use in CT, 216
Gearbox, 348
Gearbox diagnosis, 369
Gearboxes, inspection of, 388
Gearbox, NDE methods for checking, 196
Geiger counter, 129
Ghost component, 348
Glow discharge equipment, 65

Halogen leak detection, 154
Halogen torch, detection of F-gas, 155
Hardness measurement, 279
Hardness testing, 103-119
Harwell, 204
HAZ, 280
Heat affected zone, 280
Heat exchangers, 48,52,77
Helium testing, 157
HI, 202
High frequency components, 352
High frequency probes, 46
High voltage testing, pinholes, 254
Holiday detector, 251
Hologram, 209
Holographic interferometry, 209
Honeycomb panel, adhesive bonding of, 337
Hounsfield, 214

Hunter husky, 313
HV value, 106
Hydraulic systems, inspection of, 385
Hydrogen
　cell, 120-127
　cracks, 111
　generation, 120
　permeation, 121
Hydrogenation, 333, 335
Hysteresis whirl, 346

ICP, 288
Impact test, 353
Inductively coupled plasma emission
　spectrometer, 288
Inhibitor monitoring, 163
Insect damage, detection with CT, 216
Inside calipers, 189
Inspection board, 402
Inspection, visual, 379
Internal bobbin probes, 46,47
ISO 2372, 355
ISO 3057, 284
ISO 3741, 230
ISO 3742, 230
ISO 3743, 230
ISO 3744, 230
ISO 3745, 230
ISO 3746, 230
ISO 3879, 39
ISO 4964, 114,118
ISO 6506, 104,118
ISO 6507, 107,118
ISO 6508, 118
Isotope control, 399
Isotope techniques, 128-143

Jet engines, wear, 90
Joint monitoring, 136
Journal bearings, 343,346

Kaiser effect, 18
Kaiser, Joseph, 18
Kelvin, 44
Kelvin bridge, 82
Knoop method, hardness test, 109
Knots, detection in wood with CT, 216
Krautkrämer microdur, 110,116
Krautkrämer probes, 330
Kunth-Winterfeldt, E., 278

Laser distance measurements, 144-147
Laser weld, 410
Lawrence Livermore Laboratory, 214
Leakage field, 171
Leak localization, foam generators, 152
Leaks
　acoustic detection, 9
　checking for, 148

detection, 9,21,22,321
detection probes, 156
detection, with radiation, 129
rate, definition, 148
Leak testing, 148-158
Line scanners, 310
Liquid pressure testing, setup, 258
Listening apparatus, 321
Lixiscope, 141
Longitudinal magnetization, 174
Loose particles, 21
Low frequency components, 345
Low frequency probes, 46
Low voltage testing, pinholes, 254
LPR probe, 159-164
Lubricating oil, 95
Lubricating system, checking, 163

Machine vibrations, 340
Magnetic
　attraction, 24,25
　flux leakage, 198
　induction, 25
　liquid, checking, 184
　particle examination, 171-188
　plugs, 165-170
　plugs, automatic, 166
　powder, types of, 178
Magnetostatics, 26
Magnetizing, 174
Manual scanners, 242
Marine gas turbines, 99
Material condition monitoring, 198
Material strength, 408
Maxwell, James C., 44
MCM, 202
Mechanical calibration, 189-194
Mechanical polishing, 276
Medical applications, 317
Metal fatigue, 219
Metalographic method, 276
Metascope analysis, 70
Metascope equipment, 61
Meyer hardness test, 108
Micro-cracks, detection, 215
Microfocus X-ray tubes, 264
Micrometer tools, 192,382
Microstructure examinations, 279
Middle frequency region, 347
Mixing time, 133
Moiré contour mapping, 205
Moisture checking, soil, 137
Moisture content, measurement with CT, 216
Monitoring, 19
Motor console, 300
Motors, 77
Multi-channel LPR instruments, 162
Multi-channel ultrasonic equipment, 330
Multiplexer modules, 83

Index　　　415

Napier-Gazelle gearbox, 98
Narrow-band spectra, 359
NDE method combination, 195-201
NDE methods under development, 202-222
Nippon Steel Co., 98
NMK vacuum boxes, 153
Noise, 223

Noise
  levels, 224
  mapping, 226
  measurements, 223-230
  meter, 223
Non-metalic coatings, 51
Non-water based wet developers 36
Nordtest, 21,274
Nordtest
  build 123, 137
  diploma, 338
  system, 339

Octave analysis, 361
Octave band analysis, 228
Offshore installations, 85
Oil cleaning, 167
Oil industry, 83
Oil sample requirements, 97
Oil whirl, 346
Ometron, 203
On-line inspection, 19
Operational defects, 280
Optical emission spectroscopy, 64
Optical fibre light guide, 73
Optical pattern recognition, 231-239
Optocator, 144
Original defects, 280
Oscillations in smokestack, 299
Outside calipers, 189
Ozalide paper, leak checking with, 154

P scan, 240-250,331
P-scan examinations, 248
P-scan images, 245
PA, 202,218
PAIR, 159
Particle identifications, 95
Peltier element, 315
Pencil point probe, 53
Penetrant, 32
Penetrant flaw detection (see dye penetrant examination)
Phase measurements, 366
Pig, for pipeline checking, 198
Pigs, 136
Pinhole detection, 251-257
Pinholes, 251
Pipe, checking X-ray pictures, 238

Pipe leaks, detection, 324
Pipes, 52,77

Pipes
  corrosion, 85
  creep cracks in, 280
  leak detection, 11
  thickness measurement, 266
Piping thickness with P-scan, 248
Pixel, 232
Plug gauges, 193
Point measurement instruments, 310
Poisson ratio, 405
Polarization resistance, 159
Poldy hammer, 105
Positron annihilation, 218
Positron detectors, 213
Positronium, 219
Positrons, 218
Post-emulsifiable penetrant 35
Predictive NDE programmes, 199
Pressure difference, monitoring, 150
Pressure drops, checking, 152
Pressure tanks
  CO_2, testing, 138
  holography measurements, 212
Pressure testing, 258-262
Preventive maintenance, 84
Preventive maintenance programme, 316
Primary field, 42
Probes, for pinhole testing, 256
Process monitoring, 140
Production monitoring, 19
Profile measurements, 279
Projection scanning, 244
Propagation velocity, 10
Pulse-echo method, 326
Pulsed video thermography, 204
PVT, 202
Pyrometers, 310

Quality control, 150
Quantovac, 67

Radiation, measurement of, 129
Radioactive
  decay, 128
  isotope techniques, 154
Radiographic
  examination, 263
  on-stream inspection, 265
Radiography, 202,263-275
Radiometers, 310
Radiometry, 128,130
Reactors, 77
Reference electrode, 159
Reference masks, 361
Reference spectrum, 361
Refinery piping system, examination, 334

Replica technique, 276-285
Residence time measurement, 130,132
Residual stresses, 404
Rivet joint, 53
Rms/peak factor measurement, 354
Road surfaces, checking, 146
Robot inspection, 146
Rockwell B hardness test, 108
Rockwell C method, 107
Rockwell N hardness test, 108
Rockwell superficial, 108, 115
Rods, symmetry checking, 146
Rolled plates, examination, 332
Roller bearings, 343
Röntgen, Wilhelm, 264
ROSI, 265
Rot, detection in wood with CT, 216
Rotary particle dispenser (RPD), 93
RPD apparatus, 94
Ruge, 296

Safety, radiation, 129
Sampling, SOAP, 290
Scanners, automatic, 331
Scanning, 328
Scanning electron microscope, 70,166,278
Scientific Measurement Systems (SMS), 216
Scintillation screen, 142
Scleroscope hardness test, 109
Secondary field, 42
SEM, 70,166,278
Ship gearbox, 99
Shore method, hardness test, 109
Silhouette images, 233
Simmons, 296
Simple imaging instruments, 310
Simultaneous method, 210
Ski-lifts, 52
Smoke detection, 139
Smokestack, 299
Sniffer probe, 155
SOAP, 167,286-294
SOAP programme, 291
Soil
  compression checking, 137
  moisture checking, 137
Solvent removable penetrant, 35
Sound, 223
Sound waves, 17
Spark patterns, 383
Spark stream table, 383
SPATE, 202
Spectroscopic oil analysis program, 286-294
Spectrum synthesis, 360
Speed compensation, 363
Spherical particles, 97
Spring calipers, 190
Sprite detector, 315
Swedish Standard, 59

Star wheel control, 399
Static imbalance, 366
Stereo radiography, 268
Strain gauge measurements, 407
Strain gauge technology, 295-301
Stress constant, 407
Stress pattern analysis by thermal emission, 203
Stress testing, 18
Strobe light illumination, 235
Stroboscope, 302
Stroboscopy, 302-305
Struers K/S, 278
Sulphur hexafluoride detector, 156
Sulphur sticks, leak checking with, 154
Surface fatigue, 96
Surface probes, 46,50,53
Sutherland bottle, 184
Svejsecentralen, 200
Synchronous time averaging, 372

Takasaki, 206
Tangential imaging, 266
Tank leaks, detection, 324
Tebo-dorn, 193
Telebrineller, 105
Temper embrittlement, monitoring, 197
Tempera colours, use of, 92
Tensile strength, 103
Test coupons, 306-309
Texas nuclear alloy analyser, 69
Thermoelectric cooler, 315
Thermography, 310-320
Thickness gauge, 190
Thickness measurements, P-scan, 247
Thin layer activation, 134
Three dimensional measurement, 235
Time averaging HI, 210
Time averaging technique, 351
TLA, 134
Tolerance forks, 193
Tomogram, 213
Tomography, 214
Towing cables, 52
Tracer techniques, 128
Tracers, leak detection with, 154
Train wheels, checking, 146
Transportation worm, inspection of, 392
TRD 508, 284
Trend curve analysis, 375
Turbine blades
  holographic examination, 212
  profile checking, 146
  ultrasonic examination, 333
Turbines, 77

UCI hardness measurement, 110
Ultrasonic
  leak detection, 321-325
  pulser-receiver, 241

Index 417

Ultrasonics, 152,202,326-339
Ultraviolet illumination, 77,185
Ultraviolet radiation, 37
University of Marburg, 218
University of Tokyo, 217

Vacuum testing, 153,260
VDI 2056, 355
Vernier gauge, 191,382
Vibration analysis, 343
Vibration monitoring, 340-378
Vickers hardness measurements, 115
Vickers measurement, 106
Video camera, OCR, 231
Video documentation, 76
Visual inspection, 379-396
Voids, detection, 220
von Laue, 407

Water based wet developers, 36
Water pump, inspection of, 394
Water washable penetrants, 35
Wear
  examinations, 279
  index, 88
  measurements, 134
  mechanisms, 96
  metals, 286
  particle transducer, 166
  profile, 286
  rate, 292
Weld
  electron beam, 410
  laser, 410
Welding defects, 280
Welds
  checking heat treatment, 407
  examination with P-scan, 246
  inspection of, 395
  pipe, ultrasonic examination 332
  ultrasonic examination, 327
Wheatstone bridge, 295
Windmill blades, examination, 336
Wire, examination of, 52,58
Wire, fabrication measurement, 146
Working electrode, 159

X-radiation, safety with, 409
X-ray
  checking of concrete, 137
  crawlers, 397-403
  diffraction, 66,404-411
  examination, 263,397
  fluorescence, 128,138
  fluorescence analysis, 64
  photography, 128
  tomography, 217

X-ray crawler
  battery powered, 398
  cable operated, 398
  petrol powered, 398
  push-rod, 397
  retriever, 402

Young's modulus, 103,405

Zoom spectrum, 370
Zwick, 105

Ørsted, Hans C., 44